The Languages of Edison's Light

Inside Technology
edited by Wiebe E. Bijker, W. Bernard Carlson, and Trevor Pinch

Janet Abbate, *Inventing the Internet*

Charles Bazerman, *The Languages of Edison's Light*

Marc Berg, *Rationalizing Medical Work: Decision-Support Techniques and Medical Practices*

Wiebe E. Bijker, *Of Bicycles, Bakelites, and Bulbs: Toward a Theory of Sociotechnical Change*

Wiebe E. Bijker and John Law, editors, *Shaping Technology/Building Society: Studies in Sociotechnical Change*

Stuart S. Blume, *Insight and Industry: On the Dynamics of Technological Change in Medicine*

Geoffrey C. Bowker, *Science on the Run: Information Management and Industrial Geophysics at Schlumberger, 1920–1940*

Louis L. Bucciarelli, *Designing Engineers*

H. M. Collins, *Artificial Experts: Social Knowledge and Intelligent Machines*

Paul N. Edwards, *The Closed World: Computers and the Politics of Discourse in Cold War America*

Herbert Gottweis, *Governing Molecules: The Discursive Politics of Genetic Engineering in Europe and the United States*

Gabrielle Hecht, *The Radiance of France: Nuclear Power and National Identity after World War II*

Kathryn Henderson, *On Line and On Paper: Visual Representations, Visual Culture, and Computer Graphics in Design Engineering*

Eda Kranakis, *Constructing a Bridge: An Exploration of Engineering Culture, Design, and Research in Nineteenth-Century France and America*

Pamela E. Mack, *Viewing the Earth: The Social Construction of the Landsat Satellite System*

Donald MacKenzie, *Inventing Accuracy: A Historical Sociology of Nuclear Missile Guidance*

Donald MacKenzie, *Knowing Machines: Essays on Technical Change*

Susanne K. Schmidt and Raymund Werle, *Coordinating Technology: Studies in the International Standardization of Telecommunications*

The Languages of Edison's Light

Charles Bazerman

The MIT Press
Cambridge, Massachusetts
London, England

147848

Set in New Baskerville by The MIT Press.
Printed and bound in the United States of America.

Library of Congress Cataloging-in-Publication Data

Bazerman, Charles.
  The languages of Edison's light / Charles Bazerman
  p.   cm. — (Inside technology)
  Includes bibliographical references and index.
  ISBN 0-262-02456-X (hc. : alk. paper)
1. Electric lighting—history. 2. Technology—Social aspects.
3. English language—Rhetoric. I. Title. II. Series.
TK4131.B39   1999   98-51881
303.48'3—dc21   CIP

Contents

Acknowledgments   *ix*
How the Edison Papers Are Cited   *xi*
Credit for Illustrations   *xiii*

Introduction   *1*

I

The Opening Scene

1

Edison's Front-Page Story   9

II

Establishing Meanings in Evolving Systems

2

The Public Stage of News   23

3

Finances for Technological Enterprises   39

4

Menlo Park: The Place of Invention   47

5

Patents as Speech Acts and Legal Objects   85

6

Professional Presence: Edison in the Technical Press   111

7

A Place in the Market   141

III

Making It Real: The Rhetoric of Material Presence

8

Boasts, Deceptions, and Promises   159

9

The Menlo Park Demonstrations   179

10

Fairs and Exhibitions: Museums of the Future   199

11

Lighting New York: Urban Politics and Pedestrian Appearances   219

IV

Establishing Enduring Values

12

Patent Realities: Legal Stabilization of Indeterminate Texts   237

13

Charisma and Communication in Edison's Organizations   259

14

Rhetoric of Capital Investment: Solvency, Profits, and Dividends   291

15

The Language of Flowers: Domesticating Electric Light   313

Conclusion   333

Notes   351
Bibliography   397
Index   411

Acknowledgments

This project has been supported by funding and released time from the National Endowment for the Humanities, Baruch College of City University of New York, the Georgia Institute of Technology, the University of Louisville, and the University of California.

While working on this project I haunted many libraries, archives, and other institutions, including the following: Burndy Library, Dibner Institute, Bakken Collection, Smithsonian Archives, American Institute of Physics, Edison National Historical Site and Edison Papers, Hagley Collection, American Philosophical Society, Smithsonian Museum of American History, Edisonia Collection at Greenfield Village, Hammer Collection, New York Historical Society, Archives of the City of New York, Boston Public Library, Detroit Public Library, Georgia Institute of Technology, University of Louisville, City University of New York, University of California at Santa Barbara, University of California at Los Angeles, Institute of Electrical and Electronics Engineers, Center for History of Electrical Engineering, University of Pennsylvania, University of Minnesota, Popular Culture Collection at Bowling Green University, Popular Culture Collection at Case Western University, and Filson Collection. I am indebted to the endless kindness of the librarians and the archivists. I particular wish to thank George Tselos at the Edison Archives, who from the very beginning pointed me in useful directions and who was generous with his time and with access to materials.

Over the years of searching out materials and trying to make sense of them, I have bent the ears of uncountable students, colleagues, and friends, including Larry Badash, Carol Berkenkotter, James Britton, Bernard Carlson, Aviva Freedman, Paul Geller, Lisa Gitelman, Anita Guerrini, John Gumperz, Britt-Louise Gunnarson, Mary Ann Hellrigel, Paul Hernadi, Paul Israel, Debra Journet, Gene Lerner, Ken Knoespel, Peter Medway, Carolyn Miller, Greg Myers, Mike Osborne, Paul Prior,

David Russell, Patrick Sharp, John Swales, Sandy Thompson, Spencer Weart, George Yater, Audrey Young, students in technology and culture courses at Georgia Tech and UCSB, and the many participants at seminars and conferences where I delivered versions of these chapters. My special thanks go to David Nye and Keith Nier for their comments on the entire manuscript.

Finally, I would like to thank my partner, Shirley Geok-lin Lim, and my son, Gershom Kean Bazerman, for enduring my obsession with micro-films, with late-nineteenth-century industry and technology, and with old Tom. I think they will be happy to see me get onto something new.

How the Edison Papers Are Cited

The Edison papers, housed at the Edison Historical Site in West Orange, New Jersey, are published in two formats: an extensive selection in microfilm and a narrower selection in print. Both editions were begun under the general editorship of Reese V. Jenkins.

Thomas A. Edison Papers: A Selective Microfilm Edition, published by University Publications of America, is projected to consist of six parts, three of which have been released as of this writing: part I (1850–1878), part II (1879–1886), and part III (1887–1898). A list of libraries holding the microfilm collection and other information about the edition can be found by consulting the Edison Papers web site (http://edison.rutgers.edu). I cite the microfilm edition (which reproduces almost all the documents for the period examined in this book) whenever a document may be found within it. The microfilm edition is cited by reel and frame numbers—e.g., "(17: 979–981)." In most cases this identification is preceded by a description of a document, such as "letter from George Barker to Edison, 23 October 1878."

A much more selective print edition is being issued by The Johns Hopkins University Press under the title *The Papers of Thomas A. Edison.* As of this writing, four volumes have been issued, subtitled *The Making of an Inventor (February 1847–June 1873), From Workshop to Laboratory (June 1873–March 1876), Menlo Park: The Early Years (April 1876–December 1877),* and *The Wizard of Menlo Park (1878).* The documents published in this edition are largely selected from the same material reproduced in the microfilm edition; however, extended commentary and occasional documents are new to the print edition. For material not to be found in microfilm, I cite the print edition. I identify material from the print edition by the abbreviation *PTAE* and by volume, document, and page numbers—for example, "(*PTAE* III, 1118: 628)." Citations are usually accompanied by descriptions—e.g., "letter from George Barker to TAE, 20 November, 1877."

Credit for Illustrations

The Edison National Historic Site, which is under the auspices of the National Park Service of the U.S. Department of the Interior, is the source of most of the figures in this book. Figures 14.1 and 14.2 are from the Burndy Library of the Dibner Institute for the History of Science and Technology. Figures 15.5, 15.6, and 15.8 are from the Smithsonian Institution Libraries. Permission to reprint figures is gratefully acknowledged.

The Languages of Edison's Light

The Man Who Moves the World, New York **Daily Graphic,** *October 26, 1878 (94: 398).*

Introduction
Rhetorical and Material Action on the World Stage

Thomas Edison is an American myth, an exemplary life encapsulating cultural beliefs: the achievement of the self-made man; the creativity of technological America; the rewards of hard work wedded to imagination, daring, and technological entrepreneurship. For more than 100 years he has been considered the quintessential individual of capitalist society.

Through the historical record, we know that Edison was a major social force in his own time—a public figure who hobnobbed with the other movers and shakers of industrial America as the country moved from the Civil War into the twentieth century.

But we know Edison most through the transformative technologies with which he is so strongly associated: the phonograph, the movies, and (most fundamentally) electric light and power. Utility companies around the world bear his name.

The beginning of electrification has been admirably documented from several perspectives—the emergence of technology (Robert Friedel and Paul Israel's book *Edison's Electric Light*), the interrelatedness of multiple technologies within a system (Thomas Hughes's *Networks of Power*), the relationship between emergent corporate and financial structures and the delivery of power (Harold Passer's book *The Electrical Manufacturers*), and the response of the public imagination to the new electrified world (Carolyn Marvin's *When Old Technologies Were New*, David Nye's *American Technological Sublime* and *Electrifying America*), and of course Edison's comings, goings, and doings (countless biographies, memoirs, and Edisonia). Throughout this book I will rely heavily on the work of other scholars. But I will tell a different kind of story—a story of communication and symbols, a story of how Edison and the people around him represented light and power to themselves and to others as the technology emerged from the ideas of a few pioneers into a daily fact of life for us all. I will look at the languages in which Edison and his colleagues spoke to us, giving electric light and power meaning and value and incorporating it into our existing cultural systems of meaning. These representations may be viewed by some as little more than charming curiosities—quaint period costumes in which the important work of invention and industry building was accidentally clad. But the accounts I will give in this book argue that these representations, which created meanings and accommodated novelties to existing sets of beliefs and social institutions, influenced people to understand and participate in the electrification of all domains of life. It is the representations that go between minds, creating meanings and accommodating novelties to existing sets of beliefs and social institutions.

Certainly, tangible objects (light bulbs and power plants) and the tangible benefits of clean, safe light provided individuals with material experience of the technology. But even the tangible is experienced within social meanings, institutions, and activities. Electrical light and power emerged within and were supported by a social matrix. Light could not just appear through the mute work of a few mute technologists. It had to emerge as part of the drama of human meanings.

The drama of the construction of electrical power and light was both prominent and significant, both highly visible and highly consequential. Some significant actions require only the cooperation of a small number of people and can be carried out in a quiet corner of the stage until their consequences start proliferating, transforming the very nature of the stage. But other actions, from early on, require the involvement and cooperation of many actors, mediated through multiple channels of communication. Because of the many communications that are entailed, such enterprises become socially visible early on, in many different locations and many different ways. The creation of a system of light and power required visibility in order to harness the support and the force of the legal, financial, corporate, technological, public, and civic systems.

Edison and the people around him necessarily and willfully spoke to the discourses around them to create presence, meaning, and value for the emergent technology in the laboratory, in patent offices and law courts, in financial markets, in the boardrooms of newly created companies, in city halls, in newspapers, and in the consumer marketplace. These venues are saturated with language. Newspapers are the daily world gathered into words and pictures. Government and law are notorious for their verbosity. Finances and corporations do their business over meeting tables and on paper. Consumers are wooed through pamphlets, advertisements, publicity, and fast-talking salespeople. Even the laboratory and the shop floor are socially bound through diagrams, instructions, plans, reports, questions, orders, confirmations, and records.

The material technology of electrical power and light, once produced, had persuasive force and compelling priorities of its own. The night lit up at the flick of a switch argued for itself, electrocution of beast and man signified electricity's terrifying power, and regular delivery of light was one means of persuading consumers to pay their monthly electric bills. Edison was savvy enough as a rhetorician to use all these material arguments. But, as a walk through the technology section of a library or the files of an electric company will reveal, the production of these physical arguments required the cooperation of many people, aligned through words and pictures. Within those representations the physical technology takes on meanings and force that it would not have on its own, mutely. Lightning may terrify a herd of wild animals, but those animals do not turn the lightning into an instrument of the gods to foster social order, nor do they understand the lightning to be related to chaff clinging to rubbed sulfur, nor will they invent machines to produce charge and send the charge through wires into every home so that they may read books at night and watch television.

Technologies emerge into the social configurations of their times and are represented through the contemporary communicative media. This book examines the emergence of many of the social and communicative arrangements that shaped and transformed the world in which Edison and his colleagues acted: the American patent system, newly large cities, large-circulation newspapers, technological professions, transformed universities, national markets, large corporations, financial investment, and commercial display.

In Edison's day, even more than today, talk was a major medium of sharing information. Many of the social systems I will be considering were still relatively small and local, and much work was of the face-to-face kind. But the late nineteenth century was also was a period in which more extensive systems grew and began to leave substantial documentary trails. The rise of professions brought specialized literatures and training documents. The development of government bureaucracies meant burgeoning files. Broad investment in financial markets made financial news a valuable commodity. Big corporations left increasing numbers of letters, reports, memos, and forms. During this period, so many actions were carried out on paper that we can examine the documents not just as fragmentary remains of activities (like potsherds of ancient civilizations), or as documentary records of non-literary events (like medieval chronicles of royal doings) but as the actual media of social action. An exchange of letters was indeed the interaction itself. More of the work was carried out on paper in some areas (such as the legal system) than in others (such as the laboratory). In some cases, the vanished spoken words leave us with questions: What meanings and values were exchanged between the Edison Electric Illuminating Company of New York and the New York City Board of Aldermen over dinner one night in December 1880, for example. But for the most part the documentary record provides a deep picture of the communicative interactions through which Edison and his colleagues gave electrification meaning and value.

I have already referred to Edison's colleagues and collaborators. He may have started as a freelance inventor working after hours in a telegraph office, but he rapidly developed a series of networks, collaborators, employees, agents, and allies who stood at the intersection of many forms of communication. Thomas Alva Edison became Edison, Incorporated. Many individuals were involved in the nurturing and the spread of the representations of Edison's project. In some instances it is possible to separate the communicative work of Edison the individual from those of the larger enterprise, but in others it is not; and sometimes there is, for the purposes of this book, no point in making such distinctions.

Many of the communicative actions examined herein show clear signs of careful forethought, strategic planning, and art. Others seem spontaneous, more the results of momentary circumstances than of design. That is the way with all communication and representation. We communicate responsively and reactively. Even when we write with planning and reflection, our range of conscious attention is limited and our writing is directed toward a communicative landscape we have only partly articulated. Indeed, the purpose of studies such as mine is to bring more of the dynamics of communication to consciousness.

I will not always try to discern what people thought they were doing, or whether they knew what they were doing or had just done. Evidence of reflections is rare in the Edison papers, and there is limited profit in reading the minds of actors long dead. I will keep my attention focused on what the texts accomplished in the environments to which they seem to have been responsive. To contain the task, I will focus primarily on material directly related to the introduction of light in the period 1878–1882. Rather than attempt comprehensive coverage of all the material, I have chosen representative texts that reveal the rhetorical activity of the discourse.

In chapter 1 I recount the immediate reactions of various publics to Edison's announcement of incandescent lighting in order to establish the story and to reveal some of its multidimensionality. In chapters 2–7 I explore how Edison established meaning and value for his project in the several meaning systems that would bear on its success: news, finances, laboratory, patents, engineering, and the market. These chapters are organized as historically parallel, each reaching back to the formation of the relevant discursive systems and continuing forward to examine Edison's intervention. In chapters 8–11 I examine the tension between what Edison produced materially and how he represented the technology symbolically; together these chapters provide a chronological narrative of invention and development in the period 1878–1882. In chapters 12–15 I consider how the new material and social realities of incandescent light created new and enduring meanings in several domains, and I follow a number of separate but parallel trails into the ensuing years.

While researching and writing the book, I grappled with many ideas about rhetoric, technology, and society. In the text, I have not put these theories in the foreground; rather, I have tried to use them to give shape to the story. Nonetheless, I think the story has some important consequences for theories of rhetoric, social organization, and technology studies; I try to make these visible in the concluding chapter.

I

The Opening Scene

1

Edison's Front-Page Story

Edison's first announcement of interest in electric lighting was a front-page newspaper story, and electric light became the most prominent technological story of a technological age. Many other specialized dramas were played out in other discursive venues—finances, law, corporations, the laboratory, the technological press—but these were energized and supported by public attention in the press and were then reflected back onto the journalistic stage—until the United States saw itself as a society powered by electricity, and electricity flowed through all activities of daily life.

Electric Light before Edison

By 1878, when Edison entered the field of electric lighting, arc lighting (produced by a spark in the gap between two carbon electrodes) was already an established technology. Humphrey Davy had demonstrated the possibility of electric arc lighting in 1801, and in the 1840s means were invented to maintain appropriate distance between quickly consumed electrodes. In 1862 the Dungeness Lighthouse in Kent, England, was converted to arc lighting in the first successful application of the technology. By the late 1870s Paul Jablochkoff in Europe and Charles Brush in the United States had produced commercially viable arc systems for lighting streets and large public spaces. In 1878 arc lighting illuminated Paris and London streets and John Wanamaker's department store in Philadelphia. In 1880 arc lights were mounted on large towers to illuminate the entire city of Wabash, Indiana. Arc lighting, however, was too bright for homes or offices. The gentler incandescent electric lighting, produced by a glowing filament, promised to compete with gas lighting, which was in place in almost every city in Europe and the United States in the 1870s. No viable system of incandescent lighting existed, however, before Edison took interest in the problem, despite more than 30 patents by various inventors dating back to 1841.[1] Finding a filament that would reach sufficient temperature to glow without burning or melting was the main challenge.

Edison's Entry into Lighting

Until August of 1878, Edison had taken no more than passing interest in arc or incandescent lighting. Only a few passing experiments were recorded in his notebooks, among many other preliminary explorations.[2] Edison's early career as an inventor was primarily devoted to telegraphy, but in 1876 his interests turned to telephony. The telephonic investigations led serendipitously to the invention of the phonograph in late 1877. This startling invention brought Edison fame, news attention, and a hectic demonstration schedule.

In July of 1878, Edison took a break from his work to accompany a scientific expedition to the Rocky Mountains to measure a solar eclipse. On August 26 he returned to his laboratory at Menlo Park, New Jersey, talking of producing electricity from the great western falls. The next day he began preliminary investigations into lighting while still pursuing other projects. On September 8 he took a day trip to Ansonia, Connecticut, to see William Wallace's new arc lighting system, with its powerful generator. According to a September 10 account, which appeared in both the *Sun* and the *New York Mail*, he was "enraptured."[3]

Back at Menlo Park on Monday, September 9, Edison excitedly began a series of experiments. He immediately wrote his first electrical lighting patent caveat[4] and wired Wallace to send one of his generators.[5] Shortly thereafter he granted an interview to a *Sun* reporter in which he claimed to have solved the problem of the incandescent lamp. The story ran on Monday, September 16, under the headline "Edison's Newest Marvel. Sending Cheap Light, Heat, and Power by Electricity."[6] By Tuesday, Edison's lawyer, Grosvenor Lowrey, had begun financial negotiations for what was to be the Edison Electric Light Company. A preliminary agreement, reached within a month, granted Edison a $50,000 advance for research and development. Gas stocks took a tumble, and newspaper stories about Edison's marvels proliferated. Edison's mail over the next few months was filled with letters that took his claims as a fact.

All this belief and all this activity hung on one man's premature and optimistic projection of success for an improvement on the unsuccessful work of many others. Although at this point Edison had likely conceived an overall approach that would lead to incandescent lighting and central power,[7]

T. A. EDISON.

Menlo Park, N. J., *Sept 13* 187 8

William Wallace
            Ansonia Conn

Hurry up the Machine
I have struck a big
bonanza   T. A. Edison

10 DN

Telegram, TAE to William Wallace, September 13, 1878 (17: 925).

887

The Sun.

MONDAY, SEPTEMBER 16, 1878.

EDISON'S NEWEST MARVEL.

SENDING CHEAP LIGHT, HEAT, AND POWER BY ELECTRICITY.

Illuminating Gas to be Superseded—Edison Solving the Problem of Dividing the Too Great Brilliancy from an Electric Machine.

Mr. Edison says that he has discovered how to make electricity a cheap and practicable substitute for illuminating gas. Many scientific men have worked assiduously in that direction, but with little success. A powerful electric light was the result of these experiments, but the problem of its division into many small lights was a puzzler. Gramme, Siemens, Brush, Wallace, and others produced at most ten lights from a single machine, but a single one of them was found to be impracticable for lighting aught save large foundries, mills, and workshops. It has been reserved for Mr. Edison to solve the difficult problem desired. This, he says, he has done within a few days. His experience with the telephone, however, has taught him to be cautious, and he is exerting himself to protect the new scientific marvel, which, he says, will make the use of gas for illumination a thing of the past.

Mr. Edison, besides his power of origination, has the faculty for developing the ideas and mechanical constructions of others. He visited the Roosevelt pianoforte factory in this city, and, while examining the component parts of the instruments, made four suggestions so valuable that they have been patented. While in the mining district of the West, recently, he devised a means of determining the presence of gold below the surface without resorting to costly and laborious boring and blasting. While on a visit to William Wallace, the electrical machine manufacturer, in Ansonia, Conn., he was shown the lately perfected dynamo-electric machine for transmitting power by electricity. When power is applied to this machine, it will not only reproduce it, but will turn it into light. Although said by Edison to be more powerful than any other machine of the kind known, it will divide the light of the electricity produced into but ten separate lights. These lights being equal in power to 4,000 candles, their impracticability for general purposes is apparent. Each of these lights is in a substantial metal frame, capable of holding in a horizontal position two carbon plates, each twelve inches long, two and a half wide, and one-half thick. The upper and lower parts of the frame are insulated from each other, and one of the conducting wires is connected with each carbon. In the centre, and above the upper carbon, is an electro magnet in the circuit, with an armature, by means of which the upper carbon is separated from the lower as far as desired. Wires from the source of electricity are placed in the binding posts. The carbons being together, the circuit is closed, the electro magnet acts, raising and lowering the upper carbon enough to give a bright light. The light moves toward the opposite end from which it starts, then changes and goes back, always moving toward the place where the carbons are nearest together. If from any cause the light goes out, the circuit is broken, and the electric magnet ceases to act. Instantly the upper magnet falls, the circuit is closed, it relights, and separates the carbon again.

Edison on returning home after his visit to Ansonia, studied and experimented with electric lights. On Friday last his efforts were crowned with success, and the project that has filled the minds of many scientific men for years was developed.

"I have it now!" he said, on Saturday, while vigorously turning the handle of a Ritchie inductive coil in his laboratory at Menlo Park, "and, singularly enough, I have obtained it through an entirely different process than that from which scientific men have ever sought to secure it. They have all been working in the same groove, and when it is known how I have accomplished my object, everybody will wonder why they have never thought of it, it is so simple. When ten lights have been produced by a single electric machine, it has been thought to be a great triumph of scientific skill. With the process I have just discovered, I can produce a thousand—aye, ten thousand—from one machine. Indeed, the number may be said to be infinite. When the brilliancy and cheapness of the lights are made known to the public—which will be in a few weeks, or just as soon as I can thoroughly protect the process—illumination by carbureted hydrogen gas will be discarded. With fifteen or twenty of these dynamo-electric machines recently perfected by Mr. Wallace I can light the entire lower part of New York city, using a 500 horse power engine. I purpose to establish one of these light centres in Nassau street, whence wires can be run up town as far as the Cooper Institute, down to the Battery, and across to both rivers. These wires must be insulated, and laid in the ground in the same manner as gas pipes. I also propose to utilize the gas burners and chandeliers now in use. In each house I can place a light meter, whence these wires will pass through the house, tapping small metallic contrivances that may be placed over each burner. Then housekeepers may turn off their gas, and send the meters back to the companies whence they came. Whenever it is desired to light a jet, it will only be necessary to touch a little spring near it. No matches are required.

"Again, the same wire that brings the light to you," Mr. Edison continued, "will also bring power and heat. With the power you can run an elevator, a sewing machine or any other mechanical contrivance that requires a motor, and by means of the heat you may cook your food. To utilize the heat, it will only be necessary to have the ovens or stoves properly arranged for its reception. This can be done at trifling cost. The dynamo-electric machine, called a telemachon, and which has already been described in The Sun, may be run by water or steam power at a distance. When used in a large city the machine would of necessity be run by steam power. I have computed the relative cost of the light power and heat generated by the electricity transmitted to the telemachon to be but a fraction of the cost where obtained in the ordinary way. By a battery or steam power it is forty-six times cheaper, and by water power probably 95 per cent. cheaper. It has been computed that by Edison's process the same amount of light that is given by 1,000 cubic feet of the carbureted hydrogen gas now used in this city, and for which from $2.50 to $3 is paid, may be obtained for from twelve to fifteen cents. Edison will soon give a public exhibition of his new invention.

888

"Edison's Newest Marvel," New York Sun, *September 16, 1878 (94: 354).*

he would soon abandon the specific approach of thermal regulation of the lamp upon which he had based his immediate claim. It was to be more than a year before he had a working light, and an additional year before a full system was ready. Yet Edison's announcement was taken as credible. His correspondence from this period shows how the meanings people attributed to him were embedded in specific and well-developed systems of communication that made his work seem credible.[8]

Scientific Friends and the Public Culture of Science

The earliest correspondence concerning Edison's new interest in light and power[9] was an exchange with George Barker dated September 5 and 6, the purpose of which was to arrange the aforementioned visit to Ansonia.[10] Barker was a physicist friend who had been part of the summers' solar-eclipse expedition.

Edison's announcement of his perceived breakthrough changed Barker's relationship to him. Barker, who previously had written as a peer requesting aid or discussing possibilities, became a supplicant. On September 16 Barker enclosed a clipping of the *Sun* article in a letter asking whether Edison had any new items available to display at a lecture Barker was to give in January.[11] On October 23 Barker wrote expressing disappointment and upset at not being able to exhibit lamps, as though he assumed the lamps were already working and available for exhibit.[12] In early November Barker renewed the request for the lamps and offered to postpone the lecture if he could have them.[13] He later wrote Edison with an account of the lecture, which he said had gone well despite the lack of the demonstration of the Edison light.[14] This sequence of letters bespeaks the existence of a well-established genre of public lecture, dating back to colonial times.[15] These lectures as education and entertainment depended on the exhibit of the latest wonders. In the nineteenth century there was some move to institutionalize such lectures and to ensure that speakers were legitimate authorities. Barker, a professor at the University of Pennsylvania and a public figure, was clearly part of this system of public edification.[16] Lecturers had to keep in touch with inventors in order to gather material on the wonders that held public attention.

The dependent relationship between the lecturer and the inventor is revealed in Barker's letter of October 23, 1878. Barker begins by addressing Edison as Chevalier de la Legion d'Honneur (referring to Edison's recent award) and thanking him for some information. Then he describes

Barker
Ansd nov 3, 78

3909 Locust St.
Philada. Oct. 23d. 1878.
My dear Dr. Edison: —
(Chevalier de la Legion d'Honneur)
          Your letter of the 21st
I received this morning). I am
glad to know what you write
concerning the business management of the Electric Light since
now I can refer all who apply
to me to Mr. Lowrey. I gave my
card a day or two ago to Prof.
Chauvenet of St Louis, who came
to me seeking just this information.
     What you say about the use
of your new light for my lecture
quite upsets me. You will remember that before I consented to
give the lecture, I came to Menlo
and saw you about it. And that
it was upon your promise to let

Letter, George Barker to TAE, October 23, 1878 (17: 981).

his consternation at Edison's denial of a demonstration lamp. The petition for aid goes on for two handwritten pages and ends plaintively:

Not to have one of your lights there at my lecture after all I have promised, places me in a position before this community which I would rather lose my right-hand than occupy. Only two weeks ago and you wrote that you would try and come over to the lecture. Now you dont know whether you can loan me even a light for the occasion. I beg of you not to desert me now. Do let me show something that represents your new invention.[17]

Other letters expressing interest in lectures on or by Edison came from a Professor Morton of the Stevens Institute of Technology (Hoboken, New Jersey), from John Blattau, the superintendent of German Catholic schools in Philadelphia, from a Professor Charles Suley of New York, and from one David Olegar.[18]

Throughout the development of incandescent light Edison was under pressure from the press, lecturers, and financial backers to have something demonstrable, and the Christmas 1879 spectacular light show, illuminating the entire grounds of the Menlo Park laboratory, became a major public moment. Such demonstrations create an aura of the spectacular and celebrity surrounding inventors and their work, an aura that Edison both exploited and had to live up to.

Patents, Finances, Corporations, and Cities

Another letter dated September 16, 1878, also mentioning the *Sun* article of that day, suggests another document-circulation system that pervades Edison's correspondence: the patent system. (Patents tie the inventor's projection of a workable object into the legal system, providing the inventor ownership rights in courts and contractual actions.) This letter, from one A. B. Williams, introduces Charles A. Shaw as a potential patent manager for Edison's consideration.[19] The letter is accompanied by a publicity brochure for Shaw based largely on a biographical article that had appeared in *Frank Leslie's Magazine.* In addition to the formal legal correspondence concerning the filing of patents and caveats, Edison received a number of letters offering to buy, sell, manage, or otherwise trade in patents by Edison or the correspondent.[20]

Patents, as valuable intellectual properties have potential meaning within financial markets, which have their own networks of document circulation. A letter from Hugh Craig, a schemer and developer who Edison knew from the telegraph industry, provides a window on this world, where

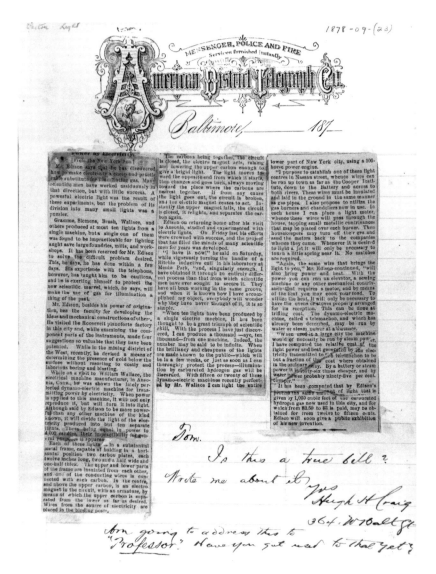

Hugh Craig's note asking for Edison's confirmation of the invention, September 23, 1878 (17: 935).

any suggestion of a new Edison invention promised wealth. Craig's letter, dated September 23, 1878, consists only of a clipping of the *Sun* article and this note:

Tom, Is this a true bill? Write me about it.[21]

This was the first of many letters Edison received (including one from the broker J. G. Kidder and one from the telegraph executive George Walker[22]) asking about investment or directly making offers. Another, from one W. C. Miller, describes a meeting Miller had with "several men of ample means" at which "a desire was expressed on the part of two or three of the best of them that they might know at an early day as to your business plans on the light."[23] N. Stucker and Gerrit Smith, whom Edison knew from the telegraphy business, wonder "Do you think telegraphers will be given the inside track?"[24] By this time, however, negotiations had been almost completed with the group of favored and high-powered investors who had been given the inside track the day after the first announcement in the *Sun*.

This intense activity betokens the financial markets and network of financial information that had been developing over the nineteenth century, as well as the particular link that had been made between financial markets and technological developments in the latter half of the century, beginning with the development of railroads and gas companies. These enterprises had the added characteristic of being large and dispersed endeavors, creating new kinds of corporate management and massive financial backing with tremendous rewards to be earned. Communications within the financial markets reflected these new realities.

One poignant example from Edison's correspondence of the way technological intelligence interacted with distant and widespread financial markets was a letter from a self-described "elderly woman in feeble health." She wrote on November 5, after gas stock prices had fallen in response to Edison's confidence in electricity. After a number of financial reverses, this woman had invested her small inheritance, upon which she depended, in gas. In desperation she implores: "I write this letter to beg you will write to me immediately, that you are sure, if you are sure of the electric light superseding the use of Gas and consequently lessening the value of Gas stock. So that I can dispose of my stock."[25] As in the financial markets of our time, everyone wanted the inside word.

The new, large corporations also meant that anyone wishing to be part of industrial activity had to become attached to a large organization. Edison received letters from small businessmen who offered to become local representatives or managers for electrical power franchises in Clarksville, Tennessee; Savannah, Georgia; Cheyenne, Wyoming; Buffalo,

New York; and Chelmsford, Massachusetts.[26] They wished to step out of the limited and weak networks of small business and into the powerful network they assumed Edison would form. Clinton Ball, a former clothes wringer manufacturer from Troy, New York, wrote a six-page letter on his bankrupt company's stationery. Ball had made a special trip to Menlo Park to offer his services, but Edison had been busy, so Ball had returned home and decided to tell his story in a letter. Having been cheated by his former partner, he was now out of work, and he wished to become Edison's agent.[27]

The new technology also had important meanings for municipalities that wanted to be in the forefront of development. Within two months of the *Sun* article, Edison had inquiries from the Louisville's board of aldermen,[28] from the mayor of San Francisco,[29] and from the city fathers of Innsbruck.[30] An inquiry about the possible use of electric light in mines as a safer alternative to gas was in a similar vein.[31]

Narrative Heroism and Material Accomplishment

Although the generally brief letters to Edison largely conform to the conventions of letter writing in nineteenth-century America, they suggest networks of texts to which the writers hope to connect Edison and his new invention. In each of those networks Edison is projected as a potentially larger-than-life figure who will shape the future. The writer hopes to gain some benefit from playing an intermediary role, or even from merely touching the hems of greatness.[32] One larger textual cultural context for all these attempted linkages is the narrative of heroic achievement, particularly scientific heroic achievement, especially as advanced in newspapers, periodicals and books of the period.[33] Indeed, several amusing parodies of this heroic story appeared in the press, as Edison was at various times proclaimed to have invented cures for most of the world's ills. A note from George Barker included a clipping from a California newspaper reporting that Edison was "said to be at work on an invention to keep Pantaloons from bagging."[34]

Edison understood how to publicize his projects in the changing press of the day—that is why the *Sun* story created such a splash. Edison's fame was not an accident; it was the consequence of many historical forces, which Edison had come to appreciate. In later chapters I will examine other cultural stories, the document systems by which they were fostered, and the human networks through which they were circulated, so we can gain a detailed view of how representations spread from one domain to another and translate power within one network to power in another and of how Edison and his colleagues actively fostered the meanings that would make incandescent light and central power everyday realities. Ultimately, Edison's ability to take an enduring place within each of these social narratives was dependent on his being able to produce a viable material technology, thereby giving solidity and functional support to the discourse. But during the years when he was working to make good on his promises, Edison's presence in the relevant discursive systems won him the credit, the support, and the leeway he needed to bring the work to completion. Edison's communications also began to build the social structures and relationships that would bring incandescent light into being, even as those same relationships made Edison responsible and accountable for producing the technology. Promises, hopes, projections of desire, and pressures for fulfillment all depend on imaginable narratives framed within recognizable social meanings. Edison's promises made a lot of sense because people had the terms to attribute sense to those words and to attribute to Edison the man the power to make things happen.

II

Establishing Meanings in Evolving Systems

2

The Public Stage of News

In the second half of the nineteenth century the American newspaper underwent a major transformation.[1] Responding to the telegraph, the railroad, the rotary press, urban growth, and the emergence of a national economy, journalism turned from small circulation partisan commentary to mass-circulation retelling of the happenings of the world. It became a new kind of stage upon which business and finance, local and international politics, urban life and technological advance, heart-rending feature stories and storybook adventures, were played out. Events that happened anywhere (from distant continents to the back rooms of Tammany Hall) or everywhere (like a climate of business and economy) or nowhere (like the promise of a new invention) could be made local and visible to millions of readers every day.

Thomas Edison grew up into this emerging world of news, just as he grew up into the emerging technological world of railroads and telegraphs. From his early teenage years, he participated in the communication industries, came to know their power, and learned how they worked as material technologies and as social forces. He understood the opportunities the new technologies offered for spreading knowledge, creating a presence, and making things happen. Using that knowledge, Edison took advantage of the opportunities the press afforded him. Particularly when the invention of the phonograph brought attention to him and his invention factory at Menlo Park, he parleyed one series of headlines and interviews into another, and then another, so that when he announced his interest in the electric light he had tremendous "media capital" to draw on.

Networks of Information

An anecdote Edison later recounted indicates how well, as a youth, he took advantage the transformative power of the new technologies of the middle of the nineteenth century (Lathrop 1890, pp. 425–435). This "luck and pluck" story of his earliest adventures in capitalism was likely embellished over 30 years; nonetheless, it suggests the adult Edison's understanding of the communications world he learned to navigate.

By 1861, young Edison was hawking newspapers and snacks on the Grand Trunk Railroad, which had just recently been extended from Detroit to his home town of Port Huron. The railroad and the telegraph had given the news business a tremendous boost, allowing the instantaneous transmission of news reports across the country and the marketing of daily newspapers over an extended region. The railway newsboy was on the cutting edge of a new information age.

Edison experienced the power of fast-breaking headlines to sell newspapers to information-hungry consumers over a geographically dispersed market. Selling papers at each stop and on the trains, Edison noticed that sales were directly related to the drama presented in the latest Civil War battlefield reports. He soon learned to gauge how many copies of a paper to stock by visiting the composing room to read the galleys of the front page. Upon previewing the particularly gruesome first reports of the battle of Shiloh that were to appear in the afternoon edition of the *Detroit Free Press* on April 9, 1882, Edison arranged credit with Wilbur F. Storey, the editor and owner of the *Free Press,* sufficient to purchase 1500 copies instead of his usual 200.[2] He then struck a deal with a telegrapher to wire news of the battle to be posted at each major stop on the train run. Increasing crowds gathered at each station to await these reports. Responding to the demand, the newsboy raised the price of the nickel paper to ten cents, then to 25. He cleared $150 in one day. The adult Edison, summing up his appreciation of the social and economic power of the new technology as a boy, commented that on that day he had decided to become a telegrapher.

Edison as an Apprentice to the News Industry

Through his early experiences on the railroad and then in telegraphy, Edison learned the ways of newspaper journalism. As a newsboy he talked with people at all levels of the *Detroit Free Press,* from the editor down to the compositor. He learned the business so well that he even published his

Tom Edison, newsboy and entrepreneur, in 1861, at the age of 14.

YESTERDAY'S

AFTERNOON EDITION.

THE GREAT BAT-TLE ON THE TEN-NESSEE.

The Fight Lasted Two Full Days.

ALBERT SIDNEY JOHNSTON KILLED.

BEAUREGARD'S ARM SHOT OFF

Gen. Prentiss, of Illinois, Taken Prisoner.

GEN. W. H. WALLACE KILLED.

GEN. W. T. SHERMAN WOUNDED.

COL. DAVID STUART WOUNDED.

The Merrimac Expected Down the First Favorable Day

COM. FOOTE'S OFFICIAL REPORT.

&c.,   &c.,   &c.

THE BATTLE ON THE TENNESSEE.

NEW YORK, April 9.

A special to the *Herald*, Pittsburgh *via* Fort Henry, 9th, 3:30 A. M., says: One of the greatest and bloodiest battles of modern days has just closed, resulting in the complete rout of the enemy, who attacked us at daybreak Sunday. The battle lasted without intermission during the entire day, and was again re-

A newspaper story that Edison marketed (Detroit Free **Press,** *April 10, 1862).*

own newspaper, the *Grand Trunk Herald,* printed on the train and carrying news of rail and telegraph workers up and down the line. He sold a few hundred copies each week, and an entire edition was sent to England after it came to the notice of the son of the great English railroad engineer George Stephenson.[3]

Later, as he developed into one of the most accomplished telegraphers of the time, Edison regularly transmitted news stories, and made the acquaintance of journalists. At age 19, while working as a telegrapher in Louisville, he sat in on late-night discussions between George Prentice, editor of the *Louisville Courier,* and a Mr. Tyler, who ran the local Associated Press office (Dyer and Martin 1910, volume I, p. 83). Two years later, living in Boston, he had his friend Milton Adams write up his minor improvements in telegraphy as a major breakthrough. After this story was published in the June 1868 *Journal of the Telegraph,* Edison was able to raise a little capital, which led him to give up his job as a telegrapher and spend full time inventing. When he set up shop as an independent inventor, he used the emerging trade press to publicize his services.

In the early 1870s, having moved to New York to invent for the major telegraphic and stock ticker companies, Edison gained the patronage of the major financiers of the telegraph industry. No longer dependent on spreading his fame through the press, he nonetheless publicized his inventions of the repeating and recording telegraphs and of duplex and quadruplex transmission systems in the technical and industrial press.

Edison, moreover, gained a taste of celebrity when on the evening of November 22, 1875, in the course of an experiment involving a vibrator magnet, he came across unexpected sparks with no measurable charge. He labeled the phenomenon the "etheric force." We now say that he observed electromagnetic radiation, but his measuring instruments were not sensitive enough to measure the high frequency of the charge's oscillation. Unaware that the phenomena had previously been observed by Joseph Henry but accurately predicting the potential for wireless transmission of signals, Edison went almost immediately to the press. The first stories appeared seven days after his initial observations, even as his experiments continued. Numerous stories in late November and December of 1875 lauded Edison's discovery of a new form of electricity, calling it "remarkable," and "wonderful" and discussing the findings in some detail, although a number of articles showed some caution by qualifying the results as only alleged.[4] The coverage was limited to experiments and claims, but some writers attributed special talent to Edison as a leading inventor.[5] Scientific opposition, however, soon led to Edison's work being discredited as misguided and overblown. Edison later regretted that he did not pursue this topic, for it might have lead him to include radio transmission among his many inventions. Nonetheless, the incident must have taught him something about the kind of publicity a startling announcement might evoke, and about the incredulity and ridicule that might result if public claims failed.

Publicity for the Phonograph

After the embarrassment of the "etheric effect" publicity, Edison was more careful in publicizing the invention of the phonograph in 1877. This remarkable device was to bring him true celebrity rapidly once it was announced, but the announcement had to wait half a year until Edison had a fully working prototype. He experimented on reproducing human speech as early as June, but he kept the work secret until the middle of November, although rumors of his invention reached the New York press. Only when *Scientific American* (November 3, 1877) suggested that the work of others might lead to recording and transmission of speech, did Edison's assistant Edward Johnson write a letter to the journal detailing Edison's invention.[6] The letter appeared in the November 17 issue, accompanied by a remarkably laudatory editorial calling the phonograph a "wonderful invention" and beginning in the most grandiose terms:

It has been said that Science is never sensational, that it is intellectual not emotional, but certainly nothing that can be conceived would be more likely to create the profoundest of sensations, to arouse the liveliest of human emotions, than once more to hear the familiar voices of the dead. Yet Science now announces that this is possible, and can be done.... The possibility is simply startling. A strip of indented paper travels through a little machine, the sounds of the latter are magnified, and our great grandchildren or posterity centuries hence hear us as plainly as if we were present. Speech has become, as it were, immortal.[7]

Two weeks later, around the time a *Scientific American* editorial praising Edison's phonograph mentioned two competitive claims of similar inventions,[8] Edison demonstrated a working model in the office of the journal. An account of the demonstration appeared in the December 22 issue, accompanied by further speculation about the possible uses of the device to record singers and legal testimony.[9]

Stories based on the *Scientific American* articles appeared in newspapers around the world.[10] A demonstration of the phonograph at the Cooper Union on January 17, 1878, kept the phonograph before the New York press, as did a public concert at the Fifth Avenue Baptist Church in which a phonograph and a telephone were used to transmit vocal music and dramatic orations.[11]

As he had with the "etheric effect," Edison worked actively to present his invention advantageously, but here the strategy included withholding information until an opportune moment, to avoid the potential embarrassment of premature announcement and the danger of leaking secrets to competitors. On the other hand, the necessity of claiming priority in the face of competition drove Edison to rapid and complete disclosure and demonstration at the right moment. Once he had gained public notice, Edison then sought to extend the publicity through demonstrations, especially the unusual media events of phonographic concerts.

From the first coverage of the phonograph, the press showed a heightened interest in Edison, in the technical detail of his accomplishment, in the effect of the invention on daily life, and in spectacular demonstrations of the technology. Edison's inventing the phonograph was reported as an exciting and life-changing event. The next round of publicity, which began in late January, brought a further personalization of the coverage. The livelier daily newspapers and popular periodicals began to show interest in Edison himself and to treat Menlo Park as a site of pilgrimage. Edison soon learned how to capitalize on these new publicity opportunities by providing the press with what they desired. As a result, he became the genial and loquacious Wizard of Menlo Park.

An Odd Man, Peculiarly American

On January 12, 1878, the *New York World* printed a different kind of article about Edison: an interview conducted at Menlo Park. This article exhibits elements that were to run through others published in the next few months, although it does not reach the same heights of hyperbole. It begins with a story of a pilgrimage to an obscure corner of the world:

> Menlo Park, N.J., January 11—I came here yesterday with a party of gentlemen interested in telegraphy and spent the entire day in the workshop and laboratory of Mr. Thomas Edison, inventor of. . . . Menlo Park is a queer-looking place, rather it would be queer looking if it were not in New Jersey.[12]

Edison leads the party through his laboratory, explaining and answering questions about all his marvelous inventions. The reporter shows interest not only in the inventions but also in the inventor:

> Mr Edison is in many ways an odd man and in every way peculiarly American. To him nothing appears impossible and in attempting to solve any given problem he

puts the discoveries of science into every conceivable combination in his effort to obtain the result sought for.

These sets of contraries—remarkable yet practical, scientific yet down to earth, a dreamer yet hard working and full of perseverance—set the themes that ran throughout Edison's portraiture as he was elevated from an odd man to a wizard.

Edison spent much time with curious visitors. Even the uninvited received tours and tolerance. Article after article boasts, often as a headline, "An Hour with . . . ," "Four Hours with . . . ," "An Afternoon with . . . ," "An Evening with . . . ," or "Edison 'At Home.' A Ninety Minutes Interview by Telephone."[13] In the middle of May, stories of Menlo Park's being overrun by visitors and bores began to appear—for example:

Mr Edison is a pleasant, genial, self-sacrificing gentleman, never wishing or willing to say no to anyone, but it must be borne in mind that he is the greatest of our present inventors, and that his brain is constantly filled with as yet unsolved problems. . . . To keep him hours each day shouting and singing into a phonograph for the amusement of unscientific people is the refinement of cruelty and a gross injustice to science as well as him.[14]

In the middle of January, about when Edison gave an interview to the reporter from the *World,* he opened himself up to even greater revelations to a writer from *The Phrenological Journal,* which was to publish a five-page biographical sketch of Edison in the February issue.[15] A phrenological examination was said to have revealed Edison as earnest, intense, vigorous, ambitious, and reflective, and filled with a sense of duty:

The mental element predominates greatly in the physical constitution, and so contributes to his energy of thought and facility of action. His brain is broad between the ears, indicating he possesses force of character in a high degree, which is exhibited in his disposition to be doing, to find opportunities of the employment of his hands or mind or both.

Edison then provided extensive biographical information which established the details for this article and many biographical sketches that were to soon follow. Words, phrases, and facts from this article were to echo not only through the press but also through numerous "life and invention" books.[16]

An interview in the February 22 *Sun* introduced and developed themes that were to characterize subsequent stories.[17] A subhead calls Edison "a man of thirty one revolutionizing the whole world." The theme of a pilgrimage to a strange place is elaborated with descriptions of the lightning rods and telegraph poles standing sentry about the laboratory. Traveling

352

THE PHRENOLOGICAL JOURNAL

AND

LIFE ILLUSTRATED.

VOL. 66.   1878.

NUMBER 2.]   *February, 1878.*   [WHOLE No. 470.

THOMAS A. EDISON,

THE ELECTRICIAN AND INVENTOR.

THIS portrait indicates an organization of remarkable activity. The mental element predominates greatly in the physical constitution, and so contributes to his energy of thought and facility of action. His brain is broad between the ears, indicating that he possesses force of character in a high degree, which is exhibited in his disposition to be doing, to find opportunities for the employment of his hands or mind or both. He has courage to work out his plans and purposes when obstacles present themselves. Earnestness characterizes his efforts in any chosen direction; and this, coupled with his mental intens'ty, renders him very thorough-going. Whatever engages his attention, so far as to make him a worker either as principal or assistant in its

"Thomas Edison," **Phrenological Journal,** *February 1878 (94: 105–108).*

past the front office and up some stairs, the reporter finds "an immense laboratory, filled with electrical instruments. A Thousand jars of chemicals were ranged against the walls. A Circle of Kerosene lamps was smoking on an empty black forge. The chimneys were the essence of blackness. . . ." In this strange land the reporter saw "Professor Edison," who "looked like anything but a professor, and reminded me of a boy apprentice to an iron molder." At the heart of the mystery is a plain, open, simple, garrulous, playful fellow—an extraordinary common man. He demonstrates the phonograph, first with gravely intoned versions of nursery rhymes, then with whistling, coughing, and sneezing. He demonstrates how one voice can be recorded on top of another. He then slips into the comic and the fantastic, playing the sounds of a "mad dog" in reverse and scheming about giving the Statue of Liberty a voice that could broadcast advertisements to Manhattan. He concludes the interview by telling a self-deprecating story about the discovery of the phonographic principle and playing a recording of a comical ditty.

In this story, versions of which were reprinted in newspapers throughout the United States and England, Edison emerges as a remarkable magician dabbling in mysteries beyond the comprehension of most mortals, but also as a down-to-earth practical man who does not take himself or his work too seriously. He appears a solid American type, easily understandable and likable. The playfulness introduced into the interview by Edison was to infuse much of the subsequent publicity, making the man, his inventions, and their fantastic promise accessible to ordinary readers of the new mass publications. His garrulousness and openness gave the reporters material with which to feed the fascination with his character.

Napoleon, Wizard, Genius

Over the next few months, as further details of Edison's life and character were supplied by his associates and acquaintances, reprints, rewrites, and pastiches of earlier articles made him a major presence in newspapers throughout the English-speaking world. The hyperbole escalated rapidly. On March 10, 1878, the *Sun* called him "the Napoleon of Science" (27: 748–749; 94: 119). On March 22 the *Paterson Daily Press* ran the phonograph story as "This Beats Fairy Tales" (94: 141). On March 29, the *New York World* headlined "That Wonderful Edison" (94: 147). By April, Edison was regularly described in supernatural terms. That month, articles appeared under the headlines "Edison the Magician,"[18] "The Wizard of Menlo Park,"[19] "A Wonderful Genius,"[20] and "The Inventor of the Age."[21] Late in the same month, when Edison appeared in Washington before a meeting of the National Academy of Science and then before both houses of Congress, the Washington and Baltimore papers reported his demonstration and carried interviews under such headlines as "The Man Who Invents,"[22] "Genius Before Science,"[23] and "Astonished Congressmen. Edison, The Modern Magician, Unfolds the Mysteries of the Phonograph."[24]

The Humor of Genius

The comic playfulness Edison injected into interviews fostered even more headlines backhandedly praising his accomplishments. On March 27, 1878, New York's *Daily Graphic* ran a front-page cartoon illustrating "Awful Possibilities of the New Speaking Phonograph,"[25] which took its scenarios from the February 22 *Sun* interview. On April 1 the *Daily Graphic* ran a long April Fool's Day story headlined "The Food Creator" and including the subheads "Edison Invents a Machine That Will Feed The Human Race" and "Manufacturing Biscuit, Meat, Vegetables, and Wine out of Air, Water, and Common Earth."[26] The next day, the *Graphic* published a slightly less hagiographic interview, headlined "The Papa of the Phonograph."[27] On April 16 the *New York Evening Post* published a short story called "The Phonograph" (94: 166), and on April 20 the *New York Weekly* published another called "The Phonograph as Detective" (27: 796–797). Both fictions have similar plots in which a deceived wife discovers her husband's duplicity through an inadvertent recording. In May, *Frank Leslie's Illustrated Weekly* ran a page full of cartoons having to do with the phonograph,[28] and the *Graphic* published several anecdotes about Edison.[29] The *Graphic* reported that a young savant had approached Edison with ideas for chemical inventions for the spirit that would help develop consciousness in sleep, allowing us to use our dreams to profit; Edison was said to have been mildly encouraging but to have warned the young man to keep a doctor nearby as he experimented on himself.[30]

Although these and many similar stories treated Edison with awe (or made sport of that awe), reports of interviews all indicate that Edison was open, congenial, expansive, and even lighthearted with the interviewers. He discussed grand possibilities for his inventions in music, business, and education; he played tricks and jokes by making comic recordings and juxtaposing incongruous pairs of recordings; he discussed fishing, told anecdotes, invented devices to rock cradles automatically when the baby cried,[31] and endlessly demonstrated his machines.

Edison invited the publicity, doing everything he could to provide good stories for reporters. He never seemed to turn a reporter away, and he went out of his way to express gratitude for the publicity he was granted. After a bit of cajolery, he complied with William Croffut's request for a testimonial letter, which was printed in a facsimile of Edison's handwriting in the *Daily Graphic* on May 16. After Edison was slow in drafting the letter, Croffut supplied a text for him to transcribe[32]:

I feel an inclination to thank you for the pleasant things you have said about me and the Phonograph in the Graphic. Your words and pictures have gratified me the more, because I had long since come to look upon your paper with pride and to regard such an illustrator of daily events as one of the marvels of the age. I am able to report to you that I am constantly increasing the sensibility and power of the Phonograph. I feel certain that it will soon justify all the hopes of its friends. By the way, Croffut's April-first hoax concerning my alleged food machine has brought in a flood of letters from all parts of the country. It was very ingenious.[33]

Edison had become a public personality in the newspapers. Dry reports of his inventions had given way to presentations of the man—a man of per-

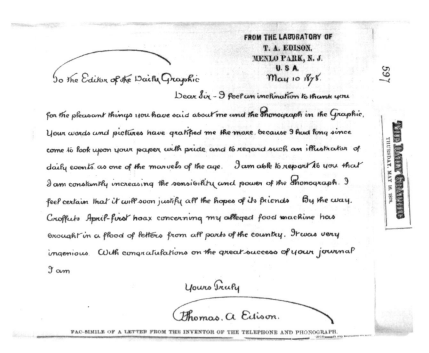

Testimonial from Edison solicited and drafted by the reporter William S. Croffut (New York Daily Graphic, May 16, 1878 (94: 197)).

sonality, vision, stories, and good humor. He was someone worth listening to, someone who could tell us about the future, someone trustworthy in his down-home judgment. He was someone who grabbed headlines and appeared as a prophet.

Changes in Journalism

Changes in journalism over the past few decades had created opportunities for such stories and for the emergence of popular heroes in the press. By 1860, steam and rotary printing allowed 25,000 impressions to be made per hour, whereas in 1800 no more than a few hundred could be made in that time. Growing urban populations provided readership for cheap, mass-produced daily newspapers. Telegraph and rail allowed the communication of distant news to fill the pages.[34] And rail distributed the papers over economically expanding regions, centered on the growing cities. Boosted by the hunger for news generated by the Civil War, newspapers became substantial businesses, independent of the political parties that had previously provided support and readership. In 1830 there were only 65 daily newspapers in the United States, with a total daily circulation of perhaps 100,000. By 1870 the number of dailies had increased to 387, with a total daily circulation of 3.5 million. By 1900 there were more than 2300 dailies, with a circulation of 15 million. Early in the century the best-selling paper sold perhaps 2000 copies daily. By the 1870s the top New York papers were selling more than 100,000 a day, and by the turn of the century more than half a million.

To attract this widening readership, newspapers invented new kinds of journalism. The New York press led the way. The papers that gave Edison the most play—the *Tribune*, the *Sun*,[35] the *Herald*,[36] and the *Daily Graphic*—were the most innovative. The *Tribune*, first under the editorship of Horace Greeley and then under Whitelaw Reid, helped pioneer the popular "penny press" in the 1830s. The *Sun*, under the editorship of Charles Dana, opened journalism up to colorful human interest stories. The only rules Dana imposed were that writing ought to be lively and interesting. The *Graphic*, edited by David Crowley, devoted almost half of its pages to illustrations and sought out subjects that lent themselves to visual drama and emotional appeal. The *Herald*, under James Gordon Bennett Sr. and then under James Gordon Bennett Jr., developed exotic, adventurous stories that often stretched over many months, taking the readers to unusual places and introducing them to unusual people.[37] At times these adventures were directly sponsored by the newspaper to produce circulation-building stories, such as Henry Morton Stanley's expedition to find David Livingstone.

A New Kind of Hero

The new journalism found in Edison good material, and the papers made a special kind of hero out of him. His laboratory became an exotic locale where the great man carried out strange adventures into the unknown. Story after story began with a reporter journeying to the exotic locale, much like an adventurer entering into a strange culture.[38] At the same time, Edison could be portrayed as an interesting, homey, informal, loquacious man, willing to share amusing anecdotes and great ideas with other Americans excited by the new technological future. And he was represented as so great a man, doing such important and exotic things, that he did not have to take himself seriously, even as he seemed almost to dabble in the occult. His accomplishments and his promises seemed so outrageous, yet so believable, that they invited a kind of humor—a humor in which he participated. As an uncommon common man—informal and democratic, yet partaking of magical genius—he was the perfect media star of his era of democratic progress. And he played the part very well.

Edison's vatic voice had such a public presence that what he thought and what he was working on became matters of great public interest. His opinions on all kinds of things (even his idea of a good diet) seemed worth reporting.

In May and June of 1878, new stories and interviews publicized Edison's light-measuring device, the tasimeter. Under ordinary conditions the tasimeter likely would not have gained much public notice, and it was rapidly forgotten after. But as the latest production of the newly proclaimed Wizard, it seemed very important, especially when Edison used it on an expedition to Pike's Peak to observe a solar eclipse. Stories of this expedition kept Edison in the public eye throughout the summer, under such headlines as "Edison Whips Old Sol"[39] and "The Gains of Genius."[40] On returning to Menlo Park, Edison used his favored position in the press to advance his new plans for central power and incandescent lighting, hatched on the journey.

Publicity and Finances

In order to beat to market the many competitors who had been working on arc and incandescent lighting, Edison needed to mount a major effort

The World.

NEW YORK, TUESDAY, AUGUST 27, 1878.

TOM EDISON BACK AGAIN.

THE MARVELLOUS MAN RETURNS TO HIS WORKSHOP FULL OF PRAIRIE BREEZES.

HOW HE TRIED THE TASIMETER IN A HURRICANE-BESET HEN-HOUSE AND KILLED A DEAD JACK RABBIT.

The wonderful Edison has returned to his workshop in Menlo Park, after an absence of over a month. He went to the West for scientific experiment, and to learn if the corona of the sun in eclipse gave forth heat. He has come back to talk about animals that stand still to be shot at, trout that leap out of the water to bite at an empty hook, broncho ponies that "buck" and mules that kick. How long he staid in his house when he reached Menlo Park yesterday no one knows, but early in the afternoon a WORLD reporter found him sitting in the office of his workshop, head over ears in business. It took a long time for the reporter to wake him up, and then Mr. Edison laughed and handing him a letter said, "Read that."

The letter was dated from Cincinnati and began, "Much respected sir." It was from the inventor of a fluid resistance neutralizer and of a gravi-motor. After the inventor had described his apparatus he wrote:

How are these for High, Low, Jack and the G. ? I shall be greatly disappointed if I shall not hear a favorable reply from you at your earliest poss. convenience. I have been financially Kicked and Cuffed about this burgh during 2½ yrs. to get these before the world. But I "never give up the Ship." God is my Help and my exceeding great joy. I hope He is the same to you. He will, without fail, be so to you, if you will seek the Lord, while He may be found, and call upon him while He is near. As I am suffering for the necessaries of life, please lend me $10 00. It shall be returned to you by God, through me, more than a thousandfold and that shortly. You will greatly oblige, for I have a family

YOURS IN CHRIST.

"There could be an awful good story made out of the letters I get," said Mr. Edison, meditatively; and while he meditated the reporter looked at him to see what change his Western trip has made in him. His face was tanned, he had got a new straw hat, his shirt was tolerably white, he had an old, long alpaca coat with bulgy pockets, his hands had lost the grime of the workshop and had taken on the bronze of the Western sun, he had a pair of new pantaloons, and he hadn't combed his hair any more than he had before he left home.

"Mr. Edison, how did you like your trip?" asked the reporter at last.

"It was bang-up," he answered quickly, lapsing into the unscientific slang common to all sections of the country. "It was bully. I never saw such a country in all my life. That's the place to go to. What with following trails, and tumbling down precipices, and riding over alkali deserts and keeping cool at 125 degrees in the shade, a person couldn't help enjoying himself."

"But what about your scientific experiments?"

"After we got through with our scientific experiments we went hunting. I killed ever so many deer, and jack rabbits, and sage hens, and other things."

"How many deer did you kill?"

"I killed one and shot at another. There are lots of antelopes there, but you can't get a shot at them under five hundred yards."

"How did you travel when hunting?"

"I rode a broncho pony and had a rifle slung at my back. Those broncho ponies are wonderful animals. I was jogging along one day when the head pony of the train started on a run, and all the others started after him. Mine kicked and reared and got himself all in a heap somehow while I held the reins. So I let the reins go and hung on to every part of him I could get a hold on, and away he went like a shot. I didn't have time to enjoy any of the scenery w— travelled through for some time."

Just then Mr. Edison's eye caught a scientific periodical that had been placed on his desk, and he was immediately lost in its contents. A moment afterwards a handsome little girl dressed in lilac silk danced into the office, and slapped him familiarly on the shoulder. He looked at her for a moment as if trying to recollect who she was and then exclaiming, "Why, Dot, is that you?" clasped her in his arms. It was the first time he had seen his little girl since his return for, as she said, she had been at Whitestone. Then after kissing her half a dozen times he slid her from his knees and took to his book again.

Dot turned to the reporter for company. "I don't care," said she, "if I do come here with this dress on. It isn't my best one. I've got one dress so covered with embroidery that you can't see the muslin, and you can't see where it is sewed on. Sharpen this pencil, please, I want to write." She bent herself over a piece of paper, and shortly afterwards entered young Tom Edison—about three years young. He seated himself in a chair and taking a clay pipe from a table began smoking to "draw" upon it.

Mr. Edison, having finished his article, looked up and, after admonishing his son not to swallow the tobacco extract, said to the reporter:

"I went one day to a telegraph station out there and asked the operator—he was a funny fellow—if there were any jack rabbits about. 'Oh, yes,' said he, 'plenty of them.' And, coming out of the station, he looked around, with his hand up to his eyes, and at last said, 'There's one, off there.' There was one, sure enough. I unslung my rifle and blazed away at him, but missed. Then I fired again and again and again, but without use. So I went up to the rabbit to hit him on the head and I found he was stuffed."

"But that wasn't as bad," continued Mr. Edison, "as the trick that was played on a Herald correspondent. He and I had been firing at tin cans—you know emigrants on the Union Pacific eat canned meat, and the road is littered with the cans, and we had used up all our cartridges. So a man gave the correspondent his rifle. It was a big old Springfield musket turned into a breech-loader. The newspaper man fired at a plate I had set up, but he didn't hit it; so he tried again until he was so sore in the shoulder that the piece knocked him over. Then it was found that the barrel of the rifle was twisted and that he had been shooting around a corner all the time."

"But the fishing was the thing," went on Mr. Edison, rising and taking a bite from a plug of Rocky Mountain tobacco. "You could throw a crumb of bread into the water, and the river was in a froth right away with trout after that crumb. If you held a hook without any bait a foot above the water the trout would leap at it. Beauties they were, too."

"That must have been glorious sport," said the reporter.

"Oh, it was too monotonous," answered Mr. Edison. "It was too much fish. We used to eat the best parts and throw the rest away."

"What do you think is the best way of riding out there?" asked the reporter.

"Well, the best way is to strap a board under you and have some one throw you on the pony's back. But the mule-packing beat me. They would load a mule until he had a burden four times the size of himself. He could stand it well enough, but if the branch of a tree struck the pack over went mule and all."

"Did you get any new ideas out there, Mr.
E——— ?"

supported by major funding. Arc lighting already was being installed on the streets and large public buildings of a number of cities. Moreover, incandescent lighting, once the problem of the filament burning out had been solved, still needed to go up against the rapidly expanding and highly profitable gaslight industry.

To get funding, Edison used the opportunity of the series of interviews granted to him on returning East. Within two days of his return to New Jersey, major interviews were run in New York's *Tribune, Herald, Graphic, Sun,* and *World*.[41] In these interviews he was expansive, telling tales of his adventures out west and of his plans for new projects. A few days later, when he traveled to William Wallace's laboratory to examine an electrical generator, he brought a reporter from the *Sun* along with him, who reported: "Mr Edison was enraptured. He fairly gloated over it. He ran from the instruments to the lights, and from the lights back to the instrument. He sprawled over a table with the simplicity of a child, and made all kinds of calculations."[42] Six days later, Edison announced to the *Sun*'s reporter that he had solved the problem of incandescent light.[43] This is the story that excited the outpouring of mail mentioned in chapter 1.

Edison's use of the public stage to gain public attention for his inventions culminated when he announced the perfection of the incandescent light in such a way that it seemed the fulfillment of many social needs and dreams. As we shall see, he was to continue to hold this stage throughout the development of the light, gaining needed support at crucial moments. While Edison had to represent himself in many arenas, the public stage of the press drew all his representations together in a single presence of the great Edison who could command resources, credit, and credibility while he developed the technology. Then, once the technology was viable and demonstrated, the press amplified local demonstrations into international events. As light and power became everyday facts of life, the press let the story sink into the ordinariness of a non-story.

3

Finances for Technological Enterprises

During Edison's life, technology was changing America's way of doing business. The railroad led the way, requiring new forms of organization, communication, and financing; then came the steamship, the telegraph, and the telephone.[1] Transportation and communication technologies provided the means for other industries to spawn large national corporations serving national markets. As the corporations grew, monopoly control beckoned, and one of the strongest tools to gain monopoly was patent control of the technologies that made the businesses possible. Thus the emergence of a new technological industry entirely controlled by fresh patents would be a tempting prospect to the major investors from whom Edison sought backing.

Changing Industry and Changing Finances

Previously businesses served local communities. Most communication with clientele and suppliers was carried on face to face. Written documents provided only first introductions and skeletal records. Even advertisement was minimal—mostly informational reminders of local businesses already familiar to the consumer. More flamboyant advertising was only for the transient, such as the purveyors of patent medicines.

Within these earlier small businesses, coordination, information exchange, negotiation, and problem solving were accomplished in person, often under the direction of a single owner-manager or a small group of partners. However, as the railroads created possibilities of national market, national goods, and national media, businesses grew in size and geographic range. Writing became more important and regularized to provide information, standardized and controlled procedures, and coordination of efforts of many different departments. The new needs of large organizations developed coordinately with new modes of communication and information production, storage, and distribution: filing systems, telegraphs and telephones, duplicating systems, typewriters. New genres of the memo, the business letter, and the report arose to regularize the communication functions within the new communication technology. (See Yates 1988.)

These new large organizations needed new forms of capitalization. Previously small businesses, depending on a few local investors, grew only as they generated their own resources. In the middle of the nineteenth century, corporations increasingly turned to the public financial markets, which themselves were changing in response to the new capital needs.

In the United States, capital markets were at first limited to government bonds funding the revolutionary war debt, and then the regional banking system (Stedman and Easton 1905; Warshow 1929; Wyckoff 1972). The only activities large and reliable enough to generate capital needs and generate sufficient trust were government, banking, and to a lesser extent insurance.

The first non-financial industry to be broadly financed on Wall Street was transportation, first with canal bonds and then railroad bonds. Railroad bonds and stocks proliferated and then consolidated as the railroads became organized into larger holding companies in the 1870s. Also in the 1870s the telegraph industry established market presence, followed by utilities in the 1880s. However, industrial equities still were thought highly speculative and were placed within an unlisted department in the New York Stock Exchange in 1885. The speculation was driven by Jay Gould and other stock manipulators, who orchestrated trust-building ventures. Frequently stocks were overcapitalized and corporations overextended in the rush to form large monopolies, with resulting price collapses, as in the National Cordage case in the early 1890s.

These expanding capital markets required new forms of communication to provide information on investments and to keep track of market news and prices. Finances became an increasing segment of newspaper reporting, and a specialized financial press emerged with the founding of the *Wall Street Journal* in 1889. Telegraphy also contributed to the new pace of financial information. The stock ticker was instituted in 1867 by the Gold and Stock Telegraph Company, which by 1871 had interlocking directorates with Western Union. (See Hotchkiss 1905.) Markets were turning from a few sleepy gentlemen sitting around in the back of a coffee house in the 1830s to the modern stock market of instantaneous communication.

It was into these new financial and corporate worlds that the electrical technology had to insert itself and gain meaning. To succeed in the new world of capital, the technology had to establish itself as a corporation, had to become an active symbol on the stock ticker, and had to produce news in the financial pages of the newspapers.

Edison and the New Technological Industries

This new world of large technology-driven corporations and finance was one that Edison had already become familiar with before he took on the project of incandescent lighting. Starting at age 16 he worked as a telegrapher on the rapidly expanding lines that stretched alongside the railroads. His tinkering on the telegraphic apparatus led to improvements that brought him into negotiation with the managers and owners of the telegraphic companies. By 1868, he became the Boston agent for the new gold ticker, transmitting the latest prices from brokers to banks and other enterprises dependent on current financial news. Soon he went into business full time as an inventor of improvements in telegraphy.

From the beginning of his career as an inventor, Edison understood the relationships among invention, production, and sales. As is often recounted in his biographies, his first invention—an electric voting machine for legislatures—found no buyers, because, as he later commented, it did not fit with the political practices at the time. (See, e.g., Lathrop 1890.) According to the biographers Dyer and Martin (1910, volume I, p. 103), he then vowed never to invent unless he was certain of "real, genuine demand."

Working in the young telegraph industry, Edison learned about electricity, tinkering, and inventing; he also learned about news, finances, stock markets, corporate manipulation, and power brokers. As an independent inventor and producer of telegraphic and telephonic equipment throughout the 1870s, Edison worked for all the competitors, including Western Union, the Atlantic and Pacific Telegraph Company, and the Automatic Telegraph Company. In the complex of litigation and mergers, he produced patents for all sides, and came to know the financiers, businessmen, and lawyers who ran the companies. By 1870, Edison had become one of the key inventors for the Gold and Stock Telegraph Company, the Western Union subsidiary that dominated the financial information industry until the rise of the Dow Jones empire at the end of the nineteenth century. By 1877, when Edison's invention of the phonograph made him an international celebrity, he already knew and was known by the leaders of the financial industry, including Jay Gould, J. P. Morgan, and the Vanderbilts.

Not only had Edison made personal contacts, he knew how to represent his work to them to be of interest and value, worth the support and investment that could turn his inventions into widely used, profitable products. He could convince them that his project was the next new technology to be backed.

Raising R&D Capital

Although Edison knew he needed in the long term to establish a place in the financial markets, he found his more immediate capital among a small group of investors whom he had known from previous contacts in the telegraph industry. They certainly had to be persuaded of the same kinds of issues that Edison was projecting before the general financial markets, relying on the current climate of financial opportunity in large technological industry. However, they also had to be convinced that the technology was producible and economically viable in the near term, and that Edison would be there first. That is, they needed to conclude that the race was soon winnable, that the race was short enough to be able to predict a winner, and that they were backing the right horse.

Through personal contact in the telegraphic and telephonic industries, Edison had established himself as an inventor who would deliver, a man on whom you could put your faith and dollars. Grosvenor Lowrey, who came to be Edison's attorney and advisor, first met Edison during an 1875 patent dispute. Lowrey was counsel for Western Union, which had accused Edison of contract violation for selling patents to Jay Gould's Atlantic & Pacific Telegraph Company. After the dispute, Edison returned to Western Union to work on a system of acoustic telephony. With Lowrey as intermediary, Edison continued for the next several years as an independent inventor on various projects for Western Union and other major companies.

As soon as Edison began work on light in earnest, he was in contact with Hamilton Twombly. Just that spring Edison had made an arrangement with Twombly and Western Union for the development of the phonograph. Twombly, a son-in-law of "Commodore" Cornelius Vanderbilt, was the business manager for the Vanderbilt family, who were major investors in Western Union and in several gas companies. Early in September of 1878, Twombly came by to discuss progress on the phonograph. Instead, Edison showed him sketches of an incandescent lamp. Twombly returned later that month with an offer to back the light project (Conot 1979, p. 123).

With Lowrey acting as negotiator for Edison, other major financial insiders were brought into their small investing group, and preliminary capital was provided to Edison by the middle of October.[2] By November 15 a final agreement was reached providing Edison $50,000 in research funds in return for one-sixth of the stock of the newly formed Edison Electric Light Company. The original board of directors included representatives of the Vanderbilt, Morgan, Western Union, Bank of New York, and Gold & Stock Telegraph Company financial empires (Josephson 1959, p. 188).

In the brief initial period when Edison was seeking financing and a corporate arrangement for the development and production of electric light and power, three related rhetorical themes emerged: the status of Edison, the state of the technology, and whether the technology would remain the property of the people who had worked on it and financed it (i.e., Edison, his colleagues, and his backers). Curiously, the usual concerns of investors as to whether there would be a market for the product and a way to deliver the product to the market were not open rhetorical issues. The success of the gas utilities and the relationship between new technologies and the rise of major industrial corporations were so well established that the backers needed no reassurance on these matters.

The best way for Edison to assure his backers that their investments would result in a workable technology was to demonstrate a working lamp. Edison knew this, and throughout the investment and development period he was promising such demonstrations, although for more than a year he had little of substance to demonstrate. Reassuring the investors of rapid substantive accomplishments, Edison acted as though his ideas would immediately be transformed into working objects. On October 5, 1878, in the midst of negotiations with backers, he promised to have five lights ready for demonstration in two days. He was not able to deliver. On October 18 he gave a brief demonstration of a platinum light to a reporter for *The Sun,* turning the power off before the light burned out.[3] On November 12, when the backers came to Menlo Park to arrange the final deal, Edison showed two models. By early December, Edison was promising a major demonstration of 2000 burning lights.

In the interim, Edison's unfulfilled promises created distrust among the backers. Until he could produce something concrete, Edison's only rhetorical resource was their trust in his reputation as a great inventor. During the period of negotiation he worked this theme of personal trust heavily.

This Agreement made the fifteenth day of November in the year one thousand eight hundred and seventy eight, between Thomas A. Edison, of Menlo Park, New Jersey, party of the first part, and The Edison Electric Light Company, a corporation created and existing under the laws of the State of New York, and hereinafter called "the Company," party of the second part.

Witnesseth:

Whereas the Company has been organized with the view of becoming the owner of and of making, using and vending and licensing others to make, use and vend within the United States and other countries or colonies hereinafter mentioned all the inventions, discoveries, improvements and devices of said Edison, made or to be made, in or pertaining to Electric Lighting, or relating in any way to the use of electricity for the purposes of power, or of illumination or heating; or relating to improvements in Electric Engines or to the developing of electric currents by machines or otherwise, for any use or purpose, except electric telegraphy.

And Whereas the said Edison is willing and desirous, in order to obtain the means to continue his investigations in the subjects above named, to transfer, upon the terms heretofore agreed upon and hereinafter fully set forth, all the right, title and interest in his said inventions, made or to be made, as herein provided, and the exclusive use thereof in the countries above named, together with all letters patent of the

Agreement, November 15, 1878 (28: 1162–1170).

In late October of 1878, when the competing inventor William Sawyer and his backer Albon Man approached Lowrey, Lowrey discussed with the group of investors the possibility of including Sawyer and his patents within the company. Edison, when informed of this through his secretary, Stockton Griffin, defined the issue now as entirely one of trust in his genius. In a letter dated November 1, 1878, Griffin reported to Lowrey that Edison "was visibly agitated and said it was the old story, that is lack of confidence—the same experience he had had with the telephone, and in fact, all his successful inventions, was being reenacted."[4] Edison reasserted his right to be trusted not only on the basis of his earlier inventions but also on the basis of the claim that his approach to incandescent lighting was "entirely original and out of the rut," unlike that of Sawyer and Man. Griffin reassured Edison of Lowrey's confidence and Lowrey shortly thereafter wrote to Edison: "My confidence in you as an infallible, certain man of science is absolute and complete."[5] Negotiations with Sawyer and Man were terminated.

In the following months, trust in Edison and his special talents was a recurring rhetorical theme, holding together a corporation that as yet had no product. For example, in a letter dated January 25, 1879, Lowrey recounts to Edison a meeting that morning at which some of the directors, in the wake of news of a Menlo Park setback, had joked about selling off their stock. According to Lowrey, the directors seemed willing to maintain their confidence in Edison if, in return, Edison were to be honest with them about the state of his work.[6]

However, as promises remained unfulfilled, backers became outspokenly skeptical of the value of their investment. To reassure the investors, Edison had to give a demonstration, despite the fact that he had not yet developed a workable high-resistance bulb. On March 22, 1879, he demonstrated sixteen low-resistance bulbs. By some accounts, they began to flicker and burn out within a few seconds.

Since $50,000 had already been committed, Edison continued, even though he had spent nearly that much already. Not until November of 1879, when the work of his laboratory had come together and when what turned out to be the key patent (223,898) was applied for, was Edison able to satisfy his backers with demonstrations of a working technology. Now Edison no longer had to play the rhetorical game of "trust me." At this point, the concerns of investors turned from the uncertainty of an unproven technology to the finances of commercial development, and thus a new set of representations of the light were needed.

Menlo Park: The Place of Invention

The scene where the light was to be produced was the Menlo Park laboratory. From early 1878 on, as we have seen, reporters regularly made the
pilgrimage to rural New Jersey to interview the wizard in his lair. Edison
repeatedly appeared as the archetypal alchemist over the forge at midnight, as in an article titled "Edison in his Workshop" that appeared in
Harper's Weekly of August 2, 1879:

It is utter, black midnight, and the stillness and awe of that lonely hour have settled upon the pleasant hills and pretty homes of the remote New Jersey village.
Only one or two windows gleam, faintly, as through dusty panes, and the traveller,
directing his stumbling steps by their light, enters a door, passes to a stairway
guarded by the shadows of strange objects, and gropes his way upward.

A single flaring gas flame flickers at one end of a long room, disclosing an infinite number of bottles of various sizes, carved and turned pieces of wood, curious
shapes of brass, and a wilderness of wires, some straight, others coiled and spiral
and kinked, the ends pinched under thumbscrews, or hidden in dirty jars, or
hanging free from invisible supports—an indiscriminate, shadowy, uncanny foreground. Picking his way circumspectly around a bluish, half translucent bulwark
of jars filled with azure liquid, and chained together by wires, a new picture meets
his bewildered eyes. At an open red brick chimney, fitfully outlined from the darkness by the light of fiercely smoking lamps, stands a roughly clothed gray-haired
man, his tall form stooping under the wooden hood which seems to confine nox-
ious gases and compel them to the flue. He is intent upon a complex arrangement of brass andiron and copper wire, assisted by magnets and vitriol jars, vials
labeled in chemical formulae, and retorts in which to form new liquid combina-
tions. His eager countenance is lighted up by the yellow glare of the unsteady
lamps, as he glances into a heavy old book lying there, while his broad shoulders
keep out the gloom that lurks in all corners and hides among the masses of
machinery. He is a fit occupant for this weird scene: a midnight workman with
supernal forces whose mysterious phenomena have taught men their largest idea
of elemental power; a modern alchemist, who finds the philosopher's stone to be
made of carbon, and with his magnetic wand changes every-day knowledge into

the pure gold of new applications and original uses. He is Thomas A. Edison, at work in his laboratory, deep in his conjuring of nature while the world sleeps.[1]

Both writers who communicated in the purple prose of late Victorian journalism and Gilded Age financiers who communicated in green dollars saw the laboratory as a place of private genius, the place where Edison worked his personal magic. Yet, as much as the press might construct the laboratory as a place of magic, and the backers as a place for the production of valuable commodities, for Edison and his colleagues it was a place of collaborative work, the work of invention carried out within "an invention factory" by a staff of about twelve (soon to double).[2]

The list of Edison's collaborators most centrally engaged in the development of electric light and power includes Edison's longtime associate Charles Batchelor (a machinist), Francis Upton (a young physicist and mathematician), John Kruesi (a seasoned machinist), Ludwig Boehm (an experienced glass blower), John and Fred Ott (mechanics), Samuel D. Mott (a draftsman), Stockton Griffin (Edison's personal secretary), the young assistants Martin Force and Francis Jehl, and the accountant William Carman.

Documents from the Laboratory

Some of the men who worked at Menlo Park kept extensive notebooks. These notebooks, supplemented by a few letters and journals kept by individual members of the team, by later interviews and memoirs, and by visitors' reports of various incidents, constitute the documentary record of what went on in Edison's laboratory. The notebooks are the residuum of the communicative acts that brought the work of the many people in the laboratory together. They are the medium by which people shared ideas, plans, and experiences. They are the designs from which the machinists made models and the records of experiments evaluated to decide on the next move. They are the crossed-out, erased, and revised transformations of a technology being puzzled through. In the laboratory, much must have been conveyed by talk, by pointing, and by manipulation of physical models, but the talk and pointing often must have been organized over the drawings and accompanying words. Moreover, the physical models and material experiments were produced in relation to the sketched plans and were recorded and compared in further drawings and words. The transient talk and manipulative experiments endure only in the plans and the records; even the models that have survived to be displayed at the Menlo

General View of Menlo Park and Edison's Laboratory, Frank Leslie's Illustrated Newspaper, *January 10, 1880 (94: 572).*

Park Restoration in Greenfield Village at Dearborn, Michigan, are interpreted through the surviving documents. The purpose of this chapter is to consider how these laboratory notebooks served within the communicative work system of the laboratory.

The Meanings of Notebooks

The documentary remains of the laboratory survive because from the beginning they were treated as legal documents, recording the state of the work at any time to substantiate patent claims and litigation. The reason why they are so useful to us in understanding laboratory communication is the same reason they were so useful in court. The work could not go forward without these communicative symbols to coordinate it, and these representations of the technology would not have been created except as part of the work of bringing the technology into being.

Yet there is a difference between the way the documents will be analyzed here and the way they have been analyzed in courts or elsewhere in technology studies. Both in court and in technology studies, laboratory notebooks indicate the state of a technology: what designs, plans, and experiments were conducted as of what date. In this spirit, the most extensive exploration of Edison's notebooks—that by Friedel and Israel in *Edison's Electric Light*—reconstructs the emergence of the technological objects and systems that Edison eventually brought to market. Friedel and Israel focus precisely on what is represented in the notebooks: the various designs of lamps and generators and wiring systems.

A secondary approach to notebooks and other documentary remains of invention has focused on the creative processes of the inventor. This work is primarily cognitive, trying to reconstruct the states of thinking of the inventors, how they moved from one step to another, and how they figured the technology out. The focus is on breakthroughs and refinements, problem formulations, and problem-solving methods. This follows a general strategy in such studies of creativity as Gruber's (1974) work on Darwin and Tweney's (1991a,b) on Faraday. Carlson and Gorman (1989, 1993) and Pershey (1989) mapped Bell's and Edison's development of designs for telephone receivers, examining how the inventors segmented problems and found trial solutions from their concrete experience in solving subproblems. This extended Hughes's (1983) pioneering work in understanding individual invention in relation to large-scale systems. An even further reconstruction is to see how thinking as represented in the symbolic forms (verbal and graphic) of notebooks interacts with construction of physical models, experiments, recording of results, and further contemplation, as Gooding (1990) did with Faraday's notebooks.

These historical reconstructions of inventions and inventive processes are appropriate and valid, for the notebooks indeed record the state of the technology and provide evidence of the plans, thinking, trials, and solutions of the people who wrote the notebooks and built the technology. However, neither approach considers the notebooks themselves as communicative documents. Prior to the question of what is represented in the notebooks is the question of why people expend the energy to compile notebooks and what role the notebooks have in individual and group activity. In considering why notebooks exist, what they do, how they work, and how they are used (long before they became subject to the retrospective scrutiny of lawyers, historians, and students of science and technology), we will be gaining evidence of the communicative and cooperative nature of invention, even where a dominating personality of great imagination takes a central role in directing the work of others through communicative acts.

The Publicness of Private Notebooks

Inventor's notebooks are a great deal more than the overflow of the private spontaneous thoughts of abstracted inventors. We can see that in Edison's earliest notebooks, when he was working as an independent, long before he had the notion or the opportunity to hire others to extend his work.

The first extant notebook in the Edison papers dates from 1867 and has entries dated through 1871.[3] Since the book opens and ends with accounts from his father's business, it is likely that Edison began using the notebook for his inventive work when he returned home to Port Huron late in 1867, without employment, at the age of 20, after five years as an itinerant telegrapher. During those years he was notorious for experimenting with the apparatus at his places of work, getting fired for it several times. However, as far as we know, such tinkering was carried out in a purely hands-on way, and we have no record of earlier plans or sketches. So this first notebook, mixed with his father's records, is as close as we get to Edison's native invention.

The notebook contains 16 pages of drawings. Some of these seem to be plans for duplex telegraphy, which Edison had been working on the previous year while a telegrapher in Louisville.

There is no indication that Edison had apparatus to work with in Port Huron, so the sketches may be memories of his previous experiments or

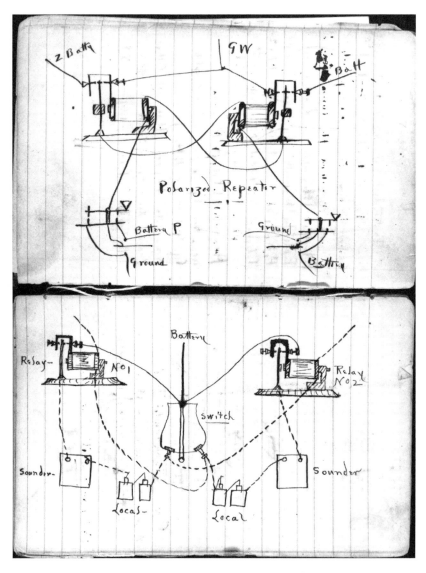

Drawings in Edison's Pocket Notebook, perhaps dating from 1867 (PN-69-08-08 (6: 767)).

plans of future ones—a way of thinking without having the objects to manipulate. Although these drawings may be symbolic surrogates for his hands-on work, they are publicly intelligible in their careful drawing, standard symbology, and labeling. They are not the sketches of someone thinking through an idea. They are drawn with solid and confident lines, well centered on the page—a conscious representation of an object that the artist knows and wishes to convey to others. Only one or two pages have the privacy of sketches, with lightly drawn, incomplete, undetailed ideas in uncentered, multiple overlapping drawings.[4] Two page of the notebook contain short lists of books on electrical and telegraphic topics, indicating that Edison was supplementing his hands-on experience with organized knowledge represented in books.

The last several pages of a second pocket notebook for 1870 reveal an even more public nature. These pages are each labeled with variations on "Drawn at L Serrells office/October 3, 1870/Constructed 2 months before."[5] Serrell was Edison's newly acquired patent attorney. The drawings and the text detail parts for a printing telegraph Edison had been working on. Having earlier that year shared some patents for a printing telegraph with then-partner Franklin Pope, Edison was now in the process of applying for his own patents, which he would receive shortly. The drawings, although crude, are detailed and labeled with cross-reference letters, and each is accompanied by about half a page of descriptive text, written in a style typical of patent specifications of the period.

These drawings appear to have been done under the supervision and advice of his patent attorney.[6] That they are legal records of prior constructions, to be used in drafting and substantiating patent applications, is confirmed by Edison's inscription on the inside back cover of the notebook: "all new inventions/ I will hereafter keep a full record!"[7]

The next notebook opens with the following inscription:

Newark, N.J. July 28, 1871

Record of ideas and inventions relating to Printing Telegraphs which work with a type wheel, all of which I am under contract with the Gold & Stock Telegraph Co of NY to give them subject to the Conditions of the Contract within the next five years from the date of Contract Reserving for myself any ideas contained in this book which I do not see fit to give said G & Stock Telegraph.

This will be a daily record, containing ideas previously formed, some of which have been tried, some that have been sketched and described, and some that have never been sketched tried or described.[8]

Drawings and descriptions, each dated and witnessed, are detailed and well labeled. These drawings then count not only as legal evidence for

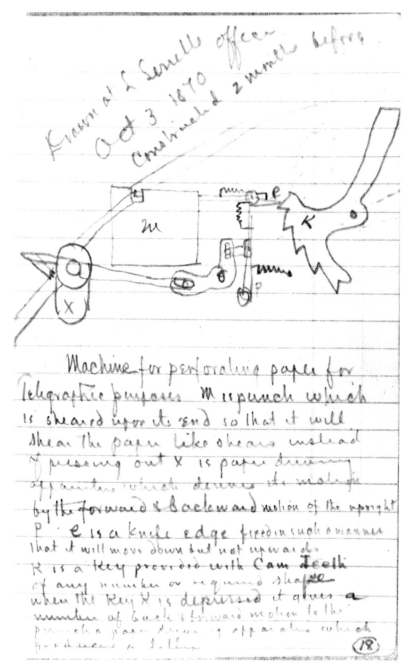

Notes prepared by Edison for patent agent, October 3, 1870 (PN-70-10-03 (6: 809)).

patent considerations but also in determining Edison's relationship with his contractor, the Gold & Stock Telegraph Company. To make the difference between contractual and personal work even more clear, the next day Edison started another notebook with a contrasting inscription:

Automatic Printing. Telegraph.   – Newark N.J. July 29, 1871
Invented by and for myself exclusively.

Daily record of ideas as they occur applicable to this system, and a record of ideas already conceived, and account of experiments-

Dot and Dash and Automatic Printing Translating System, Invented for myself exclusively, and not for any small brained capitalist.[9]

Personal Activities Enacted in Private Notebooks

Not all of Edison's notebooks were purely legal in purpose. Around this time, Edison purchased and read Michael Faraday's *Experimental Researches in Electricity*. These laboratory journals influenced Edison greatly, teaching him, as he said later, the attitude, method and mode of representing of scientific investigation. Dyer and Martin (1910, p. 101) quote Edison as recalling: "It was in Boston I bought Faraday's works. I think I must have tried about everything in those books. His explanations were simple. He used no mathematics. He was the Master Experimenter."

Faraday had a consciously adopted format for his notebooks and his more formal presentations, which presented an idealized discovery account in a format derivative of the discovery notebook.[10] Edison had access only to the more formal presentation of Faraday's *Experimental Researches in Electricity*, which were in fact re-publications of his papers in *Philosophical Transactions of the Royal Society*, but if we consider Faraday's private methods of conceiving, recording, and connecting laboratory experiences we will gain insight into his idealized representations of formal articles, and thus into what Edison might have gotten from a reading of Faraday.

In his personal notebooks, we now know, Faraday kept extensive notes both as an aid to memory and as method of thought, following his Lockean philosophy of mind. To enhance his memory and to make associative connections among the various experiences recorded in his notebooks, he developed retrieval and indexing systems for his notes, based on his conscious psychology and philosophy of discovery. His notebooks as well were recorded in the midst of experimental practice and production of results, so that the representations in them are tightly coordinated with

Cat. 1174

G-1678

Newark NJ July 28. 1871

Record of ideas and inventions relating
to Printing Telegraphs Which work with
a Type wheel. all of which I am under
Contract with the Gold & Stock Telegraph Co
of NY. to give them subject to the
Conditions of the Contract within the next
five years from the date of Contract
Reserving for myself any ideas contained
in this book which I do not see fit to give
said G & Stock Telegraph =

This will be a daily record, containing
ideas previously formed some of which
have been tried some that have been
sketched and described, and some that
have Never been sketched tried or
described

Page 1 of Newark Shop Notebook Catalog 1174 (3: 8).

Aulomatic Printing. Telegraph. = Newark N.J. July 29. 18**1** (1)

Invented by and for myself exclusively.   E-1676

Daily record of ideas as they occur applicable to this system, and a record of ideas already conceived, and account of experiments –

✓

Dot (∾) Dash (∾) aulomatic Printing Translating System, Invented for myself exclusively, and not for any small brained capitalist,

Page 1 of Newark Shop Notebook Catalog 1172 (3: 78).

planning, procedure, and results. Words and illustrations were interactive with physical experience, both shaping the production of experience and reflecting (or creating a textual surrogate for) the immediate consequences of activity.

When Faraday published his results in *Philosophical Transactions,* he followed the general prescriptions of Joseph Priestley, offered in *The History and Present State of Electricity* more than 60 years earlier. Here and elsewhere, Priestley recommended a form of presentation that shared the research experience, showing the interaction of the train of ideas, problems, experiments, and results that had led one down a path of investigation and discovery so that others might follow the path of reasoning and continue with their own investigations (Bazerman 1991). This method of exposition, which Priestley used to present his own results, contrasted with what he called the Newtonian style, which, he asserted, obscured the investigative method by placing conclusions at the beginning and presenting only a limited number of final experiments as a kind of proof. Priestley (1775, pp. 167–168) described the latter method as one by which someone reaches a great height by climbing a ladder, then kicks the ladder away hiding how he got to such a height and providing no means for anybody else to find the way up. Priestley's widely reprinted volume (five editions and two translations in his own lifetime) was influential in shaping the reports in *Philosophical Transactions* in the 1770s and the 1780s, but lost favor by 1800 as the dynamics of argument led to papers that adopted more of a Newtonian format (Bazerman 1988, chapter 3).

Faraday, in presenting his findings for *Philosophical Transactions*, returned to the Priestleyan narrative of discovery through chains of reasoning and experiences. However, he does not purport to report all the details of his turnings of thought and experiment in the exact order of occurrence, "as they were obtained"; instead he represents an idealized and simplified discovery path of reasoning and experiment "in such a manner as to give a concise view of the whole" (Faraday 1839, p. 2). He thereby avoids some of the rambling prolixity of Priestley's accounts. Nonetheless, Faraday's *Experimental Researches in Electricity* presents experiments as driven by hypotheses and puzzles, the results of which then present new hypotheses and puzzles leading to new experiments. This is exemplified early in the 1839 edition (pp. 3–4):

12. The results which I had by this time obtained with magnets led me to believe that the battery current through one wire, did, in reality, induce a similar current through the other wire, but that it continued for an instant only, and partook more of....

13. This expectation was confirmed; for on substituting a small hollow helix, formed round a glass tube....

Thus, in the published *Experimental Researches* we have, although much cleaned up and reordered, a set of investigations taken directly from laboratory experience being puzzled through by an active experimenter who is constantly tinkering with the apparatus to test new ideas.

There is no evidence and little likelihood that Edison developed a self-conscious Lockean philosophy of investigation with a well developed notion of how representation and communication entered into the reasoning and experiencing process. Nonetheless, Faraday's *Researches* provided for Edison a powerful exemplar of how researches ought to proceed and how notebooks can help record and sort out that process. Around 1873 Edison began a series of two notebooks titled Experimental Researches. These are action-oriented plans about what should be done, informally setting out tasks to be done as well as results. The books repeatedly use commands directed by Edison to himself: Try..., mix..., procure..., ascertain.[11]

Several other notebooks of this period adopt a similar style of recording personal thinking and activity and of not trying to identify inventions within legal and contractual systems. For example, two of his pocket notebooks—small pads which he used while traveling through England in 1873—contain unlabeled sketches for alternative telegraphic circuits. The few words accompanying the diagrams indicate that these sketches are plans for future physical experiments: "Try This," "Connect Thus," "or maybe," "also." These drawings are interspersed with other activity-oriented plans: "Buy 60 Bunsen or Box Grove Quantity," "Take out 4 rolls or 5-paper," "Have Wright Connect 1/2 & 1/2 Battery in quantity."[12] These notebooks helped Edison remember and elaborate his plans while away from his usual inventive environment.

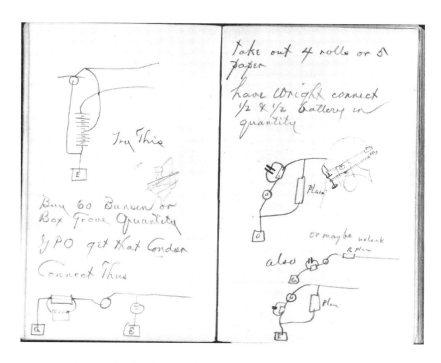

Two pages from Pocket Notebook PN 73-00-00.2 (6: 883).

Documents and the Lives of Independent Inventors

Such limited uses of graphic representation were consistent with the life led by Edison and his fellow electrical inventors in the early years of telegraphy. Edison and his colleagues were itinerant, temporarily contracted with various telegraphic and stock ticker companies to maintain and improve apparatus. Much of the work was accomplished by rearrangements or modifications of existing apparatus. A second place of employment was machine shops, which were often loosely organized, with muckers (as they were called) often under part-time, piecework, or contract employment. As independent operators, they often made separate arrangements among themselves, using others as subcontractors, or renting the use of the machine shop tools. The most important records were those affecting ownership—contractual arrangements determining who invented what, and who had a claim on patents. The conception of each project was tightly held by the originator and was close enough to the apparatus at hand so as not to require planning beyond what could be kept in the inventor's head or shared in conversation with helpers (Israel 1992; Millard 1989, 1990).

In the spring of 1870 Edison opened a small machine shop in Newark in partnership with William Unger. That winter he opened a much larger shop to produce telegraphic equipment. For that larger factory he hired the machinists Charles Batchelor, John Kruesi, and John and Fred Ott, who were to remain among his closest co-workers for many years. These machine shops were small local manufacturing facilities, requiring little of the paperwork of modern corporate business. Edison intervened directly in their operation. The primary work was manufacturing, but invention was carried out in a secondary and informal way, under the direct supervision of Edison. That few records were required beyond accounts and documents affecting patents and contracts is reflected in the large ledgers designated the Newark Shop Notebooks.[13] These four notebooks, containing entries from 1871 through 1875, record formal and explicit descriptions, witnessed and dated, in a form close to that of patent specifications, although towards the end of three of them results of some chemical trials and other experiments are recorded. In 1875 a new series of notebooks, again entitled Experimental Researches, continued the formal legal documentation of laboratory developments. Unlike the earlier series of Experimental Researches, they contain no informal records or action-oriented comments. In fact, after the move to Menlo Park, these notebooks were to be used for recopying of neatened accounts derived from the less formal Menlo Park Notebooks. This series of Experimental Researches consists of six volumes covering the period from 1875 until early 1879.[14]

Documents and the Life of an Inventive Collective

In the spring of 1876, in moving his operations to Menlo Park, Edison made a fundamental change in the nature of the work. Menlo Park was to be a full-time invention factory, where the only product would be discoveries and inventions. The inventions had to be turned out regularly and generate the money to pay the rent and payroll for this first industrial laboratory. No manufacturing business could subsidize the innovation. Edison boasted he planned to turn out "a minor invention every ten days and a big thing every six months or so" (Josephson 1959, pp. 133–134).

Full-time devotion to invention in an operation of a dozen workers and upward, depending on its efficiency for financial success, would require self-conscious attention to coordinating the work through communication. Edison, an experienced businessman, attended to setting up the necessary communicative and documentary systems, which were all in place by the time of the incandescent light project (two years later). In addition to the usual business files concerning accounts, payrolls, and business correspondence, and the legal files surrounding patents, there were primarily three kinds of notebooks: the continuation of the formal Experimental Researches (totaling six), the Menlo Park Scrapbooks (more than 150), and the working notebooks (eleven unbound notebooks, numbered from 8 through 18) and the Menlo Park Notebooks (totaling 249 in the period 1878–1882).[15] Several other notebooks were kept by individuals or in other series, including pocket notebooks and technical notebooks used in developing patent applications. Together these provided a medium for organizing the inventive work.

Scrapbooks, Collective Knowledge, Fame, and Intertextuality

The scrapbooks were begun in 1878 after Menlo Park had been in operation for more than two years. They were first kept by William Carman and Francis Upton. Carman was an accountant who had been Edison's employee from earlier endeavors and had in fact introduced Edison to the Menlo Park area. Upton was the young physicist hired in the fall of 1878, specifically for his knowledge of electricity, after having studied with Helmholtz. These scrapbooks contained extensive clippings from the technical and popular press on developments relevant to all areas of Edison's work. The clippings contained accounts of current theory, discoveries and inventions, and information about business and legal developments and about public acceptance of the various technologies. The existence of the scrapbooks suggested that Edison felt the need for himself and his employees to be fully informed about what others were doing. But he also paid attention to the business and legal and public environment within which their work would take place. The scrapbooks also contained clippings of stories about Edison's work. Part of the work of developing incandescent lighting is to draw on these knowledges, position oneself with respect to them, and place one's technology within it.

Two further kinds of notebooks suggest how Edison and his colleagues needed to position their work within the intertextual discussion of technology and technological development. First, one of Charles Batchelor's personal notebooks contained a list of journals to be purchased in lots of six to twelve and mailed out to agents, stockholders, newspaper editors, eminent industrialists, and scientists.[16] Clearly fame was something Edison would not leave to itself.[17] The second was a pair of Upton's notebooks which contained his first major assignment for Edison: notes of a search of all prior patents on incandescent light.[18] These notes recorded the patented inventions and advances in the field and pointed out useful features that Edison might draw upon. At the end of the notebooks, Upton identified features that made Edison's work distinctive and patentable.

Through these several notebooks, inventive work was collaboratively distributed. Edison was using additional eyes to extend his scanning of the technical, popular, and business literatures. Those eyes were trained in specific ways: Carman, as an accountant, was concerned with the accumulation and ordering of information; Upton, trained in the latest scientific work in electricity, examined the technical literature. Further, they then made the results of their search available to Edison and others inside the Menlo Park lab through notebooks and scrapbooks, which became reference points for the continuing work. From a modern point of view, such developments seem entirely unexceptional. However, for the first multiple-person industrial laboratory, engaged in a several simultaneous projects, these were new organizational and communicative procedures, perhaps borrowed and adapted from early usages (such as a personal secretary's scrapbook) but newly deployed and configured. It is also worth noting that the scrapbooks did not develop as a regular practice until several years into the existence of the Menlo Park laboratory, when a trained physicist was hired, suggesting the professionalization of the field and the role of journals in developing professional knowledge in this period. The need to gather together a wide array of sources had much to do with the multiplicity of publications that had only recently developed as the steam press had made cheap publication possible. Further, cheap printing made publications expendable, so they could be clipped without second thought.

Legal Documentation of Innovation: Experimental Researches

The second major kind of notebook of the period of the incandescent light are the notebooks labeled Experimental Researches. These grew out of the earlier Newark Shop Notebooks and continued the function of creating a legal record. These are after-the-fact descriptions of what was made, done, and thought in the laboratory so as to provide documentary evidence for patent applications and litigation. They are written with a formality, explicitness, and orderliness that indicate this function. What changes over the years, however, is their relation to other less formal documents. Before the development of informal notebooks, the formal legal evidentiary accounts were the first collected written record of the work. However, once the Menlo Park Notebooks were established in 1877 as a new kind of internal working notebook, the formal legal accounts became a secondary account. After that moment, the legal accounts of the Experimental Researches were abstracted from the working notebooks, gathering the parts together in a coherent narrative of invention, with cross references to the primary Menlo Park Notebooks. That is, the separation of laboratory notebooks from the major legal documentation permitted the legal accounts to become more organized and coherent while the rougher, more impromptu laboratory documents served to coordinate local activity. With the removal of the legal documentation from the laboratory, the work of creating the after-the-fact legal record could be handed over to the accountant Carman.

In a typical passage at the beginning of Experimental Researches—volume 5, describing the first development in incandescent lighting—Carman describes 21 related inventions over five pages and then draws the relevant 21 diagrams together on the next three pages.[19] This summarizing and synthesizing work was made necessary by the increasing complexity of the research program, the large numbers of people involved in the production of the technology, and the new uses to which the Menlo Park Notebooks were put.

The Working Notebooks

The Unbound Notebooks and the Menlo Park Notebooks presented a new level of text to mediate collaborative work. Immediately upon moving to Menlo Park, Edison started a series of softbound notebooks whose pages were ripped out and later reorganized according to subject. These notebooks were left in the laboratory for all to use in making working sketches for the projects they were engaged in. These drawings were signed and dated. Previously, as Edison later testified in a patent litigation case concerning the telephone, working sketches "were made on all sorts of paper and thrown into drawers"[20] although they too had been signed and dated. In 1878, Edison switched to a bound notebook. The notebooks, now designated the Menlo Park Notebooks, were largely kept intact. Early work on the incandescent light is recorded in both the Unbound Notebooks and the Menlo Park Notebooks.

Edison's testimony,[21] contemporary newspaper reports, and the contents of the notebooks all indicate that these notebooks were left out on the tables for any member of the research team to make an entry in. The entries in the earliest notebooks were in the hands of Edison, Batchelor, and Upton, but over time entries began to be made by others. The entries, which included working sketches, comments, experimental results, and orders to the workshop, created a communal record, coordinated action, and provided a site for evaluation and planning of ongoing work.

However, as the notebooks also contained documentary traces of inventive activity with legal implications, certain legal features rapidly emerged in the Menlo Park Notebooks. From the beginning the pages were almost all dated on the upper right corner. On the earliest pages of volume 1, beneath the date, were Edison's initials, TAE.[22] Some others were signed, in the same place, Charles Batchelor. Then increasingly the single signatures were countersigned by a witness, then two witnesses. The witnesses were usually individuals active in the laboratory: Edison, Batchelor, Ott, Upton, Martin Force. On the loose pages of the Unbound Notebooks, third and fourth witnesses often were added.[23] Moreover, in both the bound and the unbound notebooks, Stockton Griffin, Edison's personal secretary—who was not an inventor, and usually was located in the office downstairs and later in another building in the compound—affixed an embossed notary seal over his and the other signatures in the upper right corner, further identifying the testamentary nature of the document. On a few sheets,[24] Griffin used a rubber notary stamp and specified as witnesses Edison, Batchelor, Kruesi, and Force.

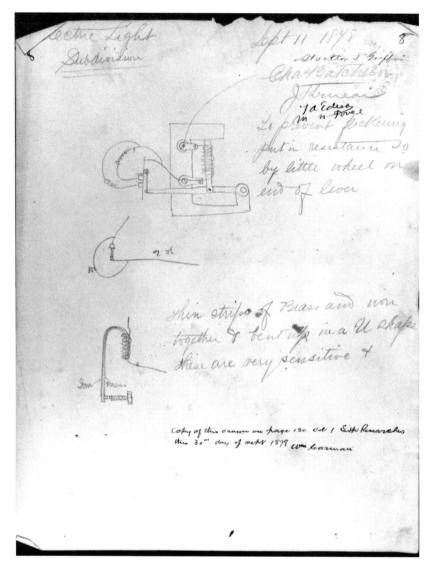

Drawings with legal and technical annotations in Unbound Notebook 16 (4: 492).

At the bottom of the page in the Menlo Park Notebooks was frequently added another annotation, usually dated from a few days to a couple of weeks later, stating the information on the page had been copied by William Carman into the Experimental Researches, with volume and page reference, indexing the raw documentary material to the transformed, formal legal record. Finally, at the bottom, selected pages are cross-referenced to later-granted patents. Thus, by means of generically typified inscriptions placed at regularized locations, these raw working documents were transformed into legal records for circulation in other communicative and documentary systems beyond the laboratory.

The Work of Invention

Ideas and Evaluation

Putting aside all the testamentary features, what can we glean from the documents themselves about how they were used in the laboratory?

Some, but far from all, seemed to be sketching out ideas, frequently containing multiple quick sketches on the same sheet, sometimes overlapping, sometimes at odd angles to one another, sometimes foregrounding a particular feature of interest that varies from one sketch to the next. This is the archetypal thinking on paper: taking one's first impulses, seeing what they look like on paper, and then filling out the idea further upon inspection. One of Edison's first sketches for the regulator light bulb was of this sort: on September 9, Edison moves from the previous day's design for regulated automatic feeding of the carbon rod for an arc lamp[25] to an automatic temperature regulator for an incandescent lamp.[26] The design of the temperature regulator seems directly derivative of the telegraph relays he had previously worked with, where the incandescent filament acts as the spring that holds the relay in contact or breaks it. Three configurations are tried out; the most developed one is at the bottom of the page. Even cruder sketches appear in the early pages of Menlo Park Notebook 1.[27]

There is, of course, an implicit evaluative component as one drawing follows on another, one plan on another. But in some cases the evaluation is made more explicit as one diagram or part of a diagram is crossed out, or an erasure mark is left on the page to indicate a change. For example, on page 35 of Unbound Notebook 16, the topmost drawing (apparently made by Batchelor) is corrected as connecting points for the wire are changed (shown by erasure marks), perhaps as the drawing emerged and the logic of the design was worked out.[28] In further details sketched below, several other arrangements of post, filament, springs, and lever are tried

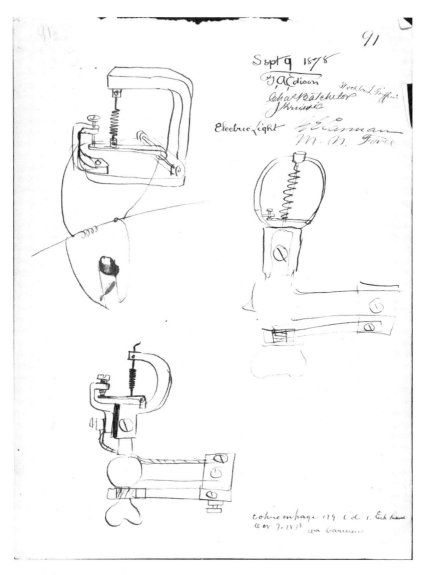

Ideas being worked out in Unbound Notebook 16 (4: 488).

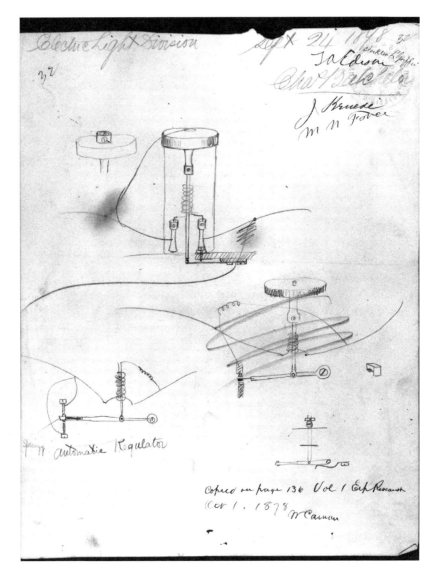

Corrected drawings in Unbound Notebook 16 (4: 0514).

out. One arrangement is entirely crossed out, and the final detail of the regulator presented in the lower right corner. This is a classic case of working out an idea by making it visible in graphic form.

Evaluation is sometimes carried out more explicitly through annotations accompanying a diagram, as when Batchelor scrawls on top of one of a series of sketches of the apparatus on page 36 of Unbound Notebook 16: "When upright wire & spiral are the same size then it is an automatic regulator for it cannot heat to burn if point is set right." He adds a verbal specification on a design feature: "Shaft slides loose."[29]

Working with Objects

Some of the evaluations and revisions seem to come from thinking about the design or the problems that emerge in drawing it; others seem to be responses to the process of trying to construct the object, or attempts to communicate the object to others who need to construct it. Page 17 of Unbound Notebook 16, for example, directs chief machinist Kruesi to make an experimental instrument for measuring temperature expansion. These rough sketches are headed by notes: "Kruesi–make 'light' instrument like this:–" and "and another with double points."[30] To the right are a couple of even rougher detail sketches of components, still in Batchelor's hand, as though Kruesi were asking for further detail. On the following page are some sketches, signed by Kruesi as first witness, that elaborate two further views of the instrument, showing specific parts to be machined.[31]

Further plans, elaborations, and evaluations seem to have resulted from actual tests with actual apparatus. On page 12, a drawing by Batchelor of a design for a temperature-regulated light is annotated "To prevent flickering put in resistance so by little wheel on end of lever." Then resistors were sketched into the circuit (but ambiguously, with cross-outs and double drawing). There is an alternate sketch of the resistor circuit just below. Similarly, the temperature resistor is accompanied by a comment on performance: "Thin strips of Brass and iron together & bent up in a U shape. These are very sensitive &" (the comment is broken off).[32] Results also may be recorded at different times as further experiments or thoughts are developed, as on page 32 of Menlo Park Notebook 1, where the first general results by Batchelor written below in a broad hand are supplemented by more detailed results from further trials, written on the side in a smaller hand and darker ink, although still perhaps by Batchelor. These later results also required new circuit lines and cross-reference letters.[33]

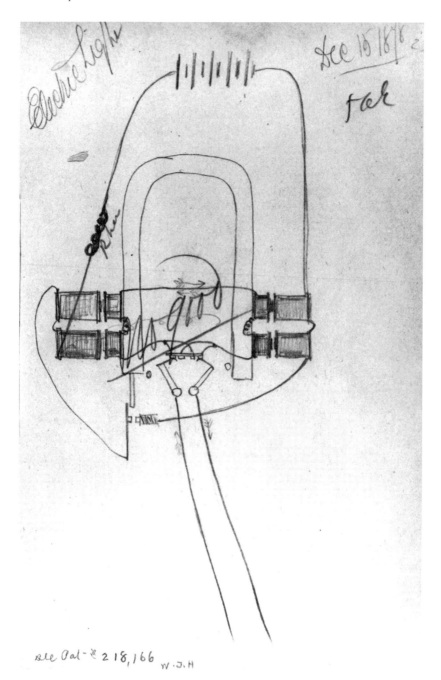

Drawing marked "no good" in Menlo Park Notebook 1 (29: 32).

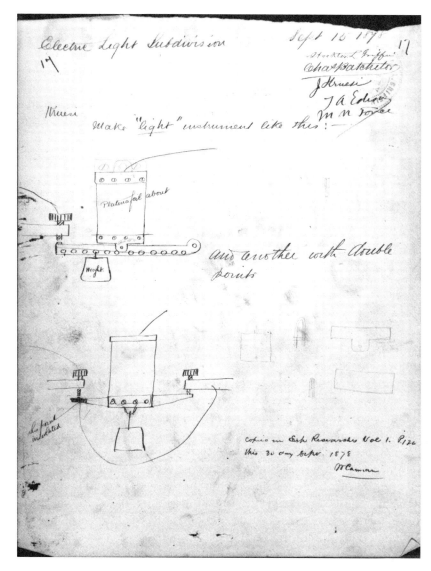

Drawings including directions that rely on local knowledge of collaborators, Unbound Notebook 16 (4: 499).

Further results of experiments sometimes lead to explicit directions to the machinists. On a sketch of a small part, for example, Kruesi directs his subordinate machinist George Jackson: "Jackson:– The little holdback for this spring at x is too far away from the points. You must put it right close up to them only 1/16 away from them. J Kruesi."[34] The use of the indicator word "this" suggests that the sketch and instructions accompanied the actual piece of apparatus when it was returned to Jackson's bench for further work.[35]

Working with Collaborators

At times, several hands collaborate on a single page. For example, page 33 of Unbound Notebook 16 consists of a central drawing apparently made at the draftsmen desk, then added to with several more impromptu sketches, differently oriented on the page, in different hands, as though several people were standing around a workbench, discussing alternatives.[36] Sometimes the drawing is in one hand and the commentary is in another, or in several, as on page 25 of Menlo Park Notebook 1, where Edison writes "no good" across the top of a careful drawing of a design.[37]

Results of experiments involving a number of trials with different substances, as with possible filaments, are often recorded in several hands, as though different trials were carried out by different individuals or many people were involved and it fell to chance who would record each result. For example, pages 47 and 48 of Unbound Notebook 16 contains observations about one experiment and then plans for another, beginning a series of tests of various chemical coatings for the platinum spiral (or filament).[38] The first several results are clearly in the same hand, but a second page of further results seems to start out in Edison's hand, then alternates with Batchelor, and then a third hand appears. Usually the name of the substance and the evaluation of the result are in the same hand, but in a few instances the results are in a different hand than the naming of the substance.

Often, on a single day or over a period of days, various individuals, such as Edison, Batchelor, and Kruesi, all draw the same or related items, each for a different purpose: Edison perhaps sketches out an idea; Batchelor works through some of the details or provides a sketch for the machinists, often with a note for Kruesi; Kruesi adds some details to direct his own work or of his subordinates; then Batchelor records some results and makes some future plans, with perhaps further evaluative comments by Edison, followed by some calculations by Upton. Moreover, they are all likely to have signed off as witnesses or added commentaries on one

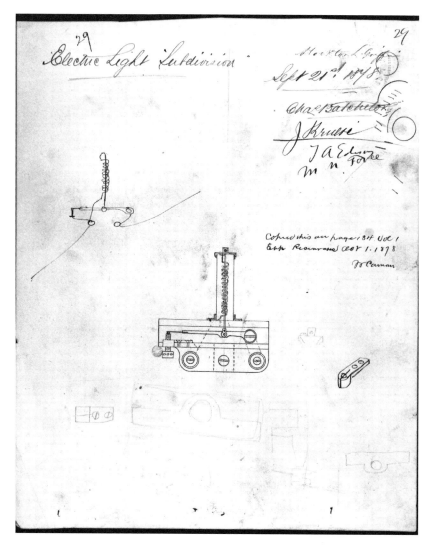

Evidence of collaborative thinking, Unbound Notebook 16 (4: 0512).

another's drawings. Each person has characteristic work, but at times they may take on surprising roles, as when Carman records some results.[39] Each level of work modifies or adds to the previous work in a collaborative process.

Representations and the Material Environment

The representations of the notebooks are embedded in the physical work of construction and testing in the laboratory, as suggested by the sparseness of instructions that are handed to the machinists. Often there is little more than a sketch with no dimensions, and a note, such as "Make instrument for testing different."[40] or "Kruesi Make one this way with air holes in the end."[41] The dimensions and details are left implicit based on Kruesi's familiarity with the apparatus and materials they have been working with. Occasionally there is even an implicit or explicit reference to a prior version, so the instruction is only for a modification of an existing piece of apparatus. Moreover, much seems to be left to Kruesi's judgment—as is evidenced by the detail sketches in his own hand. There is one other notebook in Batchelor's collection that keeps record of projects assigned to the machinists and their completion, but only occasionally (and at the beginning of the sequences) are these accompanied by careful mechanical drawings—the rest are just listed by name and date of assignment, crossed off as they are finished.[42]

Within this pattern of relying on implicit knowledge about the details of construction, the few examples of detailed explicit instructions for machinists stand out; see, for example, page 83 of Unbound Notebook 16.[43] A more common kind of explicitness is the request for a minor adjustment of an existing piece, as in Kruesi's request of Jackson (above). Why so much was left implicit is easy to understand when one realizes that the Menlo Park laboratory consisted primarily of a single large second-floor room, about 25 feet wide by 100 feet long. The library, the offices, the drafting room, and the machine shop were downstairs. Mechanical matters often could be handled largely by calling someone over to look at the physical object, supplementing the pointing and talk with a sketch to identify a new idea or modification. Of course, when mechanical information was intended to travel beyond the laboratory, as in the patent records of the Experimental Researches, more explicitness and more formality were called for.

Representations and the Evolving Practice

Because different people produce and use the notebooks at different moments in the inventive process, the various drawings and texts stand in multiple relationships to the ongoing material activity. Drawings appear beforehand as ideas, plans, and instructions for fabrication; diagrams and data record actual experiments; comments and further drawings evaluate what has come before. Some recorded results fix communal memory; others are part of evolving thought. Sometimes multiple results from extended runs are gathered in a single place to become part of communal intelligence. And occasionally some results are specifically for the record, as in Batchelor's noting the success of a certain configuration of the regulator.[44] Although Edison always had a leading role in defining directions, the members of the lab, coordinated by the drawings and comments in relation to material activity, were teaching one another how to make an incandescent lamp that worked, along with the generators, meters, and other parts of the system.

Each person's work, recorded and shared through the notebooks and the material artifacts, influenced the development of the ultimate technology, and a general understanding of the ultimate technology guided each of their work. Yet at any point no one person had the full technology in mind in all details. Clearly Edison himself set the major themes for the development in terms of the systems approach, the components of the systems, and more general strategic directions for solving technical problems. Yet many of the details were worked out by others, and other answers were in the materials themselves as they were arranged in working configurations. Edison needed the craft skills and detailed knowledge embodied in his collaborators. Edison, after all, had hired them, and not just to have more hands to share the labor.

The Laboratory as a Locus of Communication

This view of the truly collaborative nature of the work points to the necessity and the limitations of the communication. In teaching one another how to make the lamp and the system of electric light, the collaborators constantly had to be coordinating the work with one another, finding ways to document what each was doing, and articulating what they wanted of one another and of the material technology. But since no one knew exactly what would work or could fully accomplish any else's work, each could only sketch incomplete knowledge, plans, and desires, and then inspect and document the results. By creating a record, each would then constrain the work of all at the next stage of thought, representation, and action. They kept sketching parts in for one another.

Once ideas turned into representations and objects that were valued enough to be considered significant advances in the technology, the emerging ideas were drawn more fully, redrawn more often by various people, and then submitted to the patent draftsman to be turned into a public drawing for the legal process. The same drawings that served to mediate work and thought in the laboratory led directly to the drawings that served to represent the technology in the legal sphere. Many of the spirals, regulators, connectors, and other features sketched beginning on September 9, 1878, appear in Edison's first lighting patent application, filed on October 14. Notice the relationship among Edison's sketch of September 10,[45] Carman's redactive drawings (figures 1–7),[46] the detailed design of September 17,[47] Kruesi's design sketches,[48] and the final patent drawing[49]—all reproduced below.

The continuity of representation reminds us that the production of the ideas—realized in material practice but represented, recorded, and defined in symbolic terms—is sanctified as property by the patent system, to then have value in financial and commercial marketplaces. These representations, as diagrams, descriptions, and abstractions of patent numbers, are then re-represented in advertisements, publicity, and newspaper articles.

Individuals Meeting in the Laboratory

The production of materially realizable ideas through collaborative work further requires the shared attention of the many people, aligned to the shared project and to the communicative means that coordinate the work. That is, not only do the people have to find ways to communicate; they all must maintain attention on the task and the communications. Occasional doodles and other stray writings throughout the notebooks remind us that shared, focused attention is an achievement, not always to be assumed. The drawing of a wrist and hand near a similarly curved and jointed lever may suggest serious inventive analogizing, but what about the stray drawings of faces that appear occasionally, or the endemic handwriting practice that runs throughout the notebooks? The calligraphic obsession may have had its origin in young Tom Edison's development of a legible telegraphic handwriting, but it seems to have infected several members of the laboratory.

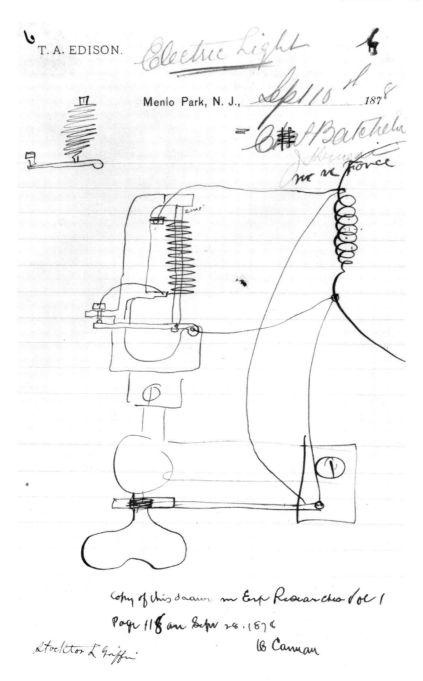

Edison's sketch of September 10, 1878 (4: 490).

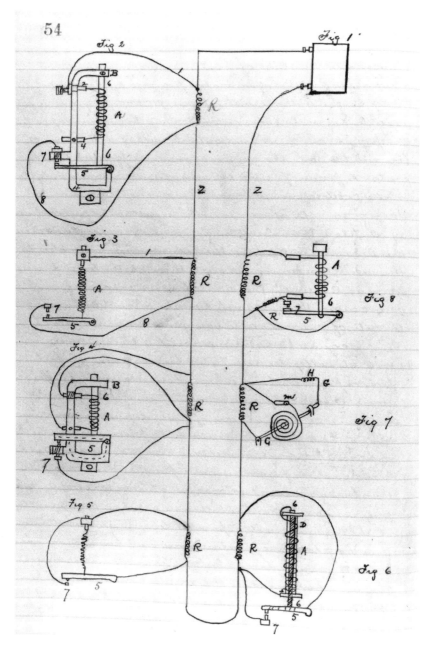

Carman's redactive drawings (3: 379).

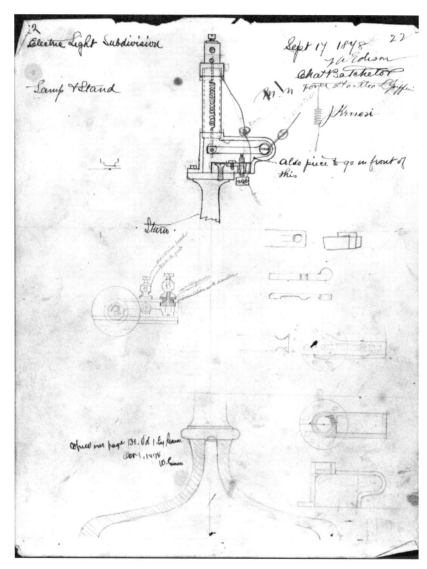

Page 26 of Unbound Notebook 16 (4: 505).

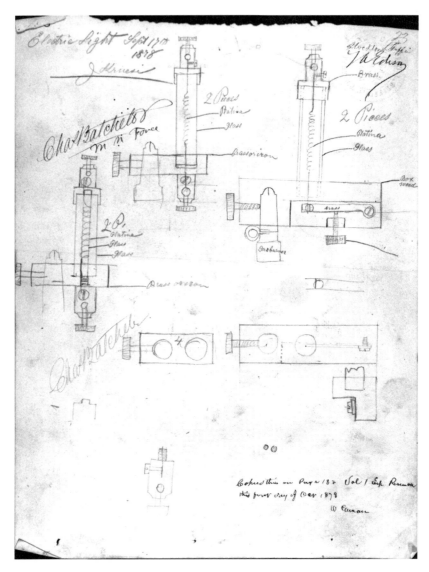

Kruesi's design sketches (4: 506).

T. A. EDISON.
Electric-Lights.

No. 214,636.   Patented **April 22, 1879.**

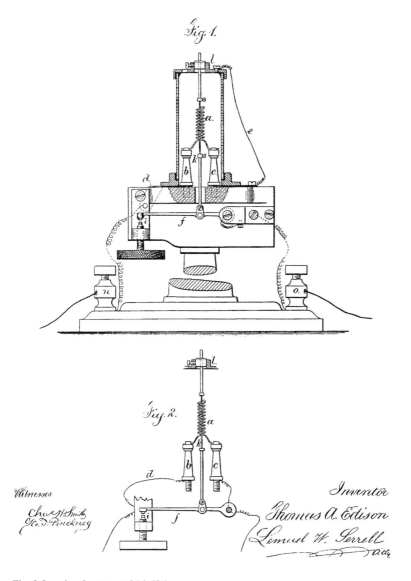

Final drawing for patent 214,636.

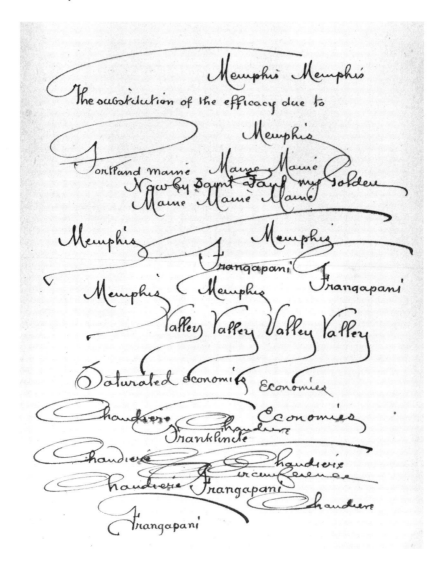

Notebook doodles (29: 0221).

But mind-wandering and doodling aside, these notebooks seemed to have been well used with commitment, focus, and communal attention. We can only speculate on the sources of this communality. Of course all the contributors to the notebooks were in Edison's employ, but the level of teamwork, shared meaning, commitment to the common project, and attention to details noted by one another went beyond the usual employee relationship.

The historians André Millard, Paul Israel, Bernard Finn, and Matthew Josephson have noted the atmosphere of mutual respect and freewheeling egalitarianism in the machine shop culture and in the community of itinerant telegraphers out of which Edison's lab grew. And a number of Edison's closest co-workers, including Batchelor, the Ott brothers, Kruesi, and Griffin, followed him from those earlier days. But in those cultures there was a sense of independent operators who might cooperate in limited ways under specific arrangements but who were out for themselves.

Yet the commitment the Menlo Park workers developed to Edison over those years suggests that they had hooked their wagons to his rising star. It was an honor to be an inventor in those days and to be working with the new genius Edison. Even more, there was the promise of enormous wealth. Several received stock in Edison enterprises. Many of Edison's early associates went on to prominence in the Edison and electrical empires. In letters to his father, Upton, in his first job and about to marry, constantly mentioned his prospects for great wealth. Batchelor and others later attested to the feeling that they were all on the verge of great wealth.

And then there was the immediate sense of reward. Each day, progress was being made on a major exciting problem. Even when results seemed uncertain, the notebooks attest to the constant flood of activity and the close collaboration.

Edison's Presence in the Laboratory Communication

Edison's personal complexities have been the subject of many biographies, and I will not begin to unpack his motives here. But I want to point out one or two things about his patterns of communicative participation. Of all the areas of discourse surrounding the production of light examined in this book, the laboratory notebooks have Edison's most direct and immediate involvement as well as the greatest part of his handwriting and drawings. Elsewhere, as with the patents, the production of text was left to his agents, surrogates, employees. Even Edison's many news interviews were mediated by reporters, although he was notorious for always having good stories to tell. In his business arrangements, he often took a more distant role, keeping his place in the laboratory. When he was pushed out of General Electric, in fact, the laboratories he created vanished with him until General Electric reinstituted them ten years later. His laboratory in West Orange remained his, organized around the multiplicity of his interests.

Numerous commentators have noticed how Edison seemed to love not only inventing but also the social atmosphere of the laboratory. He slept there, took his meals there, and stayed all night, hanging out with his boys. The lab was his home, and everywhere he moved he established a lab. His most trusted business colleagues later on were those he first got to know in the laboratory. The communication system developed in the laboratory seemed to be the most interactive, most intimate, most humanly satisfying and bond-building locale in his life.

In the interactivity of the lab we see the most intimate of the languages of light, the language of its birth. This was surrounded by the other more public languages that supported, protected, promoted, and maintained it. These other languages were thoroughly embedded in public communication systems with evolving histories, which the light and Edison had to take place in, fit into, translate themselves into. But here in the laboratory is an improvisational language, spare, creative, and spontaneous, brought into being moment by moment by people trying to work out the thing itself, using the communicative resources at hand but representing only what they needed to make the light work—then make it work better. When the light got beyond the laboratory, it became other things, to be talked about in other ways. It no longer needed people hovering over its flickering beginnings figuring out how to make it robust and steady.

5

Patents as Speech Acts and Legal Objects

In the modern capitalist world, an emergent technology is a potentially valuable property. Indeed, part of the theory of capitalism is that the desire for economic reward will foster the invention of products and production methods. As modern capitalism developed in Britain and America, special economic incentives were put into law to encourage invention and the public dissemination of inventions.[1] These incentives were in the form of "letters patent," which granted monopoly control of and advantage from the invention for a limited period. Thus a legal mechanism was created to turn ideas into property. During the period of patent protection, one could profit from the exclusive sale of a product, prohibit others from using ideas for their own profit, and sell or rent ideas to others. This new class of property, known as "intellectual property," inevitably required governmental mechanisms for certifying particular ideas as property and for protecting that property from theft.

To take advantage of monopoly in the form in which it existed in the late-nineteenth-century United States, Edison had to protect his ideas even as they were being born. The United States had developed a system of patent examination based on the novelty of a conception rather than on its proven viability, usefulness, or market value. The examination process, carried out in government offices in Washington, was based largely on submitted paper representations of an idea (picture and text), abstracted out of the particular object or product that was the realization of the idea. Neither a working prototype nor a completed product nor evidence of economic value was a part of the process of establishing that an idea was ownable and was owned by a particular person. Viability and economic value were only projected consequences—hopes that drove invention and the desire for ownership. Ownership had to come first, and the earlier the better, so that one could solicit capital and so that one could market products incorporating the invention without fear of information's getting into

hands of competitors. This system of early, prospective patenting presented puzzles for inventors and their partners as to exactly what they
wanted to protect and in what form, for they did not know how the product and the market would shape up by the time a saleable product had
been developed.

Edison's Patents

Thus, even as ideas for incandescent lighting were being developed on
paper and in prototype in Edison's laboratory, the incandescent lamp and
the system of electricity had to gain presence in another symbolic system
that circulated in hallways distant from Menlo Park. The entire research
and development enterprise needed legal protection so that Edison and
his sponsors would gain the maximum benefit from any industrial results.

On September 13, 1878, five days after visiting William Wallace, Edison
drafted his first patent caveat. In the patent system of the time, a caveat
was a first notice of intent to claim an invention. Edison's caveat was in the
form of a series of drawings illustrating regulating devices aimed at diverting current from an incandescent filament before it burned out. This idea,
as we have seen, initiated Edison's first serious line of investigation and his
public claims of having conquered the problem of incandescent lighting.
By October 14, with the help of lawyers, agents, and illustrators, he had
worked this idea into a formal patent application. This application was
approved on April 22, 1879, and patent 214,636 was granted. The same
day, a follow-up application filed on November 18, 1878, was also
approved, and patent 214,637 was granted. Two other applications concerning temperature regulation of the filament, filed on December 9,
1878, resulted in the granting of patent 218,866 (on August 26, 1879) and
patent 219,628 (on September 16, 1879). Three additional patents for regulating features were applied for on February 1, February 10, and April
21, 1879, and on May 4, 1880, patents 227,227, 227,228, and 227,229 were
granted. During this period Edison also started work on other aspects of
the system of lighting, including filaments, generators, and "subdivision
of current."[2] In ten years Edison was granted at least 245 patents concerning electric light and power—109 for lights, filaments, and their manufacture, 39 for generators, 60 for regulators and meters, and 37 for the
system and conductors.[3] Indeed, the whole technology was surrounded by
a hodge-podge of protections that made it difficult for any competitors to
enter into the terrain, even though it may never have been precisely clear
what was covered and what was not.

Patents of Ideas in the Making

To see how this thicket of patents was built out of nascent ideas, let us examine the relationship of the four early patents for thermal regulation of the filament through current regulation.

The earliest patent (214,636) is for the general principle of thermal regulation. It is not tied to any particular method, although the application illustrates an example. "Various devices for carrying my improvements into practice may be employed," Edison explained, "and I have tested a large number. I however have shown in the drawings my improvement in a convenient form, and contemplate obtaining separate patents hereafter for other and various details of construction, and I state my present invention to relate, broadly, to the combination, with an electric light produced by incandescence, of an automatic thermal regulator for the electric current."[4] The claim at the end of the patent is limited to the general idea of a regulator and a thermostatically operated shunt.

The ensuing applications identify particular mechanisms of regulation. That for patent 214,637 specifies an air or fluid expansion chamber to break the circuit, that for patent 218,866 uses a series of expansive conductive levers and springs to break the circuit as temperatures rise, and that for patent 219,628 claims the specific mechanism illustrated in the original general patent (a form of shunt that is mechanically closed when the temperature of the incandescent lamp rises). The patents of the following year also protect regulation by an improved conducting lever and spring mechanism (227,227), by combination with a generator regulator to control incoming current (227,228), and by another air expansion device obtained by surrounding a vacuum bulb with another sealed air-filled bulb (227,229). That is, Edison first patented a general approach and all the promising lines of specific development, then proceeded to investigate each of the latter. None of these mechanisms worked and none were used in the eventually marketed technology; none were even in use by the time any of the patents were granted. They were all abandoned in the time it took to process the applications, and thus all the approvals were in a sense moot. However, other features of light were being developed by the Edison team at the same time, and many of these were incorporated in various designs (some of which were patented separately—such as safety conductors,[5] which we would now see as forerunners of wire safety fuses). Other usable features appeared as parts of multiple claims in conjunction with regulating devices—for example, the use of a sealed glass vacuum

United States Patent Office.

THOMAS A. EDISON, OF MENLO PARK, NEW JERSEY.

IMPROVEMENT IN ELECTRIC LIGHTS.

Specification forming part of Letters Patent No. **214,636**, dated April 22, 1879; application filed October 14, 1878.

CASE 156.

To all whom it may concern:

Be it known that I, THOMAS A. EDISON, of Menlo Park, in the State of New Jersey, have invented an Improvement in Electric Lights, of which the following is a specification.

Electric lights have been produced by a coil or strip of platina or other metal that requires a high temperature to melt, the electric current rendering the same incandescent. In all such lights there is danger of the metal melting and destroying the apparatus, and breaking the continuity of the circuit.

My improvement is made for regulating the electric current passing through such incandescent conductor automatically, and preventing its temperature rising to the melting-point, thus producing a reliable electric light by rendering conducting substances incandescent by passing an electric current through them.

In my apparatus the heat evolved or developed is made to regulate the electric current, so that the heat cannot become too intense, because the current is lessened by the effect of the heat when certain temperatures are reached, thereby preventing injury to the incandescent substance, by keeping the heat at all times below the melting-point of the incandescent substance.

Various devices for carrying my improvement into practice may be employed, and I have tested a large number. I however have shown in the drawings my improvement in a convenient form, and contemplate obtaining separate patents hereafter for other and various details of construction, and I state my present invention to relate, broadly, to the combination, with an electric light produced by incandescence, of an automatic thermal regulator for the electric current.

Figure 1 represents the electric-light apparatus in the form in which the thermal regulator acts by the heating effect of the current itself, and Fig. 2 illustrates the same invention when the radiated heat from the incandescent conductor operates the thermal regulator.

e incandescent metal is to be platinum, ium, iridium, titanium, or any other suit-

able conductor having a high fusing-poin and the same is used in the form of a wire c thin plate or leaf.

I have shown the platinum wire *a* as a doubl spiral, the two ends terminating upon th posts *b c*, to which the conductors *d e* are con nected. The double spiral *a* is free to expan or contract by the heat, as both ends are be low the spiral.

A circuit-closing lever, *f*, is introduced i the electric circuit, the points of contact bein at *i*, and there is a platinum or similar wire, *k* connected from the lever *f* to the head-piec or other support *l*.

The current from a magneto-electric ma chine, a battery, or any other source of elec tric energy, is connected to the binding-post *n o*, and when contact at *i* is broken the cur rent passes from *o* through lever *f*, wire *k*, sup port *l*, wire *e*, post *c*, platina coil *a*, post *b*, an wire *d*, or metallic connection, to binding screw *n*. In this instance the wire *k*, bein small, is acted upon by the electric current an heated, and by its expansion the lever *f* is al lowed to close upon *i* and short-circuit th current.

The contact-point *i* is movable, and it is ad justed so that the shunt will not be closed un til the temperature of the apparatus arrives a the desired height, and, by diverting a portion or the whole of the current, the temperature o the incandescent conductor is maintained in such a manner that there will be no risk of the apparatus being injured by excessive heat o the conductor fused.

If the wire *k* is small, so as to be heated by the electricity itself, it may be placed in any convenient position relatively to the light; but if such wire is heated by radiation from the electric light, then it should be adjacent to the incandescent material.

In all instances, the expansion or contrac tion of a suitable material under changes o temperature forms a thermostatic current regulator that operates automatically, to pre vent injury to the apparatus and to the body heated by the current.

In Fig. 2 the current does not pass throug the wire *k*, and the short-circuiting lever is

2   214,636

operated by the radiated heat expanding the wire *k*. This in practice does not operate as rapidly as the device shown in Fig. 1.

The electric light may be surrounded by a glass tube or any other suitable device, such as two concentric glass tubes with the intervening space filled with alum-water or other bad conductor of heat, the object being to retain the heat of the incandescent metal and prevent loss by radiation, thus requiring less current to supply the loss by radiation.

I am aware that the electric current has been used to produce heat, and that such heat has been employed to vary the relative position of the light-giving electrodes and the length of the intervening arc. In my light there is no electric arc.

I claim as my invention—

1. In combination with an electric light having a continuous incandescent conductor, a thermostatic circuit-regulator, substantially as set forth.

2. In combination with an electric light, a thermostatically - operated shunt, substantially as set forth.

Signed by me this 5th day of October, A. D 1878.

THOMAS A. EDISON.

Witnesses:
ALFRID SWANSON,
STOCKTON L. GRIFFIN.

Text of Edison's first incandescent light patent (214,636). See page 81 for figures.

container in conjunction with patent 227,229 (a patent that includes five different claims relating to improvements in bulb design).

Edison's patent applications sought an ever-expanding set of protections around any line of development he considered promising, until the technology stabilized around an adequately protected, technologically efficient product on the market. After 1882, once a successful product reached the market the rate of patents directly connected to lamp design diminished to only a few per year until Edison Electric Light was absorbed by General Edison at the end of the decade. In the mid 1880s, the greatest emphasis was on generators and electrical distribution systems.

During the period of development of the technology, it was not clear what improvements would turn out to be truly useful. What would be considered the crucial breakthrough in retrospect was even less clear. Indeed, when everything was settled in court, in 1888, after the technology had been marketed for a number of years, only a later patent (223,898, filed November 4, 1879, and granted January 27, 1880) was considered crucial, and only one claim out of the four granted in that patent was deemed consequential (use of a high-resistance filament). The high-resistance filament did not represent Edison's main line of work, and it remained only a minor research strategy for some time; nonetheless, it turned out to be the peg on which the courts secured Edison's ownership of the incandescent light in the form that it turned out to take.

Given the peculiar nature of the patent system (which offered the lure of ownership, but which could not determine what was worth owning or what would be the crucial conceptual element of the marketed technology),

Edison and his colleagues had to represent the light in a bramble of overlapping claims, some irrelevant, some never realized, and some not even realizable. All these claims were necessarily premature, so as to create a thicket of representations that would be sure to surround the valuable product that might emerge.

To understand why Edison chose this strategy of representation within the legal system, what he had to do to create these representations, what these representations looked like, and how they functioned, we have to look in greater detail at the nature, history, and development of the patent system and at the organizational and textual forms that became incorporated in the patenting process.

Transforming Ideas into Intellectual Property

To create profit from an idea, the idea has to be transformed into an ownable piece of property assigned to an individual. The procedure for this transformation must identify an idea as an invention, establish the limits of the idea (that is, identify the size of the property), establish the period of ownership, and designate an owner (perhaps the inventor or an assignee).

This transformation process consists entirely of words and symbols. It is carried out in the legal sphere of papers circulated among officials, primarily located in the national capital, where an official record of these ideas is created, maintained, and evaluated in relation to specific criteria and instances of disagreement or contention. Secondary agents of this process are courts located in the various judicial districts of the country. Applicants or appellants to the procedure, also distributed throughout the land, are aided by their various agents and lawyers. But all must pass through what Latour (1987, chapter 6) calls "the center of calculation," which keeps the record of owned ideas.

The individual seeking ownership of an idea must define the idea in specific symbolic form, through words and pictures, and then apply for this representation to be granted status as an invention according to the criteria of the granting body. If the application is granted, the text within which the idea is represented becomes the official form of the idea, and all issues concerning the use of the idea must be referred back to that text for as long as the idea remains private property (that is, for the duration of the patent). Moreover, all questions concerning the meaning of the text must be referred through designated interpreters of that text in the patent office and in the courts.

The patented invention is itself not a specific produced object (although at times models were required to accompany an application—but these did not need to be working models, nor did they enter into the actual adjudication; they were used only for display (Dood 1983)). Thus, the granting of a patent does not require a produceable, workable material object, let alone a currently marketed, currently produced object or technology. The invention is legally not a physical entity. It is a symbolic representation—a text representing an idea. There is only a tenuous relationship between any object someone may ultimately produce and the ideas represented in a patent application; similarly, there is only a potential relationship between any product produced by a competitor and the same patent. In both cases, disputes are referred to the usual site for relating legal documents to specific concrete instances: the courts.

Because a patented idea is not the same as a produced technology and the very act of trying to produce a product risks dissemination of the idea, one may wish to seek patent protection before trying to make a workable prototype. Moreover, since the process of realizing ideas as marketable working products is likely to create substantial differences between original conceptions and final material results, one would want to obtain the widest possible patent protection for the broadest array of ideas, in order to make sure that whatever is produced is protected and that whatever competitors might wish to produce is proscribed.

The intention to get a patent is a creation of the historical development of a system. One cannot desire monetary wealth before the institution of money, because the desire is unimaginable, and one cannot desire a patent before patents have ever been granted. Moreover, once the patent system has developed, it can become the channel for the realization of other desires, such as the desire to create a large corporation or the desire to create a successful telephone system.

The History of Anglo-American Patent Laws

The patent genre, patent intentions, and the social system of patent grant developed together.[6] In Renaissance England, the letter patent was simply a designation of monopoly privilege granted by the crown for any benefit or favor to the state embodied in the monarch. For example, a king might grant a patent for the importation of salt, or the colonization and exploitation of a newly discovered or conquered piece of the Americas. Thus, the earliest patents were realized textually through the traditional forms of petition to the crown and royal grant.

Such crown privileges were, of course, open to abuse arising from the conflation of the royal pleasure and the good of the citizens. In England in 1624, out of repugnance against widespread royal abuses, all forms of state-granted monopoly were outlawed except for the single temporary monopoly granted to the inventor of a new good, under the belief that invention would advance the economic well-being of the country (Federico 1929a). A temporary monopoly was thought to encourage both invention and the sharing of knowledge to be exploited by all after the short monopoly period expired. Moreover, since invention created new value, a monopoly was not sequestering a previously open part of the economy; it was only granting temporary privilege for a value that would not have existed without the invention.[7]

Once the idea of privilege dependent on specific value to the state emerged, it became necessary to create a mechanism whereby individuals might request this privilege and present their claim to it for evaluation. In England this led a registration procedure followed by litigation in the courts when the patent was contested. This system remained in effect until 1852, when England's first comprehensive patent law was passed. In the Anglo-American colonies, patents and other monopolies were granted on an individual basis by courts and local legislatures (Federico 1929b; Bugbee 1967). The framers of the U.S. constitution were concerned to regularize and limit this practice, so they made patents and copyrights a federal responsibility under article 1, section 8, granting Congress the power "to promote the progress of useful arts by securing for limited times to authors and inventors the exclusive right to their respective writings and discoveries."

The first patent bill was signed by President Washington in April 1790, placing responsibility for approving patents on three cabinet members: the Secretary of State, the Secretary of War, and the Attorney General. They were charged with determining whether the "invention or discovery" was "sufficiently useful and important." The application was to include a specification, a drawing, and (if possible) a model. (See Dood 1983.) However, the form of the application was not further determined by the law (Federico 1936).

Because of an 1836 fire in the patent office, we only have a limited number of reconstructed files of the earliest patents. The earliest application that is currently in the patent records dates from 1790 and consists of a petitionary letter from William Pollard to Secretaries Jefferson and Howe and Attorney General Randolph requesting a patent for a spinning machine (Pollard 1790). The letter details Pollard's difficulties in obtaining a model of Arkwright's spinning machines and his failure to create a working model of it until he developed certain improvements, for which he now seeks patent monopoly. Pollard provides many financial details of the spinning industry in Britain to establish the value of the machine. The details of the machine and its operation appear to be present only in a drawing that is no longer in the patent record. Thus, in this earliest extant application the rhetorical emphasis was on the deserving character of the petitioner and the great economic value to befall the United States; the specific technical improvement was purely secondary and unargued. That is, the presentation followed the legally designated criteria of usefulness and importance rather than novelty.

The certificate for a patent granted to Francis Bailey on January 29, 1791 (*Restored U.S. Patents,* volume 1, reel 1, frame 18) bears the seal of the United States and is signed by the president and the attorney general. It looks much like a traditional diploma. The specifics of the invention are mentioned only in a single sentence, which also identifies Bailey as the inventor and attests that "the said Invention appears to be useful and important." The rest of the document includes a reference to the law, the date of issuance, testimony of the act of approval and signing, and certification. The meeting of criteria and the granting of the privilege are foremost.

During the three years this law was in operation, about 60 patents were approved,[8] but the burden of evaluating the applications was too much a drain on the time of the cabinet officers. In 1793 the law was revised to become simply a registration system with no evaluative procedures. A typical application consisted of a description of the invented object, cross-referenced to a drawing. Models (not necessarily working) were also to be provided to the patent office. The grant consisted only of official testimony that the papers were filed and the fees paid. The laxness of this law made the obtaining of patents easier, but only 9890 were issued up to 1836 under this law (Vaughan 1956). Since no check was made of prior art and since the putative inventor was not required to make a case for novelty, there were many lawsuits. Two crucial issues appear to have emerged in the litigation: the identity of the actual inventor and what exactly was being claimed as novel in the patent.

By 1830, patent applications typically included a formulaic opening statement identifying the putative inventor and a closing statement summarizing the claim.[9] These features became institutionalized in the laws that followed. At first the claim consisted only of "a listing of the important component parts of the invention" (Hantmann 1991, p. 123), but by the law of 1870 the claims marked out the boundary of the territory protected by the patent.

A new patent law passed in 1836 established a patent office with examiners.[10] The system established by this law is still in effect in the United States, with some modifications (the most important dating from 1870 and 1952). The form of the patent that was in effect in Edison's time was specified in the 1836 legislation and elaborated in practice. The procedures and criteria for examination (aimed at preventing excessive litigation) were established by both the law and the practices of the newly formed Patent Office in the Department of Commerce: novelty, invention, and utility (Bugbee 1967). These examination procedures and criteria provided a rhetorical target for applications aimed at gaining approval.

The patent reforms of 1836 initiated a period of exponential growth in patents and invention. This growth went hand in hand with the development of large industries dependent on technology and market capitalization. Throughout the remainder of the nineteenth century, the number of patents grew dramatically. In 1837, 426 patents were issued; in 1847, 495; in 1857, 2674; in 1867, 12,277; in 1877, 12,920; in 1887 20,403; and in 1897, 22,067.[11] Edison's first patent (1869) was number 90,646; his last (a posthumous one, granted in 1933) was number 1,908,830.[12]

The Typification of Patent Form

After the Civil War, the journal *Scientific American*, owned and produced by the New York patent agents Munn & Company, fostered the culture of invention by distributing promotional pamphlets that included extensive descriptions of the patent process. The *Scientific American Reference Book* of 1881, published by Munn & Company at the height of the excitement over the Edison light, gives a glimpse of the mood and the procedures surrounding patents at that time.

The 90-page section on patents opens with an upbeat assessment of invention's role in the U.S. economy (attributing to it more than $6 billion in capital investment—more than three-fourths of all capital investment in the U.S.) and of the value of individual patents. "A very large proportion of all patents prove remunerative," and useful inventions "can easily be sold for from ten to fifty thousand dollars" (p. 15). Munn & Company offer services ranging from a simple inquiry (free) through patent searches (for five dollars) and prosecution of rejections or postponements (for "very moderate" fees). The pamphlet recommends subscribing to *Scientific American* and publicizing inventions in it. It contains engravings depicting the patent office and the publisher's offices, apparently to indicate that Munn and Company equaled the mighty and intimidating splendor of the government. It warns the neophyte against cheap sale of patent rights and underhanded brokers. The pamphlet ends, as inspiration for the would-be inventor, with portraits and biographies of famous American inventors.

The main text of the pamphlet consists of advice and information on the patent process. The fees and processes for various government actions and international copyrights are presented, along with an abstract of the current laws and the full text of the latest revision. Also included are standardized forms for submission of various petitions, including the basic patent petition with cover letter and affidavit. Procedures for ensuing actions, such as rejections, appeals, interferences, and infringements, are also offered.

In the sample patent application, the opening announces the invention. The second section establishes the intention or aim of the invention. A technical description of the accompanying diagram and operations of the invention follow. The petition closes with a specification of the exact claims being made, followed by the signatures of the applicant and the witnesses. This highly regularized form is closely consistent with the format used in Edison's patents and in almost all other patents of the period. It is a form largely dictated by the 1874 law, which states in Section 4888 that the application in writing should include "a written description" of the invention and

the manner and process of making, constructing, compounding, and using it, in such full, clear, concise, and exact terms as to enable any person skilled in the art or science to which it appertains, or with which it is most nearly connected, to make, construct, compound, and use the same; and, in case of a machine, he shall explain the principle thereof, and the best mode in which he has contemplated applying that principle, so as to distinguish from other inventions; and he shall distinctly point out and distinctly claim the part, improvement, or combination which he claims as his invention or discovery. The specification shall be signed by the inventor and attested by two witnesses. (*Scientific American Reference Book*, 1881, p. 75)

A drawing might be attached. Until 1881 a model was required, but afterward was to be supplied only on request.

Consistent with these regulations, the sample patent describes an invention, identifies its inventor, and declares particular aspects of the invention as original (the claim); it further carries an official designation of the patent-granting body, a patent number, and a date from which the patent right begins.

In the late-nineteenth-century United States, a patent typically opened with one or more technical drawings signed by the inventor and two witnesses. The first page of text was headed "United States Patent Office," with subheadings identifying the inventor and the name of invention, followed by "Specification of Letters Patent xxxxx, dated xxxxx." The text, addressed "to all whom it may concern,"[13] began with a formulaic opening paragraph: "Be it known that I, xxxxxxxx, of xxxxxxxx, have invented a new and improved xxxxxxxxxxxxxx; and I do hereby declare that the following is a full and exact description thereof, reference being had to the accompanying drawing and to the letters of reference marked thereon." A general elaboration of the invention and its improvements over prior art is followed by a detailed description of the invention and its operation, typically introduced as follows: "To enable those skilled in the art to fully understand and construct my invention, I will proceed to describe it."[14] The description is usually cross-indexed to the illustration by means of reference letters. The patent then concludes with precise claims of novelty, prefaced by some such language as "I claim as new, and desire to secure by Letters Patent. . . ." The signatures of the inventor and two witnesses again appears at the end.

One obvious but unusual feature of this form as it was used in the nineteenth century is that the body of the text is in the first person in the form of a legal petitionary letter, although the patent, as indicated by the heading and opening formula, is presented as already granted. Indeed, the patent adopts the specification directly from the application, and amends it only by adding the designations of official approval by the Patent Office. Today the specification is written in the third person, without the markers of individual petition, but it is still generally the case that the language of the application is used wholesale in the grant.

Edison's first patent for lighting (patent 214,636, for the principle of thermal regulation of the incandescent conductor) follows the pattern dictated by law and custom. It begins with a drawing signed by the inventor, the patent attorney, and two witnesses. After the headings, the application is addressed "To all it may concern." The opening paragraph, consisting of one sentence, identifies the inventor and announces that he has "invented an Improvement in Electric Lights." The second paragraph specifies the problem of the metal filament's melting. The third and fourth paragraphs describe in general terms the inventor's solution to the problem (and thus his invention). The fifth paragraph generalizes the method to a variety of other devices that can accomplish the same improvement for which Edison states he is filing separate applications. The next ten paragraphs describe the specific device illustrated in the drawing, so the generalizing fifth paragraph is needed to open the door to other devices to be encompassed by the patented idea. The description concludes with a paragraph specifying that the light does not employ an arc (to distinguish it from the other existing electric light technology). The text ends with a brief summary of the claims, identifying the goal of the application to establish ownership of these specific improvements:

I claim as my invention—

1. In combination with an electric light having a continuous incandescent conductor, a thermostatic circuit regulator, substantially as set forth,

2. In combination with an electric light, a thermostatically-operated shunt, substantially as set forth.

The document is then signed by Edison and his witnesses.

Much of the non-technical text consists of "boilerplate," as is common in such legal documents, the only novelty being in the specifics of the case. In general, the aim is to gain the standard privilege and thus attempt to fulfill all the standard procedures in the safest, most traditional way.

Double Reference

Within this kind of document, as it is defined by law, practice, and Edison's example, there emerged a curious kind of double reference, the first aspect being reference to the specifics of a particular design and the second being reference to the general claims arising from the design. The design is represented by an illustration, and explication of the specifics is the centerpiece of the exposition. The specifics are surrounded by abstractions from the specific forms presented in the diagram to more general events, procedures, or objects that define the property to be owned.

Although the patent diagram and its description are continuous with the kinds of sketches that appeared in the laboratory and were conveyed to the machine shop, the claims that define the domain of ownership are more part of the discursive world of the patent system and the courts. Thus, the patent specification ties a specific configuration into a general idea, which can encompass a variety of instances and can provide a protected arena of ownership.

Edison, in his initial lighting patent, heightens this distinction by stating that he is seeking a patent for a general approach that may be realized in several different ways. The approach is to be elaborated in future patents, the details of this particular patent being only one convenient form for exposition. Claims in some other patents are more closely tied to specific configurations and designs and therefore become redundant with the accompanying illustration and description, as in claim 2 of patent 218,866 for a particular arrangement of conducting levers to act as a thermal-regulating circuit breaker:

The circuit-connection t, in combination with the levers or springs n, yoke l, light giving body i, and circuit-connections, substantially as set forth.

The cross-reference letters tie the claim to the specifics of the diagram and their elaboration in the middle paragraphs of the text.

The distinction between the specific description and the abstracted claim developed historically, as we have seen, in the processes of early litigation and later examination. However, to understand in greater depth the relationship between the concrete description and the abstract claim, we need to consider what kind of actions the text itself engages in. That is, we will consider the patent as a kind of *speech act.*[15] This analysis will also indicate the kind of transformation the light must go through as it is transformed from an idea born, represented, and explored in the laboratory into a legal entity of property.

Some Issues in Speech-Act Theory

Before proceeding with an analysis of the conditions a patent application must meet in order to be successful, we must first deal with several related difficulties concerning speech-act theory and its application to long, complex written documents.

The first difficulty is the importance of local circumstances in the identification, interpretation and realization of speech acts. For example, what I take the force of the statement "We have coffee, milk, and juice" to be depends very much on who I am, who I am speaking to, what my relationship is with them, whether I am about to go shopping, whether I have expressed thirst, and whether I am sitting at a dinner table. Simply putting such equivocal cases in the category of indirect speech acts is inadequate for several reasons: first, unless we are total strangers to a situation, we always use our knowledge of local circumstances to confirm or extend or modify our view of the explicit statement; second, most statements are not fully explicit or universally univocal in their illocutionary intent; third, there are many subtle distinctions among acts and the way acts are taken that only emerge out of the interpretation of situation; fourth, there are many kinds of acts that are conceivable only within highly defined circumstances, such as undermining the credibility of a scientific argument by mentioning a piece of apparatus used in producing the result, thereby invoking a disciplinary understanding of the inappropriateness of that apparatus to the experimental problem. Although speech acts may potentially be reduced to a few abstract categories with certain abstract guidelines, they are thereby stripped of the locally significant aspects of their meanings—the aspects that go into constructing the local event as distinctive from others and provide individuals with the subtle tools necessary to successfully respond to and negotiate events as they unfold in local circumstances.

Austin's awareness of the importance of local circumstances in the interpretation and enactment of speech acts, both locutionary and illocutionary, led him to withdraw from absolute formalizations in the closing two lectures of *How to Do Things with Words* (1962), where he qualifies his conclusions as only abstractions and cautions us to examine local circumstances. In lecture 11, in particular, Austin examines a number of examples where local factors are essential for interpretation and winds up making such statements as "Reference depends on knowledge at the time of utterance (p. 144)" and "The truth or falsity of a statement depends not merely on the meanings of words but on what act you were performing in what circumstances " (p. 145).

Searle, in *Speech Acts* (1969) and subsequent works, took on the project of pursuing the formalizations to obtain an abstract calculus of meaning that incorporated reference and illocution in a logically contained interpretive scheme. Local circumstances are included only as conditions that must be met if the completion of a speech act is to be successful. For example, a person conducting a marriage ceremony must be legally qualified to do so, and the event must be carried out in a legally appropriate place and at a legally appropriate time, between individuals legally qualified to marry, if the events are to count as a legal marriage. Here Searle helped identify some features of speech acts as they often emerge in institutionally structured settings. However, this analysis of general rules and conditions for speech acts is accomplished at the expense of suppressing analysis of the particularity of the institutional settings within which individual acts arise and at the expense of obscuring the interpretation of acts in less well defined settings.

The second difficulty is the polysemiousness of speech acts. Any speech act may be uttered and interpreted with a variety or a multiplicity of intentions and frameworks for attributing meaning. Any utterance may serve different functions for different utterers and different auditors, and these multiplicities of functions and meanings may be operating simultaneously. Moreover, the conditions of success for the utterance may become multiple, depending on the functions and meanings attributed to the utterance. In stating at a dinner party that I like vanilla ice cream, I may be placing an order, expressing delight in anticipation, revealing personal character, defending my food against the predatory habits of a dessert-loving child, making small talk, or all of the above simultaneously. The host, my child, my neighbors who served me chocolate last week, and other guests I have just met that evening may interpret the remark in a variety of ways, evaluating its effect and its effectiveness variously according to how they understand the situation and the act. A subtler example: When an intimate friend tells me of a dream, is it a personal revelation, a request for an interpretation, an invitation for commiseration, a step in the co-construction of a communal imagination, a reproach, or an invitation for me to tell my own dreams? The person telling me of the dream may have no single or clear intent and may not inform me as to what kind of response is being invited. Perhaps no particular force is attributable to the telling until the conversation has unfolded, and even then the two parties to the conversation are likely to walk away with rather different perceptions of what has happened.

The nature of a speech act or a series of speech acts is manifold and indeterminate. This indeterminacy, multiplicity, and interpretive complexity may present substantial difficulties in our closest and most spontaneous relations—difficulties we sometimes resolve only by providing some simple, determinate, and benign after-the-fact explanation that excludes some of the more troublesome interpretations triggered by the situational indeterminacy and multiplicity. Although such reductions to primary interpretations of actions and intentions may be fostered in highly institutionalized settings with highly typified actions, that still does not fully exclude multiple secondary intentions and uses packed into or pulled out of the utterances. Thus, formalizations of speech acts can at best characterize a dominant appearance in a multiple act, and can do so only in circumstances where that dominant appearance is well marked and supported in institutionalized circumstances.

The Living Complexity of Located Speech Acts

These two difficulties, which point to the power of the concept of a speech act even as they point to limitations in trying to specify the exact meaning of any particular speech act by means of a generalized understanding of speech acts, illuminate the richness of an activity embodied by utterances within circumstances. Events are alive with new forms of life that grow in the unfolding of both typified and novel utterances. Every utterance exists at the intersection of the typified and the novel as the utterance is perceived by participants coming to terms with each new moment. In Saussurean terms, speech acts exist precisely where *langue* and *parole* meet: at the alive utterance. Any attempt to reduce speech acts to a speech system removes the activity from the act and reduces complex, interpretive, intelligent, motivated human behavior to a static set of signs no longer responsive to human needs and creativity. When speech acts are reduced to a system of *langue*, the typifications—employed as resources by individuals attempting to relate through signs—are taken as the definition and the rules of the utterance. The typified speech acts then become superordinate to the activity, rather than the speech acts' being embedded parts of the overall activity. A less distorting understanding of speech acts requires constant attention to events unfolding in particular circumstances with local definition and interpretation of successful activity. Perceivable regularities in speech acts, whether perceived and acted upon by the participants or by the latecomer analyst, should be seen as historically evolved resources of typified interpretation, in relation to other social regularities and institutions that help identify the nature of each social moment as enacted by the participants.

The task of the analyst of speech activity is simplified and stabilized when the analyst looks to behaviors in highly regularized or institutional settings that help enforce recognizable and socially agreed upon characters to particular moments. Since the institutions and social understandings set the stage and define the game, it is much easier to see what is going on, and we can make plausible connections among various moments or acts if participants see and treat those moments or acts as similar. But we should not confuse a reasonably stable set of linguistic practices evolved within a particular strand of socio-historical circumstances with an absolute understanding of speech acts.

The patent process consists of a highly developed set of typified practices that surpass Searlian rigor in their mandatoriness, but that does not mean that the rigor extends beyond any particular set of typifications. Law, on the face of it, is a rigorous practice. But it is a different rigorous practice in Medieval France and in nineteenth-century America. And twentieth-century plain-language philosophy is, at least some would claim, also a rigorously typified practice, but again a different one. Each, nonetheless, evolves with novel utterances and novel moves—as do less tightly typified systems with wider ranges of freedom for novelty and multiplicity, such as contemporary literary theory (which nonetheless operates under its own set of recognizable understandings and interventions). Finally, in each of these cases, no matter how rigorous the typifications that guide the enactment any single moment may be, the dynamics of the moment grant new meaning and life to the typifications, and we must look to the dynamics of the moment to understand what is happening.

Genres as Speech Acts

The final difficulty with speech-act theory, particularly for this study, is its application to long, complex written documents. Speech acts as envisioned by Austin and Searle are short utterances carrying out single acts. For the sake of analytic clarity, Searle (1969, p. 22) explicitly excludes from consideration all but the most simple utterances.

Written texts characteristically contain more than one sentence. A text may contain many acts. Moreover, it is not clear whether sentences within extended discourses embody speech acts of specific illocutionary force in the way that isolated sentence utterances do. At best, we can imagine that a highly compulsive, closed text attempts to push a compliant reader down a certain path of reaction through a series of related acts. Nonetheless, what the sum of the various acts of a texts amount to is not clear.

However, if the text is distinctly identifiable as of a single genre, it can gain a unified force, for it is now labeled as of a single kind instantiating a recognizable social action. That is, the text effects a law (a declaration), or makes application (a directive), or contractually binds you (a commissive), or presents a scientific claim (an assertive), or conveys outrage at a governmental action (an expressive).[16] The various smaller speech acts within the larger document contribute to the larger speech act ("macro-act") of the text, and each of the sub-acts must fulfill its part within the macro-act. In fact, the expectations of generic form are such that any missing or weakly instantiated feature of the genre may weaken the text's generic force. Particularly if the genre is responsive to formal regulation, a defect in any of the sub-actions may be reason for the failure of the work of the genre. A patent application without a representation of the object, a declaration of originality, or a specification of claim is not a valid application and will not achieve the purpose of gaining approval. Thus, a defective specification will never appear as an officially approved patent, distributed in reprints by the patent office.

In a contract, many acts are fulfilled, but the overall effect is to bind parties to mutual obligations and rights, including all the stipulations agreed to in the contract. The stipulations are meaningless—both in the sense of being non-binding and in the sense of being purposeless and unmotivated and perhaps unintelligible—without the perfection of the overall contractual act. A seduction, a sale, or any other event that ended with a bottom line, a mutual agreement, and focused conjoint action among parties would have the effect of a macro-action and would give the entire proceedings the shape of the single act. Indeed, the minor actions that had gone into it would be hard to understand, hard to attribute intention to, and hard to see as effective acts without being framed by the macro-act.

Many written genres seem to resolve themselves into single acts. A patent application, a tax form, a mail order for a pair of shoes, or a final examination in English Literature 2002, once it has completed its work (to gain the patent grant or the shoes, to satisfy the Internal Revenue Service, to demonstrate competence in the subject), can be filed away purely for the record unless someone wants to call the perfection of the document or consequent actions into question. The text becomes dead and exists only in its consequences. Much of scientific writing is of this character, as articles only have a short shelf life (or citation life) and then live through their consequences or lack thereof unless someone wants to open up the dusty research. Other texts, however, must constantly be reread if they are to have force, for the texts have multiple forces that are created only by the reader's interaction with them. When we read a novel or a book of philosophy, many things are done to us. It is clearly reductionist to characterize these multiple effects under a single macro-act, such as being entertained or being enlightened. We recall a poem or a work of philosophy not just as having a single overall force (as scientific citations sometimes become symbols for single concepts), but as a collection of moments and gestures as well as an overall structure of arguments or feelings or imaginative moves. These texts live not in any sense of unified consequences, but in their multiplicity of effects on readers' minds, arising from the complex of actions realized through the texts. Left in a file or on a dusty library shelf, such texts would not do their work, nor would they do so if we were to reduce them to simple slogans which we were to carry around in our heads. Yet, even though multiplicity of action remains in these texts, attribution of genre helps to limit the domain and focus the character of the multiplicities offered by, or to be read out of, the texts—that is, genre recognition usually limits interpretive flexibility.

Transforming an Idea into a Legal Entity

Any speech act, to be successfully completed, must meet a series of conditions appropriate to that act. (Austin (1962) calls these "felicity conditions.") Accordingly, to succeed, an application for a patent must meet the conditions by which it is judged in the particular institutional conditions of patent examinations. When all these conditions are met and the patent is granted, an inventive idea is transformed into a legal property. The various parts or acts within the text contribute to the success of the overall act by addressing particular conditions.

To obtain a patent, one must have an idea for an object or a process. This object or process must be useful. It must be novel. The applicant must have invented it. All these items must be asserted in the specification. The text of a patent from Edison's time opens with an identification of the inventor and with an assertion of a new and useful invention. A description of the invention, supported by an illustration, follows. Since the applicant does not yet have the patent, he or she must cast the application in the form of a petitionary letter, closing with petitionary language (e.g., "I claim as new and wish to secure by letters patent. . . ."). This petitionary format was further framed by a cover letter, a standard form of which appears in the *Scientific American* pamphlet of 1881:

To the Commissioner of Patents:
Your Petitioner, a resident of ——,——, prays that letters-patent be granted to him for the invention set forth in the annexed specification.

signed

These petitionary features clearly signify that the applicant intends the document as a request, intends the receiver to understand it as a request, and desires the receiver do what is requested; that the text is communicated to the receiver, who is capable of interpreting the text; that the applicant believes that the person receiving the request (the Commissioner of Patents) is able to grant such a request; that the request is for something that the receiver would not already have done in the normal course of affairs; and that all social and psychological conditions of the sort that Searle (1969, pp. 57–61, 66) spells out for the act of promising must be met if the request is to be granted. The nature of the request, however, is that the receiver (the patent examiner) declare that the sender's representation of an object or process be considered a patent. That is, the petitioner must assert that his or her idea meets the criteria of a patent so that the receiver will then declare the representation to be a patent protecting the idea. Therefore, we must look into the propositions or representations embodied in the patent to see how they meet the examiner's criteria of adequate propositions.

Reference and Predication Acts in the Patent Application

Searle (1969, chapters 4 and 5) points out that every speech act has a propositional content, and that proposition consists of acts of reference and acts of predication. On one level, the act of reference of the application is to the commissioner's declaring a patent, and the predication is that the commissioner will do so. That would be the standard propositional content of a request: reference to a certain state of affairs and a predication that someone will accomplish it. However, the commissioner's declaration is based on an evaluation (to be performed by a patent examiner) of the object or process represented in the specification and of the claims predicated of that object or process. Thus, the key propositions concern the item for which patent status is sought. Because the examination performed by the receiver extends beyond the representation created by the petitioner, we must consider the propositional acts in two stages: as they are represented and as they are received. Further, we must examine what conditions must be met in each instance in order for the patent application to be successful.

The patent refers to the applicant's self, the act of invention, and the object or process that represents the invention. Thus, the patent opens with identification of the applicant, a representation of the act of invention, and details of the object. The largest part of the patent is given over to representation of the object in the form of illustrations, description of the parts of the object in relation to the illustrations, and a description of the object's operation, use, and/or construction. From the point of view of the writer, these representations rely on the writer's believing that they represent him, his actions in inventing, and (most important) the object or process he has conceived. The inventor need not have brought this idea to working perfection, so the reference is to an imagined construction that the inventor is in the process of bringing into physical realization. These representations share information about the idea with the patent examiner for the purposes of evaluation, but they also (after the patent is granted) share that information with others, thus allowing them to use or reproduce the idea once the period of protection is over.

The propositional act, however, consists of predication as well as reference. It is not enough to represent oneself as having invented or conceived of the object; one must also claim that the object is new, that it is useful, and that it instantiates some particular forms of useful novelty. Thus, a patent of the 1870s typically had near the beginning some language similar to the following, which appears in Edison's patent 214,636:

> Electric lights have been produced by a coil or strip of platina. . . . In all such lights there is a danger of the metal melting and destroying the apparatus, and breaking the continuity of the circuit.
>
> My improvement is made for regulating the electric current passing through such incandescent conductor automatically, and preventing its temperature rising to the melting point, thus producing a reliable electric light by rendering conducting substances incandescent by passing an electric current through them.

The word 'improvement' and the problem-and-solution format point toward the novelty and usefulness of what will be proposed in the patent. This patent, as has already been noted, closes with more precise claims as to what the useful innovation is: the combination of regulator, shunt, and incandescent light.

From the Speaker's Sincere Statement to the Examiner's Approval

The inventor, in making a patent application, represents himself or herself as having, of a certain date, the idea for a particular kind of device or process, and predicates that he or she believes that this idea is workable and useful, that it is an improvement of a substantial kind and therefore is an invention, and that the novel improvement can be characterized within specific claims. The applicant may always be in bad faith concerning any of these representations and predications; however, in forwarding the application the inventor must present himself or herself as sincere. The patent examiner passes public judgment on the validity of the statements in the application. In approved patent applications, the individual's belief about his or her ideas are transformed by public certification into a form of public knowledge.

The procedures for evaluation, whereby illocutionary force (embodying intent to obtain a patent) is converted to a state of belief on the examiner's part that will legally compel the desired perlocutionary effect (of actual issuance of that patent), are, however, specific and limited. The evaluation procedures attend to only certain aspects of the representations in the application. The inventor's representation of his name and geographical location are accepted on his oath. The date of filing is a matter of record of receipt and of oath. There is no procedure for determining whether the idea is workable, beyond obvious violations of physical laws[17]; the workability is left to future development.

If the idea is not workable, a patent will be of no financial value and will be abandoned, making the patent monopoly moot and insignificant. This is an important point. The patent is a monopoly only of a potential. The reference is only to an idea—a projection of a future product. The patent is of no meaning or value if that potential does not become realized or if it is not realizable. The patent examiner has no way of knowing and no obligation to determine the future prospects of the idea. Similarly, the question of the usefulness of the patent is left unexamined, because that is left to the marketplace. Since the patent monopoly will be moot if no one wishes to use or purchase the patent, there is no reason to examine the usefulness—nor is there any before-the-fact way of determining it.

Since a patent does not deal with actual produced objects, the representation is only of an idea. The idea is embodied in the patent description; there is no further examination of whether an idea is in fact present or whether this is the idea the inventor had. This loophole left open the possibility of submitting defective or incomplete representations of an object in order to stymie one's competitors, because the examiner would have no way of knowing the completeness of the idea. This is also the loophole that allowed a short-lived nineteenth-century practice of inventors' amending their already-issued patents to strengthen their positions against competing claimants on the ground that the inventor had had the correct or full idea but the representation on paper had not been fully clear or accurate. The only usual ground for rejecting a patent on the basis of its description of the idea is lack of clarity or specificity—i.e., that it is not clear what the idea is for which patent protection is sought.

The forms of examination in Patent Office practice are primarily intertextual. A patent's descriptions and claims are compared to the file of existing patents and to other representations of the current state of the art, such as textbooks and encyclopedias. Thus, of all the predications made about the idea, only the novelty of the claim is examined, leaving the examiners agnostic even as to whether this novelty is an improvement (for improvement is equivalent to the usefulness of the novelty). The most sensitive aspect of novelty is how broad the claim can be in relation to the object or process described and in relation to prior claims. In terms of ownership, broadness of claim is precisely the most crucial matter, for it will define the extent of the rights the inventor will own if the patent is issued.

Intentions and Intersubjectivity

Obtaining a patent monopoly requires that one fulfill the genre of application by meeting, in appropriate textual form (primarily of a representational kind), the success conditions of that speech act of request for a status. That is, the inventor has to represent the idea as patentable. The actual grant of the patent requires the intention of the examiner to fulfill his or her duty by applying appropriate examination procedures to determine the success of these representations of the idea as meeting the criteria of patentability. But only certain aspects of the representation come under systematic scrutiny—and even that is a kind of scrutiny that is contrary to the kind of scrutiny by which the patent is conceived (except, of course, that the inventor and the patent agent and/or attorney try to anticipate the examination procedure by patent searches and clever formulation of claims.) That is, the inventor tries to solve problems and claim turf. The examiner is not concerned about the solution of problems, but instead examines whether the turf is already occupied and tries to limit turf claimed for the specific novelties instantiated in the representation of the idea.

What the inventor and the examiner agree on is that what is sought is patent status with its monopoly privileges. If the patent is approved, they have collaborated in the creation of the patent, and they agree on the kind of thing or the status that has been achieved. They have created new value, a new property to be owned—and that property is a license to attempt to make money from a particular technology. Intentions meet over the status created by the speech act of declaration.

What has happened is that the inventor, in representing an idea he or she has had and in making reference to specific ideas through drawings representing objects in the process of being realized, must transform these objects into successful abstractions of claims that the examiner (and later the courts) will accept as patentable ideas with clear boundaries of ownership warranted by the particulars of the specific case presented in the illustration and the description and consistent with the intertextual examination procedures of the previously owned property (already in the public domain) and the legal rules governing this kind of property and its ownership. If the would-be inventor can successfully make this transformation so as to elicit the cooperation of the examiner, he or she becomes, nominally, an inventor owning an invention.

The Character and the Limits of Legal Entities

The status of inventor and the status of invention are technical, narrow legal creations necessitated by the procedures of creating a kind of property (the invention) and assigning benefits to a particular person (the inventor). These designations become legally moot once the period of patent protection is over or once the patent turns out to be valueless. Then the law then no longer cares and the designations evaporate—that is, they have no standing. Although we then popularly associate the status of inventor with the honors of the history of technology or with the orderly story of discrete inventions by discrete inventors, this is not necessarily historically warranted, as historians of technology have been pointing out with regard to the kinds of complex and continuous interactions that go into the emergence of new technologies.[18] Nonetheless, the patent system, for economic reasons only, supports a heroic notion of the history of technology.

Edison's patent, thus, accomplishes a certain kind of work of transformation, turning pieces of the technology that might produce light into pieces of property. Patents, as I have been arguing throughout this chapter, are not the final technology; they are only legal entities that surround the emergent technology as its producers try to bring it to workable use and to establish it as valuable in the marketplace. Patents surround the technology, giving it special status in the legal system and in the financial realms regulated by the laws. Thus we see why the Edison companies had to surround this uncertain new technology with as many claims that would encompass the particulars of whatever technology emerged: they had to have control of enough legal abstractions to maintain legal control of the actual material technology emerging in the marketplace.

6

Professional Presence: Edison in the Technical Press

In the technical press, Edison's efforts in incandescent lighting never got the kind of publicity that the popular press seemed to grant enthusiastically. Throughout the development process, whether he was succeeding or failing, the popular press kept the story alive and associated incandescent lighting with Edison; other developers were seen as his competitors. The technical press granted incandescent lighting only limited presence until it was becoming a material fact in city streets. Moreover, in the limited technical coverage of incandescent lighting, Edison was presented as one of many developers (e.g., Hiram Maxim, Joseph Swan, William Sawyer). Often his efforts weren't even mentioned. Further, Edison was much less aggressive in creating presence in the technical press than in giving interviews and demonstrations for the popular press. Though he had been active as a writer in the emerging electrical and telegraphic press earlier in his career, by the time of incandescent lighting Edison wrote little about his work, and what he did write was directed as much toward the general public as toward the technical community.

Despite Edison's record as a telegraphic and telephonic inventor, the American and European technical communities treated his ambitions for electric light and power with skepticism. Nonetheless, Edison needed the approval and cooperation of the emergent electrical technology community—at first to give legitimacy to his system, later to provide the knowledge and manpower to develop his industrial endeavors.

Because of the need to enlist the support and cooperation of the technical community, Edison and his colleagues worked hard to gain legitimacy and centrality for his project among the electricians. Edison strove to change his presence from that of an ambiguous and ambitious threat to that of an accepted leader of the industrial project. Yet neither he nor any of his close associates took on the task of becoming a leader of the emerging profession of electrical engineering, either in its professional societies, in its journals, or in its academic programs.[1] Instead, Edison and his collaborators (including Charles Batchelor, Francis Upton, and Samuel Insull) devoted their attention to proprietary work and to the development of the Edison companies (such as Chicago Edison). When Edison's collaborators did gain prominence as design innovators, it was in promotion and marketing—for example, Charles Hammer developed the electrical sign, and Edward Johnson and Sigmund Bergmann developed the aesthetics of lighting fixtures.

Edison and the Formation of Electrical Engineering

Edison's relationship to the development of the profession of electrical engineering had two contrary aspects.

On one hand, he was one of the exemplars of the rising field, one of the major telegraphic inventors in the period during which the groundwork for electrical engineering was laid. He also provided the primary justification for the new profession: the project of central power was the task that established electrical engineering in the United States, provided employment, and created the need for professional training and certification. And he was an early supporter of education in electrical engineering (Rosenberg 1984; Terman 1976).

Before 1882, when the Pearl Street power station opened, only six students had majored in physics at MIT in 17 years, but in 1882 MIT created a course in electrical engineering that enrolled 18 students in the first year, 30 in the next, and 105 in a single year at the end of the decade (Rosenberg 1983). Cornell University's program in electrical engineering, initiated in 1883, had an equally spectacular growth, having 218 students enrolled in 1890.

In 1886, electrical engineering was featured as a career opportunity in a three-page story in *St. Nicholas,* a magazine for young people. The article opens by characterizing electric lighting as the primary work of electrical engineers:

I suppose that most of the readers of *St. Nicholas* have seen an electric light. If there are any that have not seen the light itself, they must have seen pictures of night scenes in which the light was represented. Well, this particular method of illumination is comparatively new; it came into general use in the year 1878—just about eight years ago, —and is therefore as old as the younger readers of this magazine. When it came into use it gave employment to a new class of workers; it created a new and, all things considered, a very good profession, one that is well worth the while of our boys and young men to consider before they choose a vocation in life, —I mean, the profession of electrical engineering. (Manson 1886)

On the other hand, Edison, although he had read a considerable amount of the contemporary technical literature, was not formally trained in physics, and he always remained skeptical of too mathematical an approach to invention. He employed some trained scientists, but his approach always remained practical. He was in contact with physicists and engineers of his time, he published a few articles in existing journals, and he founded his own journal (titled *Science*); however, professional communication was not a central venue for his work. And although he played an extensive role in forming research labs and in building large corporate enterprises, he only took a passing role in the professional organizations of electrical engineering (except to receive awards and accolades).

While in some sense a parent of the field of electrical engineering, Edison was also very much its victim. Although he seems to have invented the game, he had to play by other people's rules—rules that often put him at a disadvantage. He learned to play by those rules as much as he needed to, but he never pursued the game with enthusiasm, and he regularly stepped back from professional forums once his immediate aims had been satisfied.

The Emergence of Electrical Engineering in the United States

The standard histories of electrical engineering in the United States are two volumes published by the Institute of Electrical and Electronics Engineers for its centennial in 1984: the more popular *Engineers and Electrons* by John Ryder and Donald Fink and the more scholarly *The Making of a Profession* by A. Michal McMahon. Both identify the emergence of the profession with central station power generation and the consequent founding of the American Institute of Electrical Engineering in 1884. The society itself first met at the 1884 Philadelphia Electrical Exhibition, the first showcase of electric lighting and central power in the United States.

The ostensible excuse for the founding of the American Institute of Electrical Engineering was to have an organization, parallel to those already existing in Europe, to welcome Europeans to the American exhibition. The circular calling for the formation of the society, however, pointed to a more complicated reality:

The rapidly growing art of producing and utilizing electricity has no assistance from any American national scientific society. There is no legitimate excuse for this implied absence of scientific interest, except it be the short sighted plea that every-one is too busy to give time to scientific, practical and social intercourse, which, in other professions, have been found so conducive to advancement....

It is proposed to make electrical engineers, electricians, instructors in electricity in schools and colleges, inventors and manufacturers of electrical apparatus, officers of telegraph, telephone, electric light, burglar alarm, district messenger, electric-time, and of all companies based upon electrical inventions, as well as all who are inclined to support the organization for the common interest, eligible to membership.[2]

This list indicates the heterogeneity of interests that had remained dispersed and unorganized until electric power and light catalyzed them into a new profession. In the histories of American electrical engineering, the preceding period of advance in telegraphy and telephony is treated as a pre-professional pre-history of individual entrepreneurial inventors working through informal networks and ad hoc relationships with the telegraphic companies. The inventors are presented as itinerant tinkerers working in a machine shop culture without an organized body of information, training, or regularized relations with one another. Consistent with this view of individual inventive entrepreneurs, the first organizational meeting of the AIEE, held on May 13, 1884, took up as its first piece of substantive business the sorry way in which the U.S. Patent Office was treating individual inventors and their patent agents.[3]

Machine Shop Tinkerers

Paul Israel's *From Machine Shop to Industrial Laboratory* (1992) presents the telegraphic operating room as a seedbed for itinerant tinkerers and inventors. The most active of these nomadic young men, including Edison and Pope, migrated to the urban hubs of telegraphy and telephony, where the major telegraph and telephone companies, in technological competition for efficiency and for control of patents, used small machine shops and independent inventors as suppliers of equipment. The inventors, in a climate of ad hoc product development and patent arrangements with expanding corporate enterprises, were more a part of an entrepreneurial craft than an organized profession. Their primary links being with their employers, their network with peers was informal and non-institutional, existing in random encounters and improvisatory workshop arrangements. The most substantial alliances among inventors were business partnerships, such as the brief 1869 combination of Edison, Franklin Pope, and James Ashley in a machine shop and electrical equipment manufactory.

It was out of a series of such arrangements that Edison developed the Menlo Park laboratory, which effected a major change in the working relationships of the inventors and technologists of the period, placing technologists in the employ of inventors. Important events in turning the laboratory work into a profession were the hirings of the mathematician and physicist Francis Upton (in 1878), the chemist Alfred Haid (in 1879), and the physicists Charles Clarke and Edward Nichols (in 1880). Yet Edison, relying on his "invention factory," remained an independent contractor with corporations such as Western Union.

Within this atmosphere of entrepreneurial machine shop tinkering, Edison developed his troubled professional identity. Yet before we can unpack the complexity of Edison's position within the professionalization of American electrical engineering we need to understand the earlier professionalization of electrical engineering in Britain and France. In the 1870s and the early 1880s the American technical press and engineering profession remained subordinate to those of Europe. The overseas literature, republished in American periodicals, was authoritative, and Edison needed to gain validation and acceptance for his system of lighting from the British professional world. The American electrical press and profession were industrial creatures, and ultimately industrial success would bring recognition; however, the existing British technical press and profession were already the creatures of government interests, so that gaining their professional acceptance meant addressing tests and standards.

Electrical Science in Britain

In Britain, interest in electrical science and in engineering dated back to experiments performed by Francis Hauksbee and Jean DeSaguliers early in the eighteenth century.[4] In the 1760s Joseph Priestley talked of an international circle of electricians, whose work he drew together in his *History and Present State of Electricity* (1775). Although many members of this group were affiliated with the Royal Society and *Philosophical Transactions* and with an informal network of correspondents surrounding Joseph Priestley and Benjamin Franklin, no specialized journal or organization developed to confirm and regularize this imagined community.

In the nineteenth century, natural philosophic education and discussion spread throughout many classes and urban locations of British society. Societies, academies, lectures, courses, and journals proliferated and showed an increasing tendency toward specialization. An Electrical Society formed in 1837 met until 1843, when it failed for an outstanding debt of 85 pounds. Its founding chairman, William Sturgeon, issued *The Annals of Electricity, Magnetism and Chemistry, and Guardian of Experimental Science* from

1836 to 1843 (Appleyard 1939, pp. 20–21).[5] The society issued two volumes of the *Proceedings of the London Electrical Society,* one for 1841–42 and one for 1842–43. After the demise of the society and its proceedings, a successor journal, *The Electrical Magazine,* was edited by Charles Walker, who 30 years later was to be a president of the Society of Telegraph Engineers. This journal published two volumes in London between 1843 and 1846 before going under. The society and the related publications were oriented toward natural philosophic investigation, with meetings devoted to lectures and demonstrations. *The Electrical Magazine,* for example, published articles on batteries, animal electricity, resistance, and the earth as a conductor and generator of voltaic electricity. It also reported on meetings of the Royal Institute, the Institute for Civil Engineering, and the Society of Arts, and it published abstracts of foreign journals and notices of experiments.

Earlier, engineering had emerged as a profession distinct from natural philosophy. The Society for Civil Engineering, founded in 1771, became the Institution of Civil Engineers in 1818. During the 1850s and the 1860s the problems of laying submarine telegraphic cable came to the attention of the ICE, largely as mechanical and structural issues. The electrical aspects of telegraphy were hardly touched on.

Though telegraphy was seen only as a limited engineering issue, it became an increasingly important industrial and financial force. *The Electrician: A Weekly Journal of Telegraphy, Electricity and Applied Chemistry,* published from 1861 through 1864, contained news and advertising related to the industry, but its editorial content was dominated by original papers and by reports of scientific meetings. In this respect, it stayed within the tradition of scientific and engineering societies. Something of the orientation of this journal can be gathered from an editorial in the first issue:

A Publication specially devoted to the interests of Electro-Telegraphy, and to the advancement and application of those branches of Science upon which it is founded, has long been a desideratum in this country and its Colonies. The professions of the Engineer, Mechanical Science, Railways, Photography, and Technical Chemistry, are fully represented by their own organs in English journalism. . . .

While enabling the Telegraphist to keep *au courant* with the scientific and material progress of his profession, "The Electrician" is designed to meet the requirements of the student in a field of knowledge which is every day being extended. . . .[6]

After six volumes as a weekly, *The Electrician* ran into financial difficulties and folded. It reemerged under new editorship, with much the same format in 1878, by which time the industry had developed and stabilized and interest in telephony, electro-plating, and arc lighting had developed. The new subtitle, *A Weekly Journal of Theoretical and Applied Electricity and Chemical Physics*, indicates the changing scope of inquiry.[7]

For a while after the first series of *The Electrician* folded, there was no organized communication specifically devoted to electricity. But the telegraph was becoming an important part of Britain's military and imperial expansion, drawing together its extended responsibilities and obligations throughout the world. A reliable cable to the continent was completed in 1851. A successful transatlantic cable was laid in 1866, and by 1869 a submarine cable between Aden and Bombay brought the major parts of the empire within telegraphic reach of Britain.[8] This international communication system required an extensive network of lines, the development of submarine cables, the development of standard metrics and standard knowledge, and militarily disciplined engineers trained to establish telegraphic stations in dispersed and difficult circumstances. The British military began training telegraphers at the Military Telegraph School at Chatham in 1857 (Takahashi 1986, pp. 301–313).

William Ayrtoun, the central figure in establishing electrical engineering education, also had an imperial government background. After training in physics at University College in London and Glasgow, he served in the Indian Government Telegraph Service from 1868 to 1872 and played a leading role in transforming that service. After a brief return to Britain, in 1873 he took a position in Japan at the Imperial College of Engineering. There he established the first course in the world in electrical engineering (as opposed to telegraphic training). He then brought the model home to Britain in 1878 to help create Finsbury Technical College (Brock 1981; Gooday 1991).

Government Control of Telegraphy in Britain and France

In Britain, private industry was removed from telegraphy in the 1860s because the submarine cable required trained consistency and international standardization, particularly as established in the French sponsored International Telegraph Treaty of 1865. France had already taken the lead in standardization and centralization. Its electrical telegraphic system had developed within an existing state-run system of mechanical semaphore telegraph, regulated by state civil service inspectors (Butrica 1987). Count Theodore du Moncel, who was to be a key figure in Edison's acceptance among the continental electricians, served as the chief electrical engineer of the French telegraphic system from 1869 until his death in 1884.[9]

Britain, as a party to the Telegraphic Treaty, moved to standardize service for the sake of domestic consistency and coordination with continental telegraphic systems. An 1868 British statute gave the Postmaster General authority to acquire and maintain domestic telegraphs. In 1872, William Preece, who was to become a leading electrical engineer and whose support was extremely important for Edison, left private industry for the Postal Service, where he eventually rose to Engineer-in-Chief, serving in that capacity from 1892 to 1899.

The British unification of posts and telegraphs, in turn, served as a model for the French, who determined to unify their postal and telegraph services in 1873 and actually carried it out in 1878. In France, electrical engineering education was developed to train the civil servants for the new combined service.[10]

The Society of Telegraph Engineers

The Society of Telegraph Engineers was formed in 1871 in Britain out of the telegraphic alliance of the empire-oriented government service and military (Reader 1991).[11] Of the eight founding members, five had military backgrounds: Captain P. H. Colomb, Major R. H. Stotherd, Captain C. E. Webber, Captain E. D. Malcolm, and Major Frank Bolton (Appleyard 1939, pp. 29–33). Two others, Wildman Whitehouse and Robert Henry Sabine, had been instrumental in developing submarine telegraphy. Of the 268 members on the first official list, at least 57 were from the British Postal Telegraphs, 41 from the military, and 25 from the railway and foreign telegraph departments. Only 30 were identifiably from natural science, and only 10 identifiably from industry; 105 were not identifiable (Appleyard 1939, p. 43).

In addition to many papers on cables, batteries, and other telegraphic technology, the first volume of the *Journal of the Society of Telegraphic Engineers* included a substantial number of papers showing interest in the underlying electrical science. There were studies of condensers, the duration of sparks, the production of current by various means, various applications (such as measuring temperature and detonating explosives), and the causes of natural phenomena (such as earth currents, lightning, and auroras). Over the years, the society took up telephony, arc lighting, incandescent lighting, and other emerging electrical technologies. These subjects were addressed rigorously, dispassionately, and cautiously, in contrast with the entrepreneurial enthusiasm that was driving the development of the technology in the United States.

Thus, in Britain the Society of Telegraphic Engineers became a solid and conservative forum for the evaluation of all electrical technologies, an epistemic court that claimed a corner on electrical expertise. New electrical technologies had to face inspection and evaluation according to the methods, standards, and style developed within this influential body, which in 1881 expanded its name to Society of Telegraphic Engineers and Electricians and which in 1889 became simply the Institute of Electrical Engineers. This culture of electrical expertise embodied military, imperial, and governmental values of respect, order, gravity of decision, and social responsibility. The British electricians who scoffed at Edison's claims, and before whom he had to prove the legitimacy of his enterprise at the Paris and Crystal Palace exhibitions, were among the leaders of this society. For example, the three British electricians who came to visit with the Edison representatives at the Paris exhibition on August 1, 1881 were William Preece (who had been president of the Society of Telegraphic Engineers in 1880), David Edward Hughes (who was to be president in 1886), and Sir Charles Tilson Bright (who was to be president in 1887).[12]

The membership of this society reflected the same government and military interests that established the go-slow parliamentary legislation that delayed the development of central power in Britain for almost ten years. Sir Charles Siemens[13] and other members of the society testified as expert witnesses on behalf of regulation during the parliamentary hearings on that legislation.[14]

The Development of Telegraphy in the United States

In the United States the telegraph was an instrument of domestic economy and corporate growth rather than of national power. In Britain—a geographically small and industrially developed country that already had an integrated domestic economy—the telegraph and the railroad only supplemented the existing system of internal posts, transport, and news; in the United States, they opened up and tied together vast expanses of territory. Before the railroads and the telegraph made the United States into a single economy, only a brief period of canal building had initiated commercial growth, and only the brief Pony Express experiment had attempted to bring speed of communications to the vast expanses of the west.

In the United States, telegraphy grew through private industry; attempts to bring these services and uses under government control were minimal.

The military demands on telegraphy presented by the Civil War and by westward expansion were few, sporadic, and geographically variable, so the military relied on existing commercial infrastructure and expertise, creating little of its own. Centralization and rationalization of a national system, to the extent that they occurred, were achieved through corporate mergers, buyouts, and patent control that left Western Union with a de facto monopoly by the mid 1870s.

Moreover, the technical problems of overland transmission in the United States were fewer and simpler than those posed by submarine telegraphy, which required substantial knowledge of electrical science to solve. In the United States, stringing cables, locating breaks, and repairing them were largely straightforward, as was the problem of renewing weak signals traveling long distances. It was always possible to have a human operator—or eventually a mechanical repeater—send the message on. In the United States, technological improvement was largely devoted to reducing costs through multiplexing (sending multiple signals over a single line), through the use of repeating telegraphs, by speeding up the signal, and by slowing the signal down for interpretation by operators. Efficiency-improving and cost-reducing inventions were powerful tools for obtaining competitive advantage through the control of patents. Accordingly, proprietary interests constrained the dissemination and the regularization of knowledge. Thus, American telegraphic electricians were self-taught, under private employ and contract, and engaged in small entrepreneurial innovation, rather than being part of a centralized professional system of control, training, and employment as in Britain and France.

What impetus there was for telegraphers to organize and professionalize lay in their industrial identity as telegraph operators. In the 1850s, two early industrial journals—*The American Telegraphic Magazine* and *The National Telegraphic Review and Operator's Companion*—provided commercial news, technical self-education, and industry boosterism, but each lasted less than a year.[15]

The need for news and information during the Civil War led to the founding of *The Telegrapher* in 1864. Within ten years, two other journals were founded: *Journal of the Telegraph* (1867) and *The Operator* (1874). These journals (with their various mergers, absorptions and name changes) constituted the primary media of exchange for electricians until the founding of the American Institute of Electrical Engineering in 1884. They also were the primary media by which Edison established his professional presence in the early years of his career.

The Telegrapher *and the National Telegraphic Union*

The Telegrapher, founded in 1864, was the official organ of the National Telegraphic Union, founded in 1863 for the self-improvement and self-protection of those telegraphers who identified themselves as "the intelligent and respectable class among operators" (Smith 1864). The primary interests of the union and of the journal were fraternal organization and welfare; this was revealed by the debate over a constitution at the NTU's first national congress, news of which dominated the opening issue of the journal. The most hotly debated issue was the creation of a weekly payment to operators who were sick and could not work. Subsequent issues were primarily devoted to procedural minutes and attendance lists of regional meetings.

Self-improvement was an additional theme in *The Telegrapher,* as is evident from frequent references to the sobriety of the members, accounts of the members' attempts to improve themselves, and stigmatizing stories of those who were not so sober-minded and reputable. By the third issue, self-improvement was expressed through one-page accounts of basic electrical science and recent inventions. But these bits of self-education, sometimes reprinted from European journals, always remained subservient to the more extensive stories of new telegraph systems, the expansion of the business, and the international growth of the industry. The growing status of the industry and the specific opportunities opening up were more important than the knowledge needed by the profession. Advertisements were largely for telegraphic supplies, rather than for books or journals. Thus, self-improvement was most forcefully expressed through the individual opportunities created by the advance of the industry and by individual identification with a progressive industry that was transforming the world.

Edison as a Young Telegrapher

Edison, who had become identified with this growing industry early in his career, worked to increase that identification by maintaining a presence in *The Telegrapher,* its first well-established journal. His job changes were noted in the personal column (April 4, April 25, and August 8, 1868), and his fine and rapid handwriting (an important attribute for a telegraph operator) was praised (August 1, 1868). In addition, at the end of 1867 he submitted an article (noted in the January 11, 1868 issue and published in the April 11 issue) describing his double transmitter, which allowed simultaneous transmission and reception of messages on the same telegraphic line. The description, entirely technical, is based on a schematic diagram. The fourth of the twelve paragraphs is typical of the tone of the piece:

By inspection of the diagram, it will be seen that when the instruments are at rest, there will be a constant current over the line, passing through the battery B, rheostat R, helix N, line L, helix N', rheostat R' and battery B'; but, as stated above, owing to the resistance of the rheostats RR', it will be insufficient to affect the instruments. (Edison 1868)

Here Edison is presenting his own work descriptively, assuming its value, with no technical measures or argument beyond the description of its design. The object is presented as speaking for itself, though of course the presentation is itself a form of promotion. Similar pieces by Edison appeared in the May 9 and the August 8 *Telegrapher,* and in a letter dated April 25, 1868 describing his work on an induction relay.[16]

During 1868 Edison also wrote three items for *The Telegrapher* (published in the June 2, August 15, and October 17 issues) describing the enterprises and devices of the Boston telegraphic community. The first was a letter with news about the improved wire used on a new telegraphic line and a fire at a telegraphic manufactory. The latter two were featured articles on "The Manufacture of Electrical Apparatus in Boston" and "American Compound Telegraph Wire." The original letter seemed to be a wedge for his presence in the journal, serving to make him a spokesman not only for his own work but also for the Boston community within the New York-based journal. Edison also used *The Telegrapher* to advertise his duplex telegraph.[17] His fame—due in part to his presence in this journal earlier in 1868 and in part to less formal communication among telegraphers—was sufficient at this point to support the use of his name as a prominent aspect of the advertisement, and the advertisement was bolstered by an editorial comment (by the editor of *The Telegrapher,* James Ashley) in the issue of its first appearance. In the summer of 1869, Edison's name was used in an advertisement by Franklin Pope for an Edison and Anders Magnetograph.[18]

Edison was soon to join forces with Franklin Pope (a former editor of *The Telegrapher*) and James Ashley (the current editor) to form Pope, Edison, and Company, the first electrical engineering firm in the United States. They advertised their services in *The Telegrapher* from October 2, 1869 until April 23, 1870.[19] In the issue of October 9, 1869, an anonymous editorial (probably written by editor and silent partner Ashley) praised the new enterprise and Edison and Pope individually.[20] A similar anonymous letter published in the issue of December 11, 1869 was also likely written by Ashley.[21]

By making his name and his work visible in the primary industrial technical journal, Edison was establishing a reputation. He was becoming part of a self-promoting technical community, in alliance with others gathered around the new industry's journals. He and his allies now used the Edison name as a promotional tool in signed and unsigned articles, in paid advertising, and in editorials.

Western Union's Competing Journal

The year 1867 brought a competing journal, sponsored by Western Union. First called *The Telegraphic Journal* (from March to November) and then renamed *The Journal of the Telegraph,* this journal took brief but positive note of Edison's double transmitter in its issue of April 15, 1868, calling the transmitter "interesting, simple and ingenious" but downplaying its originality and giving no technical details. The sparseness of this publicity indicates that, although Edison was then employed as a Western Union operator in Boston, he was not tied to Western Union interests, nor was the patent for this device controlled by that company. The importance of Edison's direct personal and economic connections with *The Telegrapher* and its sponsors is further indicated by a letter from Edison published in that journal's issue of April 25, 1868. This letter, describing Edison's work on induction relays, is in direct response to an article in *The Journal of the Telegraph.* Edison wrote where he had a soapbox rather than where the response would be most immediately relevant.

Soon, however, Edison began inventing for Western Union, and he received some notice in *The Journal of the Telegraph.* An article in the June 1, 1868 issue ("Automatic Telegraphy," by Edison's friend M. F. Adams) devoted its two lengthy final paragraphs to Edison's recent advance, which the Adams called "a more convenient mode of transmitting and translating." In its issue of August 2, 1869, *The Journal of the Telegraph* took note of Edison's position (which was to last only briefly) as superintendent of the Gold and Stock Telegraph Company, replacing his soon-to-be partner Franklin Pope.[22]

Stepping Back from the Telegraphic Press

Having socialized himself into the ways of the industry, made connections, and established his name, Edison stepped back for a time from publicizing his endeavors in the technical press. In 1871, Edison further distanced himself from *The Telegrapher* when he left the partnership with Pope and

Ashley. Mutual accusations followed, Pope and Ashley charging Edison with double dealing and charlatanry and Edison claiming that Pope and Ashley were unsupportive. It appears that Pope and Ashley had invested in the development of the Edison name, while Edison continued to search for new opportunities and alliances.

Ashley sought revenge by attempting to undermine Edison's reputation, which he himself had previously sought to build.[23] Particularly after Edison's invention of the quadruplex telegraphy, which allowed four signals to be sent simultaneously over the same wire, the pages of *The Telegrapher* regularly vilified Edison as the 'Professor of duplicity and quadruplicity.'[24] Exercised over Edison's close ties with Western Union and the consequent positive publicity Edison was gaining in *The Journal of the Telegraph* (*The Telegrapher*'s chief competitor), Ashley complained in an editorial that *The Journal of the Telegraph* had become "a personal organ" for Edison and George Prescott (Western Union's chief electrician) "to advance their purpose and manufacture fame, reputation, and fortune for them exclusively."[25] Ashley then addressed Edison's affiliation with another new journal, *The Operator*: "Recent issues of this publication have contained such fulsome and absurd slaverings of Edison as to make that person appear even more ridiculous than he otherwise would. It is generally believed that these are productions of his own pen...." (Of course, we should remember Ashley's own anonymous editorial and letter of 1869 in support of his silent partnership with Edison and Pope.)

The energy and vitriol of Ashley's attacks, which have the flavor of the age of P. T. Barnum and Mark Twain, are out of keeping with the sober content of the rest of his journal. The attacks reflect at least as much on the journalistic and professional environment that Ashley shared with Edison as on the moral failings imputed to Edison. For example, the following comment appeared in the issue of July 29, 1876, shortly before Ashley himself made a deal with Western Union to merge his journal with theirs and to become editor of the combined publication:

Edison About to Astonish the World Again.—Stand from Under!

The professor of duplicity and quadruplicity has been suspiciously quiet for some time. Since his great discovery of the new moonshine, which he christened "etheric force," he has apparently subsided, and except as a plaintiff or defendant in various suits and applications for injunction, now against Western Union, then against the Atlantic and Pacific Telegraph Company, with occasional legal dashes at Messrs. Prescott and Jay Gould, he has made no sign. Satisfied that some great purpose was concealed under this reticence, and determined that the world, and especially the telegraphic world, should not remain in ignorance of the doings of the most remarkable genius of this or any other age and country, *The Telegrapher* has taken the trouble to penetrate the mystery which enshrouds his purpose. It has been discovered that the professor is about to astonish the world, and confound the ignoramuses who are engaged in the improvement of telegraphic apparatus, by the production of an invention which has taxed his massive intellect and unparalleled inventive genius to the utmost, and destined to revolutionize telegraphy—which is so frequently revolutionized that the process has become an everyday occurrence.

Duplex, quadruplex, electro-harmonic and automatic are all about to be consigned to the scrap heap. Inventors who have not yet secured their reward are about to find their genius and labor come to naught, and shrink abashed before the production of Edison's genius.

The professor of duplicity and quadruplicity is engaged in perfecting a new telegraphic system, by which any number of messages can be transmitted simultaneously upon one wire! The establishment of Menlo Park has not been created for nothing. That secluded precinct is yet to become famous throughout the earth as the spot where this invention was conceived and brought to light!

The subject is so vast and overwhelming that it paralyzes our good steel pen, and the shock which this announcement will create will be so great that we must for the present forbear further comments.

N.B.—It is understood that this invention has as yet, been sold only to the Western Union Co. Applications will be favorably received from others who desire to purchase. Former "wicked parties" not excluded from this opportunity to invest.[26]

In 1877, the merger of *The Telegrapher* and *The Journal of the Telegraph*, with Ashley as editor, eliminated both long-standing journals as places for Edison to publicize his ventures.

Edison and The Operator

The Operator, founded on March 1, 1874, provided Edison with a more friendly vehicle for publicizing quadruplex telegraphy and other inventions. Edison soon affiliated with *The Operator,* and later he claimed to have founded it.[27] In fact, however, as the editors of the Edison papers point out, Edison first advertised in the issue dated July 1, 1874, and joined *The Operator* as science editor in August, several months after its founding. By the end of September, though, Edison was so strongly associated with *The Operator* that Ashley could accuse that journal of being a mouthpiece for him.

The Operator was to become the medium by which the inventive community kept in touch with one another and with the industry. Over the course of its complex history it absorbed many impulses, interests, and other journals within its changing format. In 1883, with the rise of electrical power and light, it spawned the highly successful *Electrical World. The*

Operator atrophied rapidly, while *Electrical World* flourished and absorbed many of its competitors.[28] In their increasing professionalization, *The Operator* and *Electrical World* can be seen as bellwethers of the state of the technical electrical community in the United States.

Electrical World's format in the early twentieth century as an electrical engineering journal contrasts radically with the format of the initial issues of *The Operator* a few decades earlier. The masthead of the early issues defined a much more informal mission: "*The Operator*—We Chronicle but Friendship." The first issue opened with a humorous poem on the travails of working as a telegraph operator; the rest of the first page was devoted to a description of the profession that offered details about training, salaries, and recruitment. In addition to describing the snappy new Western Union messenger uniform, the four-page issue provided news briefs, jokes, and amusing anecdotes about specific operators. The "Correspondents" column was a series of comic fictions. Much of the content seems an outgrowth of informal messages and jibes that might have traveled the wires during slack times.

The Operator's editorial statement defined its mission:

A lively little paper devoted to the operators, and to such telegraph news as would be interesting to the fraternity in general, we feel has been wanting. We ask your indulgence and support in this our humble effort in that direction. As far as it lies within our power we will endeavor to make its columns fresh, spicy and sparkling.... We shall give such personal sketches as we trust will give offense to none and will be appreciated by all.[29]

The first few issues contained sports stories about the baseball teams of local telegraph offices and lengthy letters about personal events at various offices. The journal spoke not to the improved classes of *The Telegrapher* but to the rowdy bottom of the craft, those just gaining a foothold. Nonetheless, within the first few months some stories of electrical interest, about lectures, books, extension of lines begin to appear among the personal news. By the seventh issue (June 1, 1874), advertisements appeared. Edison was one of the earliest advertisers, and for a period the sole advertiser with a device targeted toward the novice: "Edison's Ink Recorder, for Amateur Students and Private Lines. Patent Applied for. Send for Circular, Edison and Murray."

On July 15, 1874 *The Operator* took note of Edison's advance in quadruplex telegraphy, which *The Telegrapher* had derided:

Mr. Thomas A. Edison, of Newark, N.J. has for some time past been engaged in perfecting a system of "Quadruple Telegraphy," or sending two sets of messages and receiving two sets at the same time over a single wire. His apparatus being in order for experimental work, Messr. Orton, Prescott, Hunter, & Brown of the W.U. Teleg. Co, and Mr. Grant of the Montreal Teleg. Co., met in the electrician's office on Wednesday, 8th inst., where they witnessed the successful working of the Quadruplex over a circuit of about two hundred miles in length, two wires having been looped to Philadelphia and back for the purpose.

Messrs. Bogart, Phillips, Boileau, Mixer, Cook, Fullum, Gramzow and Kennedy of the night force took seats at the instruments, and for an hour exchanged newspaper specials at a rapid rate, all the instruments working nicely and promptly. Some changes in the instruments have suggested themselves to Mr. Edison, and as soon as these are made, in a few days, the experiment will be repeated on a longer circuit, it being proposed to use then a wire to Boston and back.[30]

Edison, in his capacity as scientific editor, wrote a three-part series on duplex telegraphy (September 1, October 1, and November 15, 1874). The September 15 issue contained a supplement with two further articles by Edison. Moreover, Edison's own work continued to be publicized. The issue of October 15 carried a description of his advances in quadruplex telegraphy, and that of January 15, 1875 included a reprint of his article for the *British Telegraphic Journal,* "On a New Method of Working Polarized Relays."

Once Edison had established himself in the pages of *The Operator,* he stopped writing for the journal and let others report on his doings, just as he had done with *The Telegrapher* and *The Journal of the Telegraph.* This time, he left no enemies behind him. After the end of 1874, when he gave up his editorship, he still remained a presence on *The Operator*'s pages; for example, the issue of February 1, 1875 reports that Western Union has offered him $450,000 for his quadruplex invention.[31] Over the years, Edison remained a favorite son of *The Operator,* and the journal continuing to print positive reports of his accomplishments, some even evincing a bit of the light-hearted enthusiasm Edison had gotten in the popular press. *The Operator* also carried biographical pieces about Edison, and even a comic account of how the phonograph could be used to ease the operator's burden by answering repetitive inquiries automatically.

Edison's Failure as a Scientific American

The next venue in which Edison attempted to establish professional identity, this time as a man of science, was *Scientific American,* long established as a premier organ for the new culture of invention. Combining science, invention, and industry in a vision of technological and economic development, *Scientific American* promoted an ideal of science situated in

American pragmatism and in the workshop. The 1845 prospectus at the journal's founding sought an audience of both "the intelligent and liberal working-men and those who delight in the development of the beauties of Nature." When founder Rufus Porter sold it (within a year) to Orson Desaix Munn and Alfred Ely Beach,[32] *Scientific American* became even more deeply tied to the world of invention and patents. In the Civil War period, Munn and Company became the primary patent agent in the country, opening a Washington office for that purpose (Borut 1977; Shenton 1977).

Many of the pages of *Scientific American* were devoted to patent news and technological information, describing steamships, railroads, bridges, tunnels, manufacturing processes, new forms of energy, the changing urban environment, and world's fairs and industrial expositions. Science also had its place; there were accounts of discoveries, astronomical observations, natural history, and the principles of various sciences. Though the circulation figures are unreliable, owing to promotional exaggeration, before the Civil War the journal claimed a weekly circulation of more than 25,000, and in the final three decades of the nineteenth century it claimed weekly circulations ranging from 40,000 to 50,000.

For Edison, *Scientific American*'s vision of science transforming daily life defined a noble status for the inventor: that of an individual using science to advance national industry. Edison's published letters in *Scientific American* sounded this voice of American science. In his first such letter, dated November 7, 1874, he advanced the cause of American science in his own person, claiming that he had previously invented a principle that a recent *Scientific American* had reported from the British Association for the Advancement of Science. Although exactly which of his own inventions Edison was referring to remains unclear,[33] the point here is that he presented himself, an American, as the equal of British men of science speaking before reputable British scientific societies.[34]

The following June, Edison drafted another letter to the editor of *Scientific American* (never published) describing a peculiar effect he claimed to have noted.[35] Here he presented himself as a discoverer rather than as an inventor. A few months later he announced the discovery of the "etheric force." His notebooks indicate that he first noted the phenomenon on November 22, 1875, and that he continued with active initial experiments through the middle of December. He rapidly publicized his discovery by inviting representatives of the press to his Newark laboratory for a demonstration and by giving reporters access to his notebooks. In the closing days of November, articles presenting Edison as a man of science appeared in the *Newark Daily Advertiser,* the *Newark Morning Register,* the *New York Times,* the *New York Daily Witness,* the *New York World,* and other papers.[36] The *New York Herald,* for example, portrayed Edison as a man of science "whose name promises to become famous as the discoverer of a new natural force."[37]

It is worth noting that Edison, in interviews, in notebooks, and in letters, presents himself as stumbling across this odd "etheric effect" while busy inventing. He presents his scientific discovery as a by-product of his more practical work, as he was later to claim with the Edison effect and as is consistent with his often-articulated belief that science should be based on practical experience and practical goals. Yet, as this discovery came at the end of a year when Edison was attempting to establish a presence in *Scientific American* and to adopt the mantle of a man of science, at the very least we can see that he was predisposed to be receptive to observations that might lead to discoveries even if they had no apparent immediate practical purpose.

This would-be discovery, however, was not to be accepted. A debate broke out in the popular press and moved to the scientific press, primarily *Scientific American.* By December 8 Edison had sent a letter to *Scientific American* briefly describing some of his "etheric force" experiments. Responding to the controversy over Edison's claims, *Scientific American* on December 25 published a cautiously positive editorial and a description of Edison's discovery.[38] A more skeptical editorial appeared in the next issue, and letters of criticism followed in further issues.[39] In a letter published on page 81 of the February 5 issue Edison answered these criticisms, and in a longer letter on page 101 of the February 12 issue he responded to Edwin Houston's criticism that the "etheric force" was just an example of the previously discovered induction. That criticism appeared in the *Journal of the Franklin Institute,* an elite scientific journal based in Philadelphia (Houston 1876; see PTAE II: 762). Characteristically, Edison responded to this and other criticisms in *Scientific American,* where he had established presence.

Although the "etheric force" experiments are now seen as precursors to Marconi's discovery of electromagnetic radiation, at that time Edison's arguments could not sustain his claims to have observed a previously undiscovered phenomenon. Edison was unable to establish credibility as a scientist. For the next two years he did not publish any scientific or technical work under his own name. Nor did Edison join the American Electrical Society, founded in 1874, attend any of its meetings during its existence into the early 1880s, or publish in its journal.[40] Not yet the great public figure he was to shortly become, Edison remained an inventor content with his reputation within the electrical industries.

An Inventor before the Scientists

The invention of the phonograph finally won Edison a public reputation as a man of science; however, to the technical community he remained an inventor. The next time he appeared in *Scientific American*, his purpose was to publicize an invention. This time he proceeded with great caution, withholding announcement and publication for several months while perfecting the device. When he did release information about the phonograph to *Scientific American*, it was through a letter, signed by his assistant Edward Johnson, that cast the invention as a work of science.[41] Edison followed this letter with a demonstration at the offices of the journal, described in the issue of December 22, 1877. Edison let the machine speak for itself:

Mr. Thomas A. Edison recently came into this office, placed a little machine on our desk, turned a crank, and the machine inquired as to our health, asked how we liked the phonograph, informed us that it was very well, and bid us a cordial good night....[42]

The sensation created by the phonograph gave Edison a major presence in the scientific community. On a trip to the District of Columbia in April of 1878, he presented the phonograph before the American Academy for the Advancement of Science, at the Smithsonian Institution, and before both houses of Congress and the president. The extensive press coverage of these and other phonographic demonstrations lauded Edison as a man of science. Yet Edison never published a paper on the phonograph or on related principles in the scientific literature, nor did the phonograph gain much of a presence in the scientific literature under other authorship. An article in the April 1878 *Journal of the Franklin Institute* described Edison's advances in the telephone and the phonograph (Plush 1878), and one in the May issue described microscopic examination of a phonograph cylinder (Frazer 1878). Frazer never even mentioned Edison's name. These articles present the telephone and the phonograph as great technical achievements and as curiosities for the membership of the Franklin Institute to consider, but the inventor is never portrayed as a scientist. Edison's role as inventor was reaffirmed in a September 1878 *Journal of the Franklin Institute* description of the microtasimeter[43] and in a May 1879 note containing news about the electric light.[44]

In November of 1879, Edison finally published an article in his own voice in the *Journal of the Franklin Institute:* "The Action of Heat In Vacuo in Metals." In this paper, originally read before a September 2 meeting of the American Association for the Advancement of Science, Edison presented a number of detailed observations incidental to his experiments with filaments.[45] Although the paper offered some explanations for the observations, the observations were primary and nothing was presented as a spectacular discovery or a strikingly new phenomenon. Here Edison appeared in the guise of a careful technician, working on problems with filaments, who came across information of interest to scientists:

In the course of my experiments on electric lighting I have developed some strik- ing phenomena arising from the heating of metals by the electric current, espe- cially wires of platinum, and platinum alloyed with iridium. These experiments are in progress. (Edison 1879)

The Christmas 1879 demonstration of incandescent lighting gained the scientific notice of the Franklin Institute. At the Franklin Institute's meet- ing on January 21, 1880, A. E. Outerbridge reported his observations of the Menlo Park illumination, and the ensuing discussion included com- ments of other members of the Institute who had also seen the Menlo Park illumination (Outerbridge 1880). The discussion concerned the success, the efficiency, the workings, and the economics of the technology rather than the scientific principles. Edison, at his moment of accomplishment, appeared as a system builder and not a knowledge discoverer.

In a similar way, in April of 1882, during the construction of the Pearl Street station, Edison presented to the American Society of Mechanical Engineers a "Description of the Edison Steam Dynamo" co-authored with Charles T. Porter, a noted steam engineer whose Porter-Allen Engine was adapted as part of the dynamo. That paper became the lead article in the July 1882 of the *Journal of the Franklin Institute* (Edison and Porter 1882). Edison now was seeking scientific approval of the technical feasibility of his system.

The only scientific "discovery" Edison pursued in this period was the so- called Edison effect, an electrical discharge from the negative leg of a fil- ament drawn toward the positive leg which he noticed while investigating carbon deposits in electric lamps. He was cautious in releasing these find- ings, not making any announcement until the end of 1883. Moreover, he immediately put the effect to practical use in voltage measurement, thereby inventing the first electronic tube and the first electronic instrument.

Edison's Science

Edison, though he had given up his public stance as a man of science, nonetheless understood the importance of science as a major cultural force and as a site for validation of his work. If he could not get direct acceptance as a scientist, he was still ready to risk money to support science, to profit from its advance, and to gain scientific legitimacy for incandescent lighting and other projects. In 1880 he encouraged the freelance journalist John Michels to found the first weekly scientific journal in the United States, *Science.*

Science, which lasted only from the middle of 1880 through early 1882, was an economic failure. Originally projected to have a circulation of 5000, it achieved a subscription of less than 180, perhaps with an additional 150 in other sales.[46] Although Edison withdrew his backing and the journal collapsed, it was reborn in 1883, backed by Alexander Graham Bell, to become the prestigious organ of the American Academy for the Advancement of Science.

At some point in the winter of 1879–80, John Michels broached to Edison the idea of founding a weekly modeled on the successful British journal *Nature.* Edison put him off to a later time when electric lighting would be more advanced.[47] In the spring of 1880, Michels, claiming to have additional backers lined up, returned to Edison with a prospectus suggesting five investors, each advancing $2000.[48] Edison at first was apparently agreeable and sought other backers among his friends. When Michels expanded the proposal to a $25,000 capitalization, half to be provided by a single investor if Edison would allow his name to be used prominently in the publication,[49] Edison took control and set limits. He agreed to provide full backing for 18 months, but only on a current-accounts basis. According to Michels, Edison had declared "I do not like partners."[50] And Edison did not want his name officially associated with the project, even as a stockholder; the point of this was that it would enable Michels to deny Edison's backing.[51] In view of Edison's usual desire to promote his own name, his desire to keep his role in *Science* hidden suggests the extent to which he saw the journal's value to lie in its apparent independence.

Articles in *Science* reported on such subjects as the Iroquois language, microscopic technique, evolution, anthropoid brains, and the expansion of gas solutions. The early issues contained articles by such eminent scientists as Alfred Wallace, Thomas Huxley, C. W. Siemens, and Louis Aggasiz. The opening editorial suggested that the contents would be under the supervision of recognized authorities in each of the fields.

Vol. 2.]   Whole Number 60.   AUGUST 20th, 1881.   [Price 10 Cents.

Cover page of the Edison-sponsored journal Science.

Moreover, there was extensive reporting of the meetings of various American scientific societies and the Smithsonian Institution. The American Association for the Advancement of Science received particularly detailed coverage. Extended proceedings of meetings were printed, as was the address of the outgoing president George Barker.[52] Barker was also to endorse the journal at the end of its first year.[53]

Under the rubric of general communication and advocacy of science, however, *Science* aggressively supported Edison's work. The premiere issue (July 3, 1880) contained an article by Edison's assistant Francis Upton on "Electricity as Power." The second issue (July 10) had a strong editorial supporting Edison and discounting his detractors; in the next year and a half, four more editorials lauded Edison's light[54] and one editorial praised his system of preserving foods in vacuum sealed bottles.[55] An article published in the issue of December 4, 1880 under John Michel's signature drew an analogy between the scoffing that had accompanied the introduction of gas lighting and the current skepticism toward Edison's work (Michels 1880). And the issue of October 1, 1881 included a reprint of the *New York Times* report of the Hawaiian King Kalakaua's visit to Menlo Park. Michels was quite aware that it was a function of the journal to publicize Edison's light, and in his correspondence with Edison and his secretaries he regularly called attention to these publicity efforts.[56]

By the middle of 1881, with *Science* seemingly not moving toward profitability and with the electric light gaining wide credibility, Edison began losing interest in the journal. He agreed to let Michels seek backers to form a corporation,[57] and in October, when Michels proved unsuccessful, Edison gave notice of his intent to cut off funding, which he did at the end of the year.[58] The journal vanished shortly thereafter, taking with it Edison's last attempt to define himself within the world of science.

Back among the Electricians

Even in the world of electrical invention, Edison received uneven coverage. Personal competitions, ill will, and the sense of threat that Edison seemed to evoke in some electricians led to idiosyncratic reporting.

After *The Journal of the Telegraph* absorbed *The Telegrapher* under the joint editorship of Ashley and Pope, the combined journal paid little attention to incandescent lighting until the issue of January 1, 1881, which carried a report of Henry Morton's paper on the possibilities of incandescent lighting, presented at the annual meeting of American Electrical Society.[59] Morton was among the American electricians least favorably disposed toward Edison, and his report focused on the Maxim lamp. Though it appeared after the 1879 and 1880 Menlo Park illuminations, Morton's article merely mentions that Edison was engaged in similar work.[60]

A two-part translation of an article by Theodore du Moncel from *La Lumière Electrique* on the "Progress of Electrical Science in 1880," published in the March 16 and April 1, 1881 issues of *The Journal of the Telegraph,* does not mention Edison's work at all (du Moncel 1881). An August 1 lead article on "The Future of Electricity" by the British engineer Maurice Keil does not mention Edison or incandescent lighting (Keil 1881). A half-column announcement about the Paris exhibition in the next issue (August 16), however, does contain two positive sentences about the Edison display:

Edison's display promises, when completed, to be one of the most popular in the exhibition. By Sunday night 500 Edison lights will be burning on the first floor especially fitted up for them.[61]

Nonetheless, the three-column description of the exhibition in the October 16 issue—the journal's only substantial coverage of the exhibition—barely mentions lighting, and Edison appears only in the last paragraph in the middle of a list of names of exhibitors:

Among the exhibitors of Electric lights are the following names: Maxim, Swan, The Compagnie Force et Lumière, Edison, Jablochkoff, Serrin, Brush, The Compagnie Lyonnaise, Lontin, and Siemens.[62]

The December 16 issue contains a bulletin about the awards granted to Americans but deletes mention of Edison and his Grand Diploma, despite the proud talk of Americans' excellent showing.[63] No follow-up article appears to expand upon the bulletin. An 1882 reprint of W. Grylls Adams's London lectures on "The Scientific Principles Involved in Electric Lighting," extending over several issues, devotes only a small section to incandescence and mentions Joseph Swan and St. George Lane-Fox but not Edison (Adams 1882). Finally, toward the end of 1882, by which time the Pearl Street station was already operating, Edison is mentioned in the published results of the efficiency tests from Paris and in some business and technical notices.[64]

The Operator, however, remained supportive of its favorite son. Its issue of October 15, 1878 contains an interview within which Edison expresses high hopes for incandescent lighting:

"Now that I have a machine to make the electricity, I can experiment as much as I please. I think," he added, smiling, "there is where I can beat the other inventors, as I have so many facilities here for trying experiments."[65]

An editorial in *The Operator*'s issue of December 1 minimized Edison's difficulties in producing a successful lamp and gave him an opportunity to elaborate his optimistic plans.[66]

In its issue of January 15, 1879, *The Operator* printed a satiric poem that had originally appeared in the British journal *The Electrician*. By putting the poem in a new context, *The Operator* reversed its satirical thrust. Whereas *The Electrician* had criticized Edison's overoptimistic brashness, *The Operator* reveled in Edison's expansiveness and in British discomfort, presenting Edison as an icon of American success:

I'm the greatest genius going, and the talents I possess
"Are enough to stock the world,"—Please see the New York Press;
And I give you fair warning that I claim in every land
What could, would, should, or might come of all I take in hand.

I may mention, for example, that my *newest* Telephone
You'll hear to-morrow's rumor—so let telephones alone;
And the Phonograph I'm *making* will take part in a debate;
I'm *going* to do a lot of things—so you are all too late.

They've contrived Electric Lighting, without consulting me,
On your side of the Atlantic—this is downright piracy!
But I've taken out a patent that will cover all that's known,
And—when I've *thought it over*—I shall count it all my own.

And, lastly, as a clincher, I've determined in my wrath
To patent Electricity, and Magnetism, both;
If this is not effectual to answer my intent,
I guess I'll take a patent for my sole right to invent.[67]

The Operator continued to support Edison as reports of his failure to perfect an incandescent lamp circulated, printing a biography of Edison—the journal's second—as its lead story on February 1, 1879.[68] On March 15 it printed another anecdote from Edison's early days, presenting Edison's brashness as good business practice for an ambitious, clever young man; this story was written by Edison's friend and favored journalist Edwin Marshall Fox (Fox 1879). A short interview with Edison in the April 15 issue[69] denied difficulties, and a brief editorial in the May 1 issue faulted him only for announcing his intentions in advance.[70] More anecdotes about Edison's cleverness and good humor followed in the issues of June 15 and September 1, and two stories of his progress appeared in the July 15 issue.[71]

Only in its issue of October 1 did *The Operator* waver temporarily: "Much had been expected of Prof. Edison, but as yet we have seen no Edison light."[72] The article saw great future in arc lighting but little hope for incandescence. The next issue, dated October 15, reprinted a similar brief skeptical comment from a local newspaper.[73]

But after the Christmas 1879 demonstration, *The Operator* returned to fully supporting Edison. The issue of January 1, 1880 contained an article titled "The Electric Light a Reality." After recounting the hard work, persistence, and inventive genius that had gone into the success of the incandescent light, the article moves to other Edison projects, including experiments on a medicine for headaches and other nervous disorders and a scheme to extract gold from mining tailings.[74] A January 15 article detailed the Menlo Park exhibition.[75] The February 15 issue included a column of material republished from Upton's article in *Scribner's Monthly*.[76] A March 15 article gave a positive assessment of Edison's accomplishment.[77] The August 15 issue had several columns of Edisonian anecdotes. The October 15 issue contained a column of quotation from Edison's *North American Review* article, and other stories published during this period told of Edison's work on an electric locomotive.[78] In 1881 *The Operator* published extensive positive coverage of the Paris exhibition; the January 15 issue reprinted much of Theodore du Moncel's article in *La Lumière Électrique*.[79] And *The Operator* covered the opening of the Pearl Street station in its issues dated August 1, November 11, and December 23, 1882.[80] *The Operator's* strong advocacy of Edison was very much in the spirit of the popular press. It even reprinted material directly from newspapers and popular journals.

In January 1883, with volume 14, *The Operator* became a weekly. For a brief time it appeared under the title *The Operator and Electrical World,* but by the end of April the two journals were separated. *The Operator,* with its interests narrowed to telegraphy, soon vanished. *The Electrical World* flourished and soon boasted "the largest circulation of any electrical journal in the world."[81] However, the new journal distanced itself from Edison.

Edison's Peculiar Place in **The Electrical World**

The Electrical World treated Edison with caution and even some distrust. Although it had been created to meet the expanded interest in electrical industries fostered by electric light and power, the journal took the stance of representing all the emerging electrical industries and refused to be a cheerleader for a single industry leader. The heroic portrayals of Edison in the popular press were treated with increasing irony. Fewer stories on Edison appeared, while his competitors received extensive coverage.[82]

Stories about the Edison enterprises appeared in *The Electrical World* largely as business news or within general commentaries on the advance of electric lighting.[83] In 1883 and 1884, the journal's first two years, only two features on Edison installations appeared: "The Edison System on Board Steam Vessels" and "The Edison Electrical Lighting Station, at Berlin, Germany."[84] No feature stories on Edison's success in the United States were run during this period, although there was some indirect publicity in two lead stories on his suppliers of fixtures[85] and in an interview with Edward Johnson, vice-president of the Edison Electric Light Company.[86]

Edison did get play in stories about fairs and expositions, such "The Electric Lighting of the Louisville Exposition" and "Incandescent Lighting at the [New Orleans] Exposition."[87] But his technology was treated mostly as a given, rather than news. As the business side of the industry sorted itself out, Edison's importance became apparent in his numerical superiority whenever statistics were cited. For example, the issue dated November 1, 1885 reported statistics from the recent *American Electrical Directory* indicating that Edison controlled about 80 percent of the incandescent lighting industry.[88]

Edison also had a large but ambiguous personal presence in *The Electrical World* as a commentator on the industry. Interviews with him in this capacity appeared in the issues of June 30, 1883, April 19 and November 29, 1884, and January 21, 1885.[89] But nowhere in the journal was this attention to Edison's opinions translated into explicit recognition that he had emerged as a leader of the industry. Though it did not accord such personal attention to any other like figure during the 1880s, *The Electrical World* treated Edison's pronouncements with skepticism. For example, citing Edison's repeated statement in the middle of 1883 that "there is nothing more in electric lighting to be invented or required," an editorial commented that if Edison believes this

there is a strong probability that he will awake at no very distant day and find that the Edison system has been left behind.... As we have before pointed out, however, Mr Edison's public utterances, whenever they may have any bearing upon the commercial Status of the Edison Electric Light Company, are to be received with a grain of allowance.[90]

Clearly *The Electrical World* took a broader view of the industrial possibilities than an Edison portrayed as trying to corner the market for himself.

Edison had established dominance of the industry by 1887, and in that year *The Electrical World* ran a few features on the Edison companies—but only in supplements and special numbers, sponsored by Edison for special distribution to promote new installations. The "Kansas City Number"

(July 16, 1887) was devoted entirely to Edison and Kansas City,[91] and the twelve-page "Supplement" of August 25, 1888 was devoted entirely to a single article describing the new Edison Machine Works in Schenectady.[92]

An American Industrialist, and Not a Scientific American

The Electrical World reported extensively on the electrical industry's new professional organizations, but Edison and his colleagues did not take active roles in those organizations. The initial meeting of the National Conference of Electricians, held at the Philadelphia Exposition on September 8, 1884, was attended by many of Edison's competitors, American and foreign, but not by any of Edison's collaborators. The minutes, printed in full in *The Electrical World,* do not mention of any of Edison's co-workers or any Edison installation, despite the prominent displays of Edison products and Edison's ceremonial appearances at the exposition. The first president of this organization, which was to become the American Institute of Electrical Engineering, was Henry Rowland. Rowland had been a friend of Edison and had collaborated with George Barker on one of the early published tests of the Edison system, but his inaugural speech as the first president of the AIEE was a paean to pure science—a speech, as Hounshell (1980) has pointed out,[93] countering the practical direction Edison indicated for American science. Rowland defined a profession that left Edison on the outside.

The early meetings of the National Electric Light Association, reported in detail in *The Electrical World,*[94] also seem to have lacked any participants from Edison enterprises, while all other interests in the industry were broadly represented, including many individuals and corporations with whom Edison had had fallings out (e.g., Otto Moses) or lawsuits (e.g., U.S. Electric Lighting).[95]

Professional legitimation of Edison's technology was left to friends and dependents outside his organization. Publicly distributed tests and evaluations were carried out by George Barker, Henry Rowland, C. F. Brackett, and Charles Young—American academics who had hitched their star to Edison and who (at least for a time) had strong bonds of interest and friendship with him. International legitimation was to come at the Paris and Crystal Palace expositions with the enlistment of British and French electricians, including William Preece, Joseph Swan, and Theodore du Moncel.

Edison focused his organizational attention on the Edison companies, relying on their growth for the advancement of his technology. Thus, even though he was eventually to accept many honors and honorary offices in professional organizations, even though earlier in his career he had used the professional networks of the day to learn the ropes and to gain personal presence, and even though in retrospect he is seen as one of the founders of the profession of electrical engineering, during the "lighting years" Edison increasingly sought identity as an entrepreneurial inventor and industrialist. Although he became the great American hero of science and technology, he never gained a strong voice as a scientist or even as an electrical engineer. Instead, the enduring identity of his later career placed him with his industrialist buddies Henry Ford and Harvey Firestone.

7

A Place in the Market

"Different genius," Adam Smith writes, "is not the foundation of this disposition to barter which is the cause of the division of labour. The real foundation of it is that principle to perswade which so much prevails in human nature. . . . We ought then mainly to cultivate the power of perswasion, and indeed we do so without intending it. Since a whole life is spent in the exercise of it, a ready method of bargaining with each other must undoubtedly be attained." (Smith 1978, pp. 493–494) This observation suggests that all economic transactions are rhetorical—that is, they are exchanges of value, and value is a human discursive construct. Yet the economics that developed from Smith's analysis seems to ignore reflection on persuasion (or rhetoric), treating value as that which is worked out in the marketplace. The needs and desires of the customers that drive the demand curve are viewed as unreflective givens rather than as perceptions that can be influenced. Customers are treated as knowing what they want and how much they are willing to pay for it (at least as worked out in the market process). Further, purchasers are treated as knowing not only how desirable a commodity is but also how much is available and how it compares to other desirable commodities. That is, economics assumes free and rational choice by individuals having knowledge of the marketplace, of themselves, and of commodities. Any lack of knowledge is quickly worked out in the marketplace, which then, in its behavior, provides encapsulated knowledge.

Such an understanding of economic transactions may be appropriate for corn, for cotton, and even for eighteenth-century luxury goods. In the twentieth century it may be appropriate for calculating the decision among competing heating technologies (oil, gas electricity, wood), because these technologies are available in the marketplace, well known and understood, and reliable and stable in their delivery systems and their quality. Advertisements by a local gas or Edison company may evoke some curiosity, but any reasonably diligent homeowner can get sufficient information to make something approaching an economically rational decision—within the limited terms of economic calculus.

New Technologies and New Values

But what about new technologies? When the world of opportunities is changing, both suppliers and consumers only have visions and predictions, chimeras and hopes, to go by. Those who wish to develop and sell new products must locate the seeds of desire in the consumer and then nurture that desire. Purveyors of new technologies must make hopes seem realizable, project anticipated but unproved benefits, promise plausible but uncertain long-term costs, and elevate their future systems above current real options to entice risk-taking customers to buy into the dream so as to make the dream real. They must create value by defining a place for the technology; a place for people to put their needs and desires into; a recognizable place where consumers, governments, and investors want to go. Moreover, since this place must be imaginable to consumers, it must be built on existing sites in the market. Every technology, no matter how novel, must work by analogy with existing products defined as competitors. The new technology must locate itself against a product already established in the market and then displace it. In order to displace the current niche holder, the technological entrepreneur must identify or create a dissatisfaction with the current technology so that the consumers will find change worth the trouble.

Technological change reminds us that market value is more than the result of rational economic calculation and market forces: value is driven by perceived desire and perceived benefit. When available products and market niches are stable and are understood by all actors, the value that people attribute to products, and thus their desire as expressed through the demand curve, can seem like a natural fact, as economics does not question why people value or desire the products and services for which they bid. But when tastes or values are changing, desire is being created and channeled in new directions, and producers may have some persuasive role in the social construction of demand. Demand is no longer an essential given of a rational quantitative calculus. Desire must be rhetorically elicited and directed, perceptions must be influenced, and meanings must be attributed.

Rhetoric, Marketing, Created Value, and Social Change

The rhetorical economic work of locating unmet desire and matching potential products to desire is the work of marketing. Marketing, however, is not traditionally considered a significant force in economic calculation, and marketing theory remains underdeveloped both as a practical and as an academic endeavor (van Nostrand 1997, chapter 7). However, if we wish to understand any culture enacted through any set of economic arrangements, no matter how stable, we need to consider how people have come to attribute values to goods and how they have come to demand particular products and services. It makes some difference to economy, society, and culture if people save all year in anticipation of an annual binge of gift giving, if they have complex and expensive mating rituals, or if a substantial part of the emotional life of a country is invested in an extensive network of spectator sports that has grown through corporate-controlled electronic media.

The late-nineteenth-century United States was a site of rapid socio-economic change, where the new culture of techno-industrial corporate America was being formed. The rapid economic growth took shape within the existing values and structures of American society and culture, even as it transformed those values and structures. Moreover, the dialectic shapings and transformations were accomplished through the active agency of individuals and groups of individuals engaged in their separate projects, as they built on the terrain of existing society. Thomas Edison, both as an individual and as the center of numerous collaborative corporate enterprises bearing his name, was a highly influential agent of socio-cultural-economic change, and intentionally so. As he matured from newsboy tinkerer to one of a small group of industrialists who came to stand for a new way of life, Edison came to understand that he was engaged in a socio-economic and ultimately cultural project as well as a technological one.

Edison, in developing a system of centralized power distribution, had to persuade many people attending to various discourses to place value in his projected vision, so that they would risk investments, commitments, and purchases of various sorts. This chapter examines how Edison created value for a number of audiences by positioning his new system against competing technologies. The points of competition make visible how Edison and his surrogates sought to transform existing market values into new ones. That is, the competition defines a place in the market as well as a target for creating comparative value until the product is valued sufficiently in itself to maintain its own market.

Typified Representations and Typified Technologies

In other chapters, examining the representation of incandescent lighting in the various discursive systems in which it needed to gain meaning and value, I have emphasized the typified forms of utterance, or the genres, by which system-appropriate meaning is recognizably asserted. I have also examined the historically developed structures of activity within the discursive systems that made certain tropes, topoi, and strategies effective. However, once a technology has achieved a robust social presence, it is perceived as more than a collection of representations from discursive realms; it is perceived as something of a unitary social fact; thus it aggregates its own core of meanings, which draw the various other relevant discursive systems together in a new configuration. The automobile, for example, was valuable not only from economic, industrial, personal and other perspectives; it fostered a new way of life, creating new commitments and activities. Roads were built, suburbs grew, and the automobile industry became a key component of the American economy. Even more, the separate values, although analyzable separately, cannot be acted on independently. A change in the cost of fuels cannot in itself dictate a switch to a more efficient mode of transportation; the organization of cities, commerce, and leisure must be taken into account. The automobile, as a technology, became a form of life with its own obduracy, a social type that then created place and precedent for other technologies that might be considered of its type. People want cars, not for one or another distinct reason, but because cars are a necessity of modern life—in most places, a full modern life would be unthinkable without one. If the gas-powered automobile is to be displaced, it will likely be displaced by something that seems in many respects like it—the same sort of thing, but without the "un-economical" or "un-ecological" values we now attribute to the technology—so we are more ready to seek an automobile powered by natural gas, hydrogen, or electricity than we are to reorganize our lives so as to eliminate the need for rapid, flexible, personal transport.

Because of this typification of technologies and their meanings, it would help to advance Edison's project if he could do more than just cobble together meanings in a variety of discursive system. It would be better to find an existing technology that was already accepted as a social fact, had already worked its way into numerous meaning systems, and had a unifying presence of its own—provided that Edison's new technology could be represented and perceived as more valuable than the existing technology. That is, a beatable competitor could help advance the acceptance of

Edison's novelty by making its reality seem more imaginable and by decreasing its disruptiveness.

Edison Wants to Electrify the World

Edison's underlying project was truly transformative. The modern system of centralized electrical power would carry out the work of industry and would provide essential transport for commerce and personal travel, domestic conveniences and comforts, and lighting for homes, streets, offices, and factories. The breadth of Edison's vision is evident from some of his interviews and actions just at the point when he began work on incandescent lighting.

In the September 10 *Sun* interview, under the headline: "Invention's Big Triumph: An Electric Machine that will Transmit Power by Wire,"[1] Edison attributed his interest in electrical power to his having witnessed the difficulty of the western miners' work when fast-running streams were going to waste nearby. Edison claimed to have turned to George Barker at one such site and asked "Why cannot the power of yonder river be transmitted to these men by electricity?" The story considers the great industrial possibilities of electric energy that could easily be transported hundreds of miles by wire and "that will revolutionize the motive power of the world." Light is not mentioned in the article, except for the eight arc lights used by William Wallace to demonstrate the efficiency of his generator.

On September 16, the *Sun* announced that Edison had solved the problem of incandescent lighting. The story carried the subheadline "Sending Cheap Light, Heat, and Power by Electricity." Edison was quoted as follows:

The same wire that brings the light to you will also bring power and heat. With the power you can run an elevator, a sewing machine or any other mechanical contrivance that requires a motor, and by means of the heat you may cook your food.[2]

The following day, September 17, the *Sun,* under the headline "Power Flashed By Wire," continues its discussion of Edison's perceptions of the great potential of the Wallace generator for the transmission of power for a wide range of industrial uses.[3] The *New York Telegram* of September 18 reports Edison's promises of his new invention in terms of illumination, power, and heat and specifically mentions the powering of the machine.[4]

Throughout the development of the light and power system, Edison kept the other uses of his power system in mind. In September of 1879, even before achieving a successful lamp, he was inquiring about the power needs of sewing machines in order to develop an appropriate motor for domestic and industrial use.[5]

Even earlier, in September of 1878, simultaneous with the start of his work on incandescent light, Edison had shown interest in the problems of noise and pollution from the new elevated railroads.

While Edison was reported to be offering small, immediate solutions and making no promises about electrical railroads, one of his main backers for electrical light and power was the railroad magnate Henry Villard. Supported by Villard's active interest, Edison took up electrical traction as a problem in spring 1880, just after his first successful demonstration of light, while he was still developing the system for installation. His first trial run of a primitive model was in May of 1880. By 1883 Edison had formed the Electric Railway Company of America, which demonstrated a successful passenger train at the Railway Exposition in Chicago in June of that year.[6]

Throughout the development of the illuminating system, Edison devoted more effort to the improvement of generators than to the more highly publicized light; this is evident from the number of laboratory notebook pages devoted to each. One of his biggest breakthroughs was developing a generator with an efficiency of 80 percent at a time when the efficiency of the best generators available was 55 percent. Once Edison was convinced of the possibilities of improved generation as a result of his visit to Wallace, he turned his attention to the light bulb and the surrounding system. Moreover, he and his colleagues perfected many parts of the system, including meters, fuses, switches, and connectors. The integrated system was recognized by Edison's contemporaries and by later historians as providing a major competitive advantage over other incandescent lights, but for Edison the system was from the beginning the larger product he was selling: centrally generated electric power. Light was just a means of entry into the market.

Edison's vision of the potential uses of electric power may well explain his insistence on marketing electrical light as part of a central system, rather than in small isolated systems as arc lighting was being sold. Sales of isolated Edison systems during the first few years of the production of incandescent lighting indicated that short-term opportunities were best in small, private generating units. The Edison Company for Isolated Lighting was profitable in its first year (1882), paying a 10 percent dividend,[7] whereas the Edison Electric Light Company did not turn a profit until 1886.[8] In October of 1885, the number of lamps illuminated by isolated systems totaled 132,875, while central stations illuminated 88,300.[9] Edison, however, held back development of the business of isolated systems so that such equipment would not compete against central stations and would not impede the development of the larger electrical power industry. In 1881, for example, work on isolated systems for ships and railroads was slowed while attention was turned to the development of the Pearl Street station.[10] In their 1883 report to stockholders, the directors of the Edison Light Company went so far as to state the following:

Valuable as the business of Isolated Lighting unquestionably is, there can be no doubt that Central Station Lighting will prove to be the much larger source of Revenue.[11]

But He Sold Us on Incandescent Lighting

Despite Edison's clear interest in a broad range of electric power possibilities, his public representations, his public attention, and even his relations with backers focused specifically on the incandescent light. The massive publicity that continued until the opening of the Pearl Street station focused overwhelmingly on the light.[12] Interviews with Edison were almost totally about the light, with no mention of other uses and with little concern for generators or other parts of the system. Transport of power across large distances was rarely mentioned.[13]

Demonstrations exhibited how well Edison was able to produce incandescent light. The Christmas 1879 and 1880 displays centered on the illumination of Menlo Park; little attention was paid to other uses of electric power, although motors for water pumps and sewing machines were demonstrated at the 1879 display (Jehl 1937, volume 1, p. 423).[14] During the period leading up to the opening of the Pearl Street station, the public's attention was directed almost entirely toward lighting. However, a brief comment in a short interview with Edison vice-president S. B. Eaton in the *New York Tribune* of March 28, 1881 reveals Edison's intentions:

We shall have a larger revenue there [the downtown District] from power than light. Power will be supplied for industrial and domestic purposes; for instance, for small machine shops, elevators, printing presses, lathes, and all sorts of domestic contrivances, such as sewing machines, house elevators, pumps for elevating water, forced draughts of air for ventilation, fans suspended from ceilings, etc. Besides there will be furnished an arrangement for heating water, so that a bathtub of cold-water can be warmed by electricity almost instantly and almost at no expense.[15]

The technical press too focused on illumination. Although developments in efficiency and size of generators received some attention, they were often discussed in relation to the economic competitiveness of incandescent lighting.

Indeed, although Edison clearly had larger designs for the electrification of all forms of work (which in the twentieth century was been realized beyond what even Edison could imagine), and although he had already attempted to publicize his prophetic vision, the projected technology soon was narrowly represented as incandescent illumination. Edison cooperated in presenting his total design as that of incandescent lighting delivered through centralized distribution of power, with the broader vision of a multiple-use system remaining only to guide his long-range planning and development of the system made possible by the success of the incandescent light. How and why did this happen?

Quite simply, city and town dwellers had been well prepared to understand, accept, and pay for a centrally delivered lighting technology by their experience of gas lighting, supplemented by the newly emerging electric arc lighting industry. They had already developed, articulated, and acted upon a perceived need for illumination. They were familiar with a mode of centralized delivery and organization, they recognized a workable corporate and financial structure, and they had accommodated themselves to the cost. Gas lighting was a complex fact of life that had established multiple values. Arc lighting extended the possibilities to electricity, and now incandescence could be perceived as a further extension and an improvement.

One might give a technology-determinist explanation of Edison's choice to foreground illumination as the core of his project, arguing that lighting was better developed, ready to happen, and therefore the right technology to exploit. However, there were other technologies on which an electrical power industry might have been built. Small electric motors and electric railroads, also close to exploitation, might have presented fewer difficulties and occasioned less skepticism in the expert community. Indeed, Edison attempted to develop these once he felt sure that he had overcome the major obstacles to illumination. However, they did not gain his first research attention, nor did he give this work high public profile. All the early Edison companies concerned with central power contained the word 'light', 'lighting', or 'illuminating' in their names, and the one company devoted to railroads did not carry Edison's name despite his insistence everywhere else of the importance of having his name in companies' titles. Even more curious, although the screw-in socket was invented early in the development of the system, pronged plugs were not developed until well into the twentieth century, so that for many years everything other than a light bulb could be connected to the central power system only through a direct wire or clamp connection or by means of a screw-in light socket adapter (Schroeder 1986). The other uses were not even designed into the system in a convenient manner.

The Rise of Gas Utilities and the Formation of a Market

To understand Edison's rhetorical use of the competitive analogy with gas lighting, we need to examine the position gas lighting had arrived at by the last quarter of the nineteenth century.

Around 1800, gas produced from distillation of coal began to be exploited for the lighting of factories (Stotz 1938). In the early decades of the century, in major cities in Europe and the United States, on the model of existing water utilities, companies developed to supply gas to customers through underground mains from centralized gas factories. Improvements in burners, compression, and gases led to increasing intensity in the lighting. Although there were episodes of unbounded competition, with as many as four or five gas mains of different companies on a single street, the industry moved toward consolidation, so that each gas district was typically supplied by a single company.

In New York City, where Edison was to open his first regular central station, the first gas utility was founded in 1823 and capitalized at $1 million through offerings of common stock. Gas utilities became one of the first vehicles for public investment in industry, preceding the development of the railroad and telegraph industries. The New York company was immediately profitable and paid dividends of 12.5 percent by 1830. By the end of the Civil War, three other companies shared the Manhattan franchise, each doing well. Mains were laid throughout the city, and the majority of streets were lighted by gas. Gas lighting became a regular feature of middle-class housing, of industry, and of business.

After the Civil War, the gas business boomed; dividends of the four Manhattan-based companies were typically between 20 and 35 percent. However, in the 1870s, new methods of producing gases that burned brighter and cost less brought about a period of intense competition, loss of profits, and corporate vulnerability. During this period the exploitation of natural gas (available in only some regions) brought further turmoil to the industry, though not in New York.

Fear and Instability in the Gas Lighting Market

Around 1880, despite the wide acceptance of gas use and the presence of lighting utilities (which delivered centrally produced illuminants through underground conduits, metered them, then distributed them through the walls of buildings to points of use), new products were changing the competitive market, and dissatisfaction left consumers open to new solutions. The gas technology had a number of drawbacks. The fear of explosions and gas-ignited fires was real and was reinforced by many highly publicized incidents. The warranted fear of gas poisoning was reinforced by the use of gas for suicide and by the frequent appearance of trace odors of gas. Contamination of soil, groundwater, and cisterns also was a recognized problem. Further, room air was made unpleasant and even stifling by the soot from lamps, the consumption of oxygen, and increased room temperature. Finally, gas lighting was seen as too harsh for domestic relaxation, as distorting natural colors, and as unflattering to female beauty. Thus, many drawing rooms still were lighted by tallow and oil lamps, which were perceived as more aesthetically pleasing.

The illumination market was further destabilized and opened to electrical solutions by arc lighting, which was on the verge of commercial exploitation in the late 1870s. Preliminary demonstration projects at fairs, in theaters, in lighthouses, and on urban streets had shown the technology to be powerfully effective outdoors and in large spaces. The large market that gas utilities had developed for street lighting was now threatened.

In the late 1870s, New York newspapers were filled with stories about competition in the gas industry, about changing stock prices, and about the newest wonders of arc lighting, all of which provided useful templates for the Edison stories to exploit. Investors were on the lookout for a new technology that might corner the illumination utility market. Customers were ready to buy a replacement product.

Edison Addresses the Gas Lighting Market

From the beginning of his interest in central power, Edison was aware of the need to position his technology against gas lighting. In the September 16, 1878 *Sun* article that announced his "Newest Marvel," Edison stated: "When the brilliancy and cheapness of the lights are made known to the public . . . illumination by carbureted hydrogen gas will be discarded." He further described how the insulated wires would be "laid in the ground in the same manner as gas pipes" and how he would "utilize gas burners and chandeliers" already in place. "The housekeepers," he proclaimed, "may turn off their gas and send the meters back to the companies whence they came."[16]

Edison's notebooks, from the earliest days of his work on incandescent lighting through the period of development, contained estimates of cost compared to gas light. Of course, Edison was not the only one interested in such estimates; potential backers, clients, and competitors also wanted to know. Some of the taste of the rough-and-ready competitive atmosphere is evident in this early telegraphic exchange between Edison and his London agent.

Nov 19, 1878. Please give approximate cost compared gas. Barrett, London. Paid. Nov 22, 1878. Very Much Cheaper. Collect.[17]

Calculations of competitive cost appeared in newspapers, electrical journals, and gas industry journals throughout the period, the estimates becoming more precise as the details of the system took shape.[18] Moreover, during the early years of actual service Edison companies provided lamps and actual service at a loss so as to keep the price competitive with gas until economies of scale and further development could bring the actual cost down.

Design Analogies

The incandescent lamp, as it was called,[19] was reminiscent of the glass chimney surrounding a gas lamp, which in turn followed the model of the oil lamp. Further, early incandescent fixtures were designed to resemble gas fixtures, and many installations were created by wiring previously installed gas fixtures, the wiring being run through abandoned gas lines. The early on-off switches also were made to resemble gas valves, even though rotary switches were not the most effective kind for electrical circuits.

The central system of delivery was also patterned on gas lighting. Central districts (like those around gas-producing plants) were created around power plants; these would then deliver the power through underground conduits or mains laid in the streets. In fact, companies seeking to lay cables under the streets of New York had to follow laws designed for the gas industry and had to incorporate under the Gas Statutes (Friedel and Israel 1986, p. 192; "The Edison Light," *New York Times*, 26 October 1881). Use was metered, and billing (as with gas) was based on actual use.

Even the corporate and financial structure of electric lighting was modeled on that of the publicly held gas utilities. Thus, electricity could

Nov 19 - 78

N.Y.

Edison Mo

Please give approximate
Cost Compared gas

Barrett
London

7 Paid
ma G

Telegram, William Barrett to TAE, November 19, 1878 (17: 1044).

T. A. EDISON.

Menlo Park, N. J., *Nov 22d* 1878

Barrett London

Very much Cheaper

5 Collect

Menlo Park N.J. 22d

1040

G Ma

Telegram, TAE to William Barrett, November 22, 1878 (17: 1047).

become a vehicle for investment, and the promises of incandescent lighting became financial news. The prices of electricity stocks were often compared in the newspapers with those of gas stocks, and the inverse relationship was apparent: as electricity stocks rose, gas stocks dropped.

Explanatory Analogies

Not only were the material system and the financial and organizational system modeled on gas utilities; soon the Edison project was explained to the public entirely in terms of lighting. A story in the *New York Herald* of October 12, 1878, headlined "Edison's Electric Light. What will Revolutionize the Present Method of Illumination. Cheaper and Better than Gas," is typical. The article, which is based on an interview with Edison, explicitly positions the electrical system as a replacement for gas lighting:

In the central stations will be the magneto-electric generating machines run by engines. Wires will then be run in iron pipes underground after the manner of gas pipes, connecting with dwellings, stores, theaters and other places to be lighted. The gas fixtures at present used, instead of being removed, will be utilized to encase the wire. In the place of the burner will be the invention, and meters will be used to register the quantity of electricity consumed. Their form is not yet determined upon....The amount of light can be regulated in the same way as can that from gas....No matches being used, and there being no flame, all the dangers incident to the use of gas are obviated. The light gives out no heat.[20]

The article then describes a demonstration by Edison, who sets his electric light directly side by side with a gas lamp. The reporter comments:

The electric lamp seemed much softer. A continuous view of it for three minutes did not pain the eye, whereas looking at the gas for the same length of time caused some little pain and confusion of sight. The inventor next exhibited the light turned down low. It gave mild illumination.

One of the noticeable features of the light when fully turned on was that all colors could be distinguished as readily by sunlight.

The Analogies of Competitive Advantages

These themes of parallel system, greater safety, greater comfort, and greater aesthetic pleasure run throughout Edison company literature throughout the period of development and competition. Edison actively defined the benefits of his electrical power system against the recognized faults of gas, so that it would be perceived as preferable and therefore as the next-generation solution to the same consumer need.[21]

The gas industry, at various times, dismissed the threat, comforted itself, fought back, and then repositioned itself on new ground. The meeting of the American Gaslight Association in the middle of October 1878, despite recent drops in gas stocks and some apprehension by some members, continued much as usual; the threats from Edison were largely dismissed, and Professor Henry Morton of the Stevens Institute gave a demonstration to reassure the membership that electric light would not compete with gas for the same business.[22] In the press, particularly that of the gas light industry in both the United States and Britain, many similar denials appear.[23]

As electric power expanded by taking away the lighting business, the gas utilities reorganized their business around heating and cooking, where it remains despite the attempts in the middle of the twentieth century to market the "total electrical kitchen" and the "total electrical house."

Edison was successful in positioning electrical power as a lighting technology to replace gas lighting for as long as it took to build the system. With time, it expanded to exploit wider possibilities that Edison had imagined from the beginning. Edison's success in selling the United States on light is suggested by the fact that it has taken scholars such as Thomas Hughes to remind us that Edison was developing and marketing a system of electric power and not just a light bulb (Hughes 1983).

A Good Competitive Analogy Is One Where You Come Out Better

The development of electricity in Britain offers an instructive comparison. There a strong alliance between the gas utilities, the government post office (which controlled telegraphy and therefore electrical development), and Parliament resulted in stringent regulations that hindered the development of an electrical power industry for almost ten years. Thus, although the analogy with gas was effective in the United States, where the private gas industry was vulnerable, in Britain it mobilized a strong opposing alliance.

Economics without Rhetoric? An Economy without People?

Technology competitions are extensive rhetorical exercises to identify a point of competition, to create a perception of one's new technology as belonging at that point of competition, and then to actually win. One has to establish a place for the new technology. When it has found its place, it becomes its own fact of life, creating a new landscape on which any new technology has to get a foothold. Competitive analogy seems to be one of the major rhetorical mechanisms by which markets grow to absorb and value new products, fostering the social change that comes with new opportunities for purchasing the tools of living. Those who successfully provide us with new products are taught this lesson in the rhetorical creation of value by practical economics. Although the formal study of economics may eschew marketing and rhetoric, those who wish to play the economic game must also play the rhetorical game, because markets exist only in people's perceptions of them and in people's actions within the markets they perceive. Market values are created and maintained by the choices made by people, who are the source of values.

Insofar as automated operations of markets reduce human intervention to assert complex and changing sets of values, unfortunate consequences follow. Values become residual and reduced, for there are no means for bringing new values into the market. Social change is stifled, for it is hard for new things to gain place and value without the creative play of rhetoric and human meaning. Finally, options for creating richness of life evaporate as markets become conservative, driven by only the values that have been thoroughly quantized and then enacted in their most abstracted form. Economics without humans really would be a dismal endeavor. And living in an economy without rhetoric would be even more dismal.

III

Making It Real: The Rhetoric of Material Presence

8

Boasts, Deceptions, and Promises

Electricity and incandescent light are more than words, numbers, and pictures: they are the material results of a material technology. Engines, dynamos, wires, transformers, insulators, and other components that can be seen and touched produce light that is visible and power that will shock the unwary. Edison's enterprise was built on the promise of electric lamps' burning brightly and not burning out. Even more, it was built on the promise of electricity's being delivered to homes and factories to illuminate the darkness and to power useful machines. If Edison's light and power had not become parts of daily material practice, all the technical, legal, financial, industrial, and public relations bubbles we have examined would have burst, leaving little meaning, value, or presence.

The representations examined in this book were in ongoing dialectic with the promise and the actual appearance of the material, observable, practical performance of electricity. The next four chapters examine how the material promise and the material production emerge out of the symbolic representations. In an important sense, the material bottom line warrants the meaning of the words. But the meaning, location, and form of the material are, in an equally important sense, mediated by the words. Electricity and incandescent light, when they appeared, became material representations of ideas they were realizing. They represented themselves—and what those self-representations meant depended on the discursive systems in which they signified.

Most noticeably, between 1878 and 1882 the material production of light and power took on the meaning of a fulfillment of a promise made by Edison. "I have accomplished all I promised," Edison told a reporter for the New York paper *The Sun* at the opening of the Pearl Street station on September 4, 1882.[1] However, exactly what that promise was and to whom it was made were evolving and variable matters. Accordingly, the conditions that the material production had to be seen to fulfill were also variable. Even the emergence of the promise from optimistic boasts was fluid and required some rhetorical management.

Claimed Facts: The Imagined and the Material

In the *New York Daily Sun* of September 16, 1878, Edison presents incandescent lighting as an accomplished fact, and the reporter corroborates. The article opens: "Mr Edison says that he has discovered how to make electricity a cheap and practicable substitute for illuminating gas." Later in the opening paragraph, the *Sun* reports that Edison claimed to have solved the problem of the subdivision of light "within a few days." "On Friday last [Edison's] efforts were crowned with success," the article further reports, "and the project that has filled the minds of many scientific men for years was developed: 'I have it now,' he said on Saturday...."[2] The only reason Edison provides for not giving details and displaying his achievement immediately is the need for patent protection, "which will be in a few weeks or as soon as I can thoroughly protect the process." But such was not the case. Though Edison had an overall conception of an overall system, made some experiments, drew numerous sketches, and filed a patent caveat in the days just before his announcement, he had no working light. Indeed, Wallace's "Telemachon" generator, the sight of which had excited Edison to begin serious work on incandescence the previous week, had not yet arrived at Menlo Park as of Friday, September 13.

Edison's commitment was to the system of central power, and in the ensuing months he made gradual advances on many aspects of the system. But by specifically publicizing his project as one of incandescent lighting (for reasons examined in chapter 7), he made the production of incandescent lighting the crucial public measure of his credibility. In order to maintain that credibility until financing was fully arranged for, Edison continued representing incandescent lighting as accomplished. Edison continued to use the need for patent protection as a reason not to make the details of his work public. Indicative of the deception was the article "Edison's Electric Light," which appeared in the *New York Herald* of October 12, 1878. A subheadline reads "How the Wizard of Menlo Park Guards His Secret." The article reports that, although American patents have just been granted, Edison still awaits European patents, and that "when word is cabled Mr. Edison that the patents have been granted he will throw his invention open to the public gaze, but until then he declines to make known its details." The article describes in some detail the measures Edison had taken to keep the strangers who flocked to Menlo Park from learning his secret. "After the patents are all granted," however, Edison is said to intend to "light up all the houses in Menlo Park, gratis."[3]

Edison's first flush of excitement with his conception of the project may provide some excuse for his initial optimistic claims of having solved the practical problems of light, even though there is no record in his notebook of any substantial success for at least six months, and it took more than a year to produce a lamp durable enough to be announced and demonstrated to the public. Moreover, it soon must have become clear to him that the solution was not yet at hand, even though certain paths for exploration had been marked out. Devices to regulate filament temperature continued to be explored, but none was found satisfactory. There is no record of success with any model constructed using Edison's initial idea of thermal regulation, whether regulated by expansion of metal, expansion of air, magnetism, or any other mechanism. Moreover, Edison had not settled on a material to be used in a filament. He had neither a filament that could withstand heating nor a means of regulating the heat that was burning out the filaments he was trying.

From Optimistic Boast to Deception

Yet Edison continued to maintain that the incandescent lighting had been perfected. In order to maintain such a public stance, he realized, he would have to provide at least some material token of his purported accomplishment. In the middle of October 1878 he invited reporters, one by one, to witness carefully manipulated private demonstrations, and in each case the reporter attested to the accomplishment.

The first—a reporter from the *New York Herald,* likely Edwin Marshall Fox—visited Menlo Park on the evening of October 11, and his story ran in the next day's edition as "Edison's Electric Light."[4] The reporter, having been led to the "inventor deep in experimental research," was shown a light "clearer and more brilliant" than a gas jet. Then the electric lamp was turned on for what the reporter claimed was three minutes (but he also claimed that staring at the light for this length of time did not hurt his eyes). Then the electric lamp was turned down low to show a mild illumination. Edison clearly knew the time limits on what he could demonstrate, and he controlled the demonstration within those limits. At no point did Edison rely on the proper functioning of the automatic temperature regulators, even though at the time all his experimental models were based on regulating the temperature of the filaments. When asked "Have you run across any serious difficulties in it as yet?" Edison replied will be dissipated by subdivision, so that by dividing into ten the current afforded by 400 powerful cells the aggregate of light would be far less than the concentrated effect. Mr. Farmer has very well stated these fundamental objections in the *American Gas Light Journal*, and I only add some confirmatory experimental data and observations.

The lighthouse authorities in England have shown that 290 cubic feet of gas per hour, consumed in one burner, afforded a light of nearly 3,000 candies, whereas, divided into 88 burners of 5 cubic feet each, the light of only 580 candies would be given. Mr. Hippolyte Fontaine, engineer of the Gramme Company, manufacturers of the electric light machines, says:—

The electric light does not interfere with gas light, nor with oil light, nor with candle light.

It will not revolutionize, as has often been averred, the question of lighting, destroying what is now in use, and monopolizing every industrial application, domestic and public. The electric light has its place marked out for it in many circumstances; but, far from diminishing the consumption of the other lights, it will lead to their further development by demonstrating the advantages of a more powerful and complete illumination.

Undoubtedly Mr. Edison will wind up with a round turn at the insurmountable obstacle—the enormous loss due to subdivision of the current.

One hardly knows what to believe of Edison. The reporters make him out to be a regular mechanical Munchausen of the John Keely order, but I am told he does not talk to scientific people in the same strain. Reports to newspaper reporters. No end of announcements of what Edison is going to do are daily made, the world wags on and nothing more is heard from them. This last piece of newspaper buncombe will in a few days longer, as every one hates gas bills heartily—they are even more disagreeable than the noises on the Metropolitan Railway, because they can't be avoided by moving.

I see notice from the reported interview with General Roome that it is not electric light that disturbs his dreams. It is the "schemers" who have prevented him furnishing the mild, radiant light of Manhattan gas to the poor man at $2 per thousand. That is good!

THE DAILY GRAPHIC

MONDAY, OCTOBER 21, 1878.

960

EDISON'S NEW LIGHT.

ELECTRICITY SOON TO TAKE THE PLACE OF GAS IN ALL CITIES.

"Positively No Admittance" was the discouraging greeting at the door of Edison's laboratory at Menlo Park, Saturday, when the representative of THE GRAPHIC entered and encountered the pleasant faces of Mr. Batchelor, Edison's indispensable assistant, and Mr. Griffin, secretary and correspondent.

"Now, what is that inhospitable sign for?" they were asked.

"You see," Batchelor said, "Edison is altogether more amiable than it is wise to be. He doesn't want to bar anybody out, so he lets all sorts of inquisitive people come here and occupy his time, when they are moved by no motive higher than idle curiosity. But at important seasons like this we can induce him to keep some of the people out."

"He is particularly busy, now, is he?" asked THE GRAPHIC.

"Yes," responded Batchelor; "the fact is, for several weeks now he has been at work night and day on this electric light. He has driven it into a pretty small corner at last. Here he comes and can speak for himself."

At this point Mr. Edison came clattering down the stairs, glowing with a pleasant excitement and evidently just emerged from his wizard's cave. "Hello, is it you?" he said rapidly. "In a week or two I'll have my electric light ready for you to illustrate, if you care to do so."

"You seem to be making a panic among the gas companies," said THE GRAPHIC.

"Well, yes; those old fellows know what they are about. I've got 'em, certain, and they are finding it out."

"Is there really any good cause for this sudden tumble of gas stocks?"

"It is a little precipitate, perhaps, but it was bound to come. The electric light is the light of the future—and it will be my light—unless," he added, with a conciliatory twinkle, "some other fellow gets up a better one. Still, the gas stocks need not decline. The companies can just adopt electricity instead of gas and run our wires inside of their pipes."

"The American Gaslight Association doesn't believe in you," suggested a scientific gentleman who was present.

"Those gentlemen are right from their point of view," said Edison, "but they talked in the dark. They didn't know what I had got hold of. Batchelor was at their meeting and was greatly amused."

"However, seeing is believing," continued he, "and I will show you the electric light and my methods of feeding and regulating it, only stipulating that you shall give no description of it which shall vitiate my patents in Germany, applied for, but not yet granted."

He led the way up stairs again, to the second floor of the laboratory, which has been illustrated in THE GRAPHIC, and paused before the bench where he first hit on the phonograph, and where he finished his telephone. Three small brass standards were there, six or eight inches high, each with a small glass globe or cylinder at the top, enclosing a curious nest of wires. From each standard a wire descended through the floor.

"These are the lamps," said the inventor, relighting the pipe which had gone out and laying it on the bench where it immediately perished again. He touched a lever on the bench. "Now the current is on this lamp," he explained, touching the smallest; "it is lighted, but you do not yet see it." Presently the nest of wire at the top assumed a dull crimson glow. In another minute it was scarlet; then it turned to a fierce white heat.

"Of course, there is no flame," he said; "the light is wholly from incandescence. That light is just about equal to one gas jet. I can increase or diminish it to any extent. I can regulate it with mathematical accuracy."

"What is that wire that glows?"

"That is platina."

"How long will it last?"

"Forever, almost. It will not burn. It never oxydizes."

Then he turned it down through all shades of red, till the light vanished. "You do not see it now," he said; "but it is lighted. It is invisible, and the electricity required is almost infinitely small, but it is there, and a touch will recall it—see!" and he tapped the lever and the illumination returned. "How's that for a sick room?" he asked with a broad smile of pleasure.

He connected the circuit with the two other lamps and showed their different patterns and capacities. Then he explained the peculiarity which rendered this electric light practicable and valuable, and said "if a statement of that were published it might invalidate my foreign patents." He didn't ask an "affidavy" of secrecy, but seemed satisfied with the negative affirmation of silence.

"This is exactly what you want it to be, then, is it?" asked one of the party.

"Not exactly," said Mr. Edison; "there are three points to be perfected. I am working on them now. One is an electricity meter. You see, this thing has to be invented from the very beginning."

"Where does this electricity come from?" he was asked.

"Down stairs. It is furnished by our engine. We use Wallace's machine—William Wallace, of Ansonia; a wonderfully ingenious man. We use his generator. It simply turns power into gas. In actual operation, one large engine would supply a whole town with light."

"Edison's New Light," New York Daily Graphic, October 21, 1878 (94: 0380).

"Well, no." He went on to say that he was a bit worried by how well it was operating, since his other inventions had run into numerous obstacles. He did note that a few details and improvements (such as a meter) needed work, and that he intended to make the lamp so durable it wouldn't need replacement, even once a year. In the context, he appears to imply that he had already achieved durability of a year.

On October 18, Edison gave a lengthy interview at Menlo Park to a reporter for the *Sun*—apparently Amos J. Cummings—who suggested in the resulting story that he was an intimate of Edison's.[5] The interview ended with a demonstration of the dynamos and the platinum light Edison had been working on:

There was the light, clear, cold, and beautiful. The intense brightness was gone. There was nothing irritating to the eye. The mechanism was so simple and perfect that it explained itself. The strip of platinum that acted as burner did not burn. It was incandescent. It threw off a light pure and white. It was set in a gallows-like frame, but it glowed with the phosphorescent effulgence of the star Altaire…. A turn of the screw, and its brightness became dazzling or was reduced to the faintest glimmer of a glowworm. It seemed perfect. The professor gazed at it with pride.[6]

Again the demonstration was short; the light was quickly turned down to a low glow. Again Edison pleaded that patent issues kept him from sharing his secret more fully: "It is from no purely selfish motive that I keep my secret from the public. I have no wish to do it, but it is necessary for my own protection." Edison suggested that within six weeks he would have the device in practical operation and would be lighting up Menlo Park. He mentioned as an afterthought, however, that he was still working "to keep the bugs out of the invention."

The *Sun* interview was published on Sunday, October 20. On Saturday, Edison gave another private interview and demonstration to a reporter from the *Daily Graphic*. Under the condition that the reporter (likely William Croffut) not provide any description that would vitiate his pending German patents, Edison demonstrated his lamp by gradually turning it up to a bright glow and then immediately turning it down, all the while emphasizing how well it could be adjusted: "I can regulate it with mathematical accuracy."[7] He particularly pointed out how the lamp was still on when it was not glowing, when the "electricity required is almost infinitely small." Edison obviously had his routine down by this point and seemed to enjoy his ability to manipulate the situation. He maintained this bald-faced lie:

"And how long will it last."

"Forever, almost. It will not burn. It never oxydizes."

Edison admitted that he was still working on three things, but he specified only one of these: the meter.[8]

Keeping Secrets and Closing the Deal

This deception of the public seems to have been linked to Edison's negotiations with his backers. Inquiries from potential backers had come almost immediately after the *Sun* article of September 16. The very next day, Edison received a telegram from Grosvenor Lowrey and some of his associates requesting a meeting.[9] Two days later, Tracy Edson of Western Union wrote a similar letter, referring to a discussion that had taken place on Monday (the day of the article's publication) and requesting a meeting the next Saturday, September 21.[10] Other inquiries followed,[11] but by this time negotiations with a group centered on Lowrey, Edson, and Hamilton Twombly (of the Vanderbilt family) were already underway. By the beginning of October, the outlines of a deal had been worked out by Lowrey, with Edison's approval.[12] Papers of incorporation were filed with the city and the state of New York on October 16[13] and were made public as an appendix to the *Sun*'s story of October 20. Final transfers of stock and funds were made through documents dated November 15. The twelve backers included representatives of the Vanderbilt, Western Union, Morgan Bank, and Gold & Stock Company financial empires.

Another incentive for Edison to convey optimism was a meeting of the American Gas Light Association held during the week of October 14. Executives of that association spent the week denying the threat from incandescent lighting, and their denials were mentioned in numerous stories in the *Herald* and the *Sun*.[14] At the same time, the *Graphic* was reporting that the prices of gas stocks were falling in New York and in London because of the belief that Edison's light would soon drive the gas companies out of business.[15] Edison clearly wanted to make his technology appear as complete, as ready, and as threatening as he could, at least until the financial backing was signed and sealed.[16]

Pressure as a Motivator

Edison's remarkable public presence, nurtured throughout 1878, led the public to believe that his word was as good as the deed—that for the Wizard of Menlo Park material production of electric light was only an afterthought. The financiers capitalizing the development of the technology and those investing in energy stocks acted as though the effects of Edison's pronouncement would be rapid and extreme. Correspondingly, Edison and his Menlo Park team were put under great pressure to produce the working technology—a pressure that helped them keep up the relentless pace and cooperative work in the laboratory.

The pressure helped Edison and his team to focus their attention their energies on electric lighting, so that it displaced other projects at Menlo Park and elicited the extreme commitment documented, during a slightly later part of the project, in Francis Upton's letters to his father and in Francis Jehl's much later account (mythologized though it may be). Edison himself seemed to be aware of the motivational value of putting himself and others on the spot. Upton later commented: "I have often felt that Mr. Edison got himself purposely into trouble by premature publications and otherwise, so that he would have a full incentive to get himself out of trouble." (Dyer and Martin 1910, p. 297)

Whose Trust Was Needed, and When?

Although the insiders at Menlo Park must have understood how far they were from an accomplished technology and must have been aware of the pressure they were under to produce, the outside world, including the financiers, believed that Edison only had to get his patents arranged and a few details perfected before Menlo Park and then the rest of the world would be lighted. Once Edison had backing, he had the problem of turning his fraudulent claim into a promissory note while still maintaining his credibility before the various publics that had taken him at his word.

Who, specifically, were the various people paying attention to Edison's claim, and what was their continuing stake in his success?

Edison's fellow workers at Menlo Park were close to the facts, personally acquainted with Edison's enthusiasms and public relations, and familiar with the long developmental work that went into any technology. They were deep enough believers in Edison's eventual triumph that they did not seem to have much of a loss of faith in Edison and the light, except perhaps when Edison himself began losing faith in the spring of 1879 (Dyer and Martin 1910, pp. 289–290).

Although claims of invention were made most forcefully and specifically to the patent system, the patent examiners did not have any stake in whether Edison's designs became manufactured and marketed reality, or even experimentally workable. The patent system, since it left all issues of efficacy and fulfillment to the commercial market, did not have any official way of noting whether Edison fulfilled on the promises of any of his patents. When there were valuable material consequences of a patent, the patent system did provide a repository of representations of objects and owners that could be the starting point for appeals and litigation. Then the courts would address the relationship of the actual material technology to the patent representation, but only in later years. In the short term, no one in the patent system or the courts was actively concerned with the reality of Edison's claims or the fulfillment of his promises.

Public excitement and credibility aided persuasion of financial backers, but once the business arrangements were made, the full faith and trust of the general public were no longer crucial. After all, the public would be there when incandescent lighting was ready and demonstrable. All that was important was that the public not be so disillusioned as not to pay attention when the time came. On the other hand, continuing curiosity kept them in some watchful suspense.[17] Edison's failure to produce could potentially strain their acceptance of the new technology if and when it became available, as well as their willingness to accept the general authority of Edison concerning future enterprises.

Similarly, panic among holders of gas stocks was of only short-term benefit in its capacity to excite the financiers who were to back the Edison Electric Light Company. However, strong conviction of Edison's fulfillment of his promise would not need to be renewed among stock investors until Edison was ready to make a public offering of stocks in the winter of 1880–81.

The one community whose continuing trust Edison most needed was his financial backers. This, however, was the group that was most in the position to feel betrayed by Edison, since he had taken their money deviously (even though he felt confident of his own ideas). Looking for a quick return on their investment, they soon became impatient. Within days after the signing of the preliminary papers establishing the Edison Electrical Light Company, Edison's backers repeatedly asked Grosvenor Lowrey to allow them to visit Menlo Park—for example:

... the Board of Directors would like to visit Menlo Park some day next week. Please write to Mr. Soren and let him know if you can manage to set up a few lights so that they can be exhibited, say by Friday of next week.[18]

The backers were expecting a quick return on investments in an accomplished technology. What Edison wound up offering them was a long-term investment in a technology that would have to be researched before it could be developed. But the issue was never stated so clearly. Rather, Edison kept stretching out the time he would need to produce a workable light, trying the patience and the credulousness of his increasingly restive backers.

One other important group of onlookers—whose faith was needed, but was barely forthcoming, rapidly lost, and only partly re-established after energetic efforts—consisted of members of the American and European scientific community. They were needed to lend credibility to Edison's efforts, to allow the opportunity for testing and confirmation of his accomplishments when they came, and ultimately to endorse Edison's success. Even the most friendly members of the scientific and technological community were hardly willing to stand by in ignorance, lending passive support. Professor George Barker's frequent requests for demonstration models to use in his lectures were very close to calling Edison's bluff.[19] Many of the other American electrical technologists, such as William Sawyer and Moses Farmer, as competitors, had an interest in discounting and deriding Edison's claims. Many European electricians, even those not working on the light, were ready to scoff at Edison, owing to nationalism and owing to a desire to protect their professionalism against the Americans' amateur entrepreneurship.

The Problem of Maintaining Credibility

Edison's problem was to maintain authority and credibility as someone who could be trusted to tell the truth and deliver on his promises. His failure to materially deliver on his premature claim undermined his authority as the premier inventor and his reputation as a plain-spoken, truthful man capable of sound judgment. Now he could easily be seen as a huckster, and the European technical press was ready to call him one. For example, in the initial issue of *La Lumière Electrique* his early patents were characterized as "an immense and lamentable hoax."[20]

Veiling its incredulity in arch sympathy, the British *Journal of Gas Lighting, Water Supply & Sanitary Improvement* commented:

... we believe that Mr. Edison has incautiously raised expectations which he is anxious to fulfill, but which are taxing his energies to an extent which he little anticipated. We pity the man who is thus committed to an Herculean Task, and, for, his own sake, we hope he will not drive on this harassing enterprise until we hear some day that, despite all the bouyancy of his spirit, and his extraordinary power of endurance, he has sacrificed his health and broken down at his work.[21]

Thus, as Edison's team worked away at Menlo Park, there were several sets of onlookers with different stakes and perspectives, and Edison had different needs and goals with respect to maintaining or regaining the trust of each set at various stages in the process. Until the signing of the contract to provide development funds, he needed the public, the newspapers, the general stock investors, the small group of backers, and the workers in Menlo Park to have various kinds of trust in his word. After the signing, he needed to maintain the full support of his backers so his finances wouldn't collapse, but he needed only a weakened good will among other groups. Once he had a workable product ready to bring to market, he had to reaffirm his credibility with each of these audiences, including the technical and scientific judgment-makers. His demonstrations would be directed toward these various needs at appropriate moments.

Claims and Promises as Speech Acts

Because of the strain on his credibility that his earlier false claim had created, Edison's word became weaker and weaker during 1879, and by the middle of the year it counted for little. Only actual physical demonstrations would prove his accomplishment and, retrospectively, redeem him as a keeper of promises and a man of his word.

To better understand the difficulties Edison had converting a claim into a promise and maintaining the credibility necessary to carry off the promise, it is worth examining what is entailed in the speech acts of claim and promise. Following the analyses of Austin and Searle,[22] we can identify some of the differences between the two kinds of acts and some of the difficulties that follow for a promise if a claim is found to be false or even deceptive.

Edison found it necessary to change the tense of his claim from the present perfect ("has been accomplished") to the future ("will be accomplished"). But in believing Edison's claim that incandescent lighting had been perfected, people would have assumed that Edison both believed what he was saying and had good evidence for believing the truth of his claims. When it turned out that the claim was not true in the perfect tense, Edison's authority was suspect. On the one hand, if it was taken that his original belief was sincere (that is, Edison erred only out of optimism), then his judgment would be questioned. On the other hand, if his original claim was taken to be insincere (a lie or a fraud from the very beginning), his sincerity in all future claims would be in question. With Edison's command of the evidence, his judgment to determine what is done or doable, and perhaps his sincerity called into question, his ability to make a credible promise on which others might rely was severely damaged.

A claim excites hopes, fears, or changed perceptions insofar as the hearer finds what is claimed worth attending to. Similarly, a promise implies something that would not happen in the ordinary course of events and something that would please the hearer. (If it displeased the hearer, it would be a threat.) Though electric lighting may have remained extraordinary and exciting to many hearers despite a failed claim, failed claims by Edison or others may have created disbelief in its accomplishment at any point. Indeed, the British electricians adopted this posture, arguing that Edison could not achieve what many others with better scientific credentials had already failed to achieve.

Furthermore, in making a claim, the speaker obliges himself to the hearer to be telling the truth and reveals the intention to produce in the hearer a knowledge of the obligation of truth-telling. Similarly, in making a promise, the speaker puts himself under the obligation of accomplishment and intends that the hearer be aware of that obligation. When his claim failed, Edison was tainted with not having responsibly maintained the truth and with having intended to exercise authority under false pretenses. The perception that Edison had intentionally duped his audience could hardly encourage his audience to accept the amended claim as a promise.

Suspicion of Edison's reliability as a promise-maker once he had failed as a claim-maker increased because people had pinned specific hopes, plans, and actions on his failed claim. People had worked for Edison late into the night at low pay; financiers had invested; journalists had granted him credibility; civic leaders had dreamed particular dreams; consumers had entertained possible choices. These commitments heightened the disappointment and maintained a continuing dependence on Edison within a situation these audiences no longer trusted. The pressures on Edison to produce a workable material technology were great and specific.

Tokens of Material Production and Excuses for Delay

The mid-October press demonstrations allowed Edison, through sleight of hand, to turn failed experiments into tokens of a claimed achievement, closed the deal with his financiers, and bought him time to come up with a real solution. To buy even more time, in interviews he hinted at small problems that had not yet been resolved—patents, a longer-lasting lamp, a meter, and unspecified details. By suggesting that the system was not yet complete, Edison was beginning to insert a time lag in the production of the technology, providing himself with a means of turning the claim to a promise. In the interview for a story that appeared in the *Daily Graphic* on December 28, 1878, Edison used the excuses of his optimistic congeniality and the newspapers' lack of caution to further distance himself from the claim of a completed system, stressing how much needed to be done and how hard he was working:

Have I been too ready to talk about my schemes and projects? It is possible. But I like to talk about a thing I am interested in, and I have served newspapers too long as a telegraph operator not to feel an active sympathy with their efforts to get all the news. And if a reporter comes in eager to hear something I talk to him and tell him what I have done, and sometimes what I hope to do; what I have found and what I am looking for, what I am certain of and what I have projected, and he goes off resolved to make up an entertaining story, and without at all intending to exaggerate he writes it up in a sparkling, picturesque manner and he occasionally makes the mistake of saying I have done what I only hope to do and I have found what I am still looking for.[23]

In this interview Edison's shameless denial of his own careful nurturing of the press and planting of stories was surpassed by his denial of his previous claims about the completed success of the system:

Now newspapers are shouting impatiently for the light, as if it were a mere bagatelle and could be instantly introduced. But it takes time.

The story continues with admiring details about all-night sessions in the lab and the tireless work of Edison and his compatriots. He clearly maintains that he has only made a promise, and that he is working his damndest to make good on it. The subhead is explicit about the status of the light and about Edison's efforts to deliver on the newly defined promise:

Edison's Latest Word About the Promised Electric Light—A New Telephone. He Works All Night and Has Several New Strings to his Bow.

That Edison's credibility was still high until the end of the year is evident from a good-natured story about an apocryphal Edison wonder published in the *New York Herald* on January 1, 1879 under the headline "Flight of the Airship 'Science' to the White Mountains. Three Hundred Miles in Half an Hour."[24] And the British humor magazine *Punch,* in its New Year's issue, ran a series of cartoons about "Edison's Anti-Gravity Under-Clothing."[25]

In the *Herald* of January 30, 1879, Edison again claimed that incandescent lighting was "Already Perfected, but Requiring to be Cheapened"[26]

But just a week later, the *Daily Graphic,* once one of his biggest boosters, expressed doubts in an editorial titled "The Gas Stock Scare": "The general credulity with which Mr. Edison's promises of revolutionizing domestic economy by the introduction of his electric light was remarkable."[27] Edison's public credibility, which he had parleyed into a financial arrangement, was collapsing. "It is yet possible that Mr. Edison may accomplish what he has engaged to do," the editorialist noted, "but there is really no trustworthy evidence that he will do so." The writer advised people to hold onto their gas stocks.

The Value of a Good Press and Its Cost

Edison did more to keep the press on his side than give good interviews. In January of 1878, he cemented the friendship of his two most enthusiastic journalistic supporters, Edwin Marshall Fox of the *Herald* and William Croffut of the *Graphic,* with gifts of stock in the Edison Electric Light Company—rewards for past services and attempts to see to it that they would continue to have a stake in the success of the enterprise.

According to a letter from Calvin Goddard (secretary of the Edison Electric Light Company) to Edison, dated January 18, 1879, the books of the newly incorporated Edison Electric Light Company were ready for Edison to subdivide some of his stocks, in accordance with his apparently previously expressed desire.[28] Edison almost immediately arranged for gifts of stocks to a number of his close associates, most of them working at Menlo Park: 30 shares to James Adams, 30 shares to J. H. Banker, 130 shares to Charles Batchelor, 30 shares to Tracy Edson, 20 shares to Stockton Griffin, 8 shares to Edward Johnson, and 30 shares to Uriah H. Painter. And there were to be gifts to two people not directly in Edison's employ: the reporters Edwin Marshall Fox (eight shares) and William Croffut (five shares).[29] Each of these shares had a par value of $100 dollars and represented one three-thousandth of the Edison Electric Light Company, so the gift of even a few shares was not inconsiderable. By the end of the year these shares were to trade at $4800 each.[30]

Personal notes of thanks to Edison from several of the recipients remain in the Edison papers.[31] The most interesting and effusive of these are from the reporters. On January 26, Fox wrote:

Your success shall ever, believe me, be most welcome news to me and any temporary obstruction in your great work shall cause none of your friends more pain than it shall cause me.[32]

THE EDISON ELECTRIC LIGHT CO.,
3 BROAD STREET.

NORVIN GREEN, Pres.
R. M. GALLAWAY, V. Pres.
E. P. FABBRI, Treas.
C. GODDARD, Sec'y

New York, Jany 22ᵈ 1879

Thos. A. Edison Eq
    Menlo Park

Dear Sir

        I enclose Stock Cfs
as follows:

#		shrs.
1	J R Edson	30 shrs.
2	J H Banker	30 "
3	C Batchelor	130 "
4	S L Griffin	20 "
5	E. Fox	8 "
6	W C Cruffut	5 "
7	E H Johnson	8 "
8	U H Painter	20 "
9	Jas Adams	30 "
10	T A Edison	219 "
		500 "

in place of temporary Cfs
No. 2. Cancelled.

            Yours truly

                C Goddard Secy

Letter, Calvin Goddard to TAE, January 22, 1879 (50: 227).

In a postscript, Fox also mentions that he will soon be able to repay a note on a loan from Edison. Croffut's letter, dated February 3, is even more effusive:

My dear Edison.

Yes! Bless you. Yes, of course I got the five shares of stock, & have been commercially ecstatic ever since. You are a brick. If I can do anything in the world for you at any time in every way, order me up & I'll go it alone.[33]

After expressing confidence that Edison will succeed, Croffut asks for inside information:

But, if you shouldn't be able & can foresee it, do let me know so that I can unload.

Croffut's letter ends with an inquiry about Arthur Williams—a friend or relative whom Edison has given a job, thus adding to "the numerous obligations of your trust."[34] By the end of the year, Croffut had been offered $500 for two of his shares, and he wrote to ask Edison where he could buy more.[35]

Thus, it is evident that Edison was aware that good press was of some value and that reporters could be very good friends when needed. Clearly he was willing to pay a price to ensure the loyalty of his friends in the press.[36]

Corporate Obligations

Edison's financial backers were harder to string along; their stakes were specific, and Edison was contractually bound to specific legal obligations to them. When examined closely, however, those legal obligations involved less than that of handing over a working technology—indeed, less than even promising that he would do so. The documents establishing the Edison Electric Light Company, signed on November 15, 1878, provided Edison $50,000 in return for one-sixth of the newly formed company; Edison retained ownership of the remaining five-sixths.[37] In return Edison obligated himself to assign to the company all his existing "inventions, discoveries, devices and improvements" in electric lighting, heating, power, engines, and generation (including a specified list of patent applications) and all his future "inventions, discoveries, devices and improvements" in these areas for the next 17 years. There were clauses specifying his obligations to seek future patents for inventions and several forms of compensation he would receive in various eventualities. Only one short phrase in the preamble suggested that the company would engage in "making, using, and vending" the technology, but all contractual details organized the company simply to engage in owning and licensing patents.

The Edison Electric Light Company served Thomas Edison's immediate needs by relieving him of the obligation to manufacture a working technology (either right away or at any time in the future). Edison was obligated only to seek and assign patents, which in themselves carry no obligation of a successful, practicable, or manufacturable technology. The one clause that specifically instructed Edison to pursue the technology—the fourth clause—was vague enough that it did not hold him accountable by any specific date:

The said Edison agrees on behalf of and for the benefit of the Company, to prosecute with his utmost skill and diligence further necessary investigations and experiments upon the use of electricity for the purposes described in the first article, and to endeavor to discover and devise the best and most economical means, modes and apparatus for applying electricity to the purposes above named, and for rendering the means, modes or apparatus which he may have discovered or devised, more useful, economical and convenient, and to perfect and complete all his inventions and improvements described on the first and second articles, and all such as may result from the further investigations and experiments herein provided for, as far and as fast as may be in his power.[38]

Thus, under the guise of improvement and perfection, Edison (with the aid of Grosvenor Lowrey) bought himself legal time and freed himself from any legal promise other than the vague ones of using "his utmost skill and diligence" "as far and as fast as may be in his power."

Nonetheless, the backers acted as though they still believed that Edison was expected to produce rapidly. Immediately upon coming to terms with Edison, the backers insisted on an immediate demonstration. Within a few weeks they forwarded at least four such requests to Edison through Lowrey.[39]

With no concrete progress to demonstrate, Edison resisted a visit from the financiers. They then approached William Sawyer, apparently having realized that Edison was not the only inventor around and that there was some bluster in his claims. At the end of October, Lowrey hired Howard Butler to evaluate Sawyer's claims,[40] and early in November he hired Francis Upton to begin a patent search.[41]

Serious Work as the Only Available Token

Lowrey tried to protect Edison, but his powers to do so were limited. On December 5 Lowrey wrote Stockton Griffin, Edison's secretary:

I have put them [a group of the investors] off once or twice, telling them that thus we take up time which is of great value, but I do not like to repeat this too often. Edison must therefore allow this to be added to his interruptions, and after next Monday's visit I believe everyone will have seen what is to be seen, and he will then be left free to pursue his studies without further interruption, and strongly supported by the sympathy and confidence of his friends and associates.[42]

Lowrey, recognizing that success would not be coming soon, suggested that Edison step back from the pretense of imminent success in order to maintain credibility with the backers. He suggested that seriousness of work was the best Edison could offer as a token of eventual success, and that the backers would have to satisfy themselves with the promise of eventual achievement.

The December 9 visit, according to Lowrey, reassured the investors that Edison was seriously at work: "They realize now that you are doing a man's work upon a great problem, and they think you have got the jug by the handle with a reasonable probability of carrying it safely to the well and bringing it back full."[43] However, it did not convince them that they would soon have a product to sell. Another visit from Edson and Banker later in December revealed the project in disorder. Nonetheless, the backers seemed to tolerate their disappointment with humor. Lowrey, writing to Edison, recalled: "We had a very good natured laugh over their [Edson's and Banker's] disappointment at their visit. In addition to not finding you, they say that the general dilapidation, ruin and havoc of moving, caused the electric light to look very small, and that it looked rather as if you were getting ready for an auction."[44]

During this period, while being pressured for visible results by his backers, Edison switched strategies. Rather than try to regulate the temperature of the filament, he would try to find a more robust filament material. He also began to show more concern with other elements of the system.

Over the next few months, Lowrey was able to keep the backers at bay with reports of Edison's progress, even though at times Edison had to be pressured to keep up the flow of information, as is evident from a letter from Lowrey to Edison dated January 18, 1879:

There is a director's meeting on Tuesday next and I think you should come in. It will encourage them all to see you and talk with you; it is due them to be kept informed. I advise you decidedly to come.

Goddard gives me a good account of the state of progress. I hope either you or Upton will from time to time write me what is being done, as the question is one which all parties ask me, with the impression apparently that I can answer. You will remember that I have not seen you since the visit of Mr Drexel. . . . My news is therefore rather old.[45]

From this point on, however, Lowrey, owing to his regular visits to Menlo Park and to communications up to New York from Edison, Upton, and others, seems to have had enough information to report to the backers, for we hear no further complaints of this sort. The backers seemed to have grudgingly accepted that the best token of Edison's eventual success would be word of progress filtered through Lowrey. Still, the news Lowrey transmitted back to Edison was not always encouraging. On January 25, 1879, Lowrey wrote to Edison: "I went to Mr. Fabbri's today and found that Griffin had been there this morning. They all gathered around me and, in a joking way, asked if I knew anybody who wanted to buy their stock."[46]

Some Success, But the Task Not Yet Completed

In February and March of 1879 Edison had made some progress using iron, platinum, and platinum alloy filaments.[47] On March 16, the notebooks record, a dozen lamps worked continuously from Sunday through Monday.[48] Reports indicated that Menlo Park may have been illuminated for between one and two weeks during the latter half of the month. Edison invited backers to Menlo Park for a demonstration during this period.[49] The actual events of the demonstration remain murky; however, despite stories told in later years of a disaster,[50] the demonstration seems to have been a moderate success.

A hostile contemporary report from the *London Times*, while pointing to the limitations of Edison's accomplishment, did recognize that in the middle of March Edison was able "to maintain 400 coiled iron wires in a state of incandescence."[51] A more well-disposed interview in the *New York World* quotes Edison as claiming he had the light working, was planning a demonstration at Metuchen, and did not believe he had any difficulties with his backers.[52] And the *New York Herald*, consistently an Edison backer, reported on March 27:

The first practical illustration of Edison's electric light as a system has just been given. For the past two nights his entire laboratory and machine shop have been lighted up with the new light, and the result has been eminently satisfactory.[53]

Fourteen lamps reportedly glowed at 18–20 candlepower.

Upton's description in a March 23 letter to his father was more moderate:

The electric light has been on exhibition during the past few days. It shows a fair promise of being ultimately a success, and of course then giving me all the money I may need. He has lit up his shop with 18 small lights made of platinum. Each light is about equal to a gas burner. I am fully satisfied with my prospects even if the light does not succeed.[54]

Upton's remarks indicate that, while Edison was exuding full optimism in public, within Menlo Park there was a more guarded attitude toward whatever had been accomplished. Platinum was expensive, and the life spans of lamps were still measured in hours and days; neither of these facts boded well for a fully marketable product. Recognizing that further solutions were needed before that system could be represented as complete and ready to go, Edison soon dismantled the exhibition. By March 29 Edison had replied to a request for a visit by one of the directors who had missed the meeting: "Supposed Lowrey had given it up & I have taken lamps down. I will have a new exhibition pretty soon."[55]

Upton's evaluation of the demonstration, expressed in a March 30 letter to his father, remained guarded:

Mr Edison has arranged during the past week or two to show his light as he has it now. It has made a very good impression on those who are chiefly interested in its success. There is no doubt that he can have enough money to carry on his experiments if he needs more, from many sources.

I know so many troubles ahead since I see how the lamp works experimentally that I do not expect as much for a long time to come.[56]

On April 4, Upton spelled out the experimental troubles he saw:

The light is very slowly progressing, a little slower than I should wish. It is hard to make the lamps we want though we can make very good ones. We have placed our hopes very high for we can make lamps with a few hours life which are extremely economical and there seems to be no reason why we cannot make them last many hours in time.[57]

At the end of April, Upton again wrote his father optimistically, but still he expressed doubt:

There is still hope this summer will see a public exhibition of the electric light. There are thousands of difficulties to be overcome yet before it can be given to the public and Mr. Edison will overcome them if any does.... He says he will either have what he wants or prove it impossible.[58]

Keeping Up a Bright Face

With some reason to hope for the success of his current line of investigation, Edison continued to give interviews to the New York newspapers most favorably disposed toward him: the *Herald*, the *Graphic*, and the *Sun*.[59] He planned for an exhibition at Menlo Park in the near future to precede any New York exhibition.[60] The *Herald* even took the occasion of the approval of Edison's earliest patents (no longer relevant to his work) to run a major four-column story under the spectacular headline "Subdivided Lightning."[61] This was accompanied by a lengthy editorial praising Edison's success and foreseeing great benefits, and two days later there was another editorial and another interview.[62]

Edison also turned his reliance on platinum into publicity for a search for new sources of the costly metal. The *Graphic* obliged by portraying Edison descending into a mine, wearing a wizard's hat.[63] Early in July 18 of 1879, Edison, replying to Tracy Edson,[64] set out a plan for a demonstration:

> I think it will start out on a small scale without making any great outlay. I propose to put up a dynamo machine and 30 lights when I get the standard lamp, and when a 2d dynamo is done, I will add that and 30 more lights.[65]

On September 17, Tracy Edson wrote to Edison about the possibility of an exhibition in New York. He also announced that the directors would be at Menlo Park the following week to see developments and to discuss further financing to bring the experiments to a conclusion.[66] In the meantime, Edison gained a couple of newspaper headlines for his work on lighting by delivering a paper to the American Association for the Advancement of Science on September 2.[67]

Edison also committed himself to outfit the *Jeannette,* a ship being prepared for an Arctic expedition, with a system of electric lighting. The *Jeannette* expedition (backed by James Gordon Bennett, publisher of the *New York Herald* and sponsor of a number of circulation-building expeditions) had captured the public imagination. On April 21, 1879, the expedition's leader, George Washington DeLong, had written to tell Edison that he needed a workable system by June 15, when the ship was to sail from San Francisco.[68] Edison had responded positively, perhaps because he needed to show that he could produce some lighting of value. Because his incandescent lamps were not yet durable and reliable, Edison could only outfit the ship with arc lighting (powered by one of his new-model dynamos). Although this revealed Edison's inability to produce a working incandescent system, it served as a public token of his ongoing work. (The *Jeannette* was lost in the arctic winter.)

9

The Menlo Park Demonstrations

On October 22, 1879, in the Menlo Park laboratory, as part of a series of tests of carbon filaments, Charles Batchelor illuminated a lamp with a filament made of carbonized cotton thread. The lamp burned for 14½ hours, failing only after extra power was added. This event was to be glorified and elaborated by both the papers and Edison's co-workers—particularly Francis Jehl, who later recounted that the lamp had burned for *40* hours (Jehl 1929, volume 1, pp. 351–357). But even in its less dramatic reality, the event marked a definitive turning point for those in the laboratory. They now had a clear direction for their revivified experimental work and a reason to prepare for more public demonstrations. Attention now turned with a passion to procedures for carbonizing filaments and to tests of various lamps made with them. Notebooks were filled and new patents applied for. Within two weeks Edison's lab workers were convinced that they had the solution in hand and that it required only small refinements.

The changes in emotions and plans are evident in Francis Upton's letters to his father. On October 26 Upton expressed optimism about the future but still no absolute sense of breakthrough:

This week has brought nothing very new. The electric light is looking up, for we have some very good lamps. The latest gossip that Mr. Edison has told me, is that there is talk of swallowing the old Electric Light Co. in a new one and making the capital three million of dollars.[1]

A week later, Upton informed his father that Edison was ready to demonstrate working lights:

Mr. Edison now proposes to give an exhibition of some lamps in actual operation. There is talk if he can show a number of lamps of organizing a large company with three or five millions capital to push the matter through.[2]

On November 16, Upton veritably crowed to the father, to whom he had been defending himself all year:

The wire is laid to my house and I shall light up my parlor. If you wanted to show one or two of your pictures you could lend them to me and I would hang them in the best place in my parlor. Ahem![3]

Throughout November, Edison seemed quite convinced that he had the solution in hand and began preparing to demonstrate the bulb to his backers and to the public. On November 4 he asked Western Union to send two linemen to set up wiring for a "light exhibition,"[4] and on November 7 they arrived and set to work.[5] Edison also purchased steam engines to power his generators, stating on November 17 "I expect in 3 weeks from today to have the Park lighted by electricity."[6]

Early in November, Egisto Fabbri, one of Edison's backers, traveled to Menlo Park from New York to view the recent advances on the behalf of the other investors. Despite having dragged their heels on negotiations for extra funds since the spring, on November 13 the investors were satisfied enough to assess themselves $5.00 a share to cover remaining expenses of the experiments and demonstrations.[7]

The actual production of an adequate light in the laboratory, therefore, had great symbolic significance for both the workers and the backers. Besides changing discourse, plans, and actions, it became an object of representations to others (among them Upton's father, the directors, and the suppliers).

Keeping the Public in the Dark

Rumors of success began to spread, and in November Edison received several requests from the press that suggest they had caught wind of something. On November 6 a reporter for the *New York Tribune* wrote that, despite earlier skepticism based on lack of information, from what he now had heard he would be willing to write very supportively if only Edison would provide some details of recent accomplishments.[8] And on November 20, the editor of the *Boston Journal* asked to be informed of the date of the demonstration early enough so he could send a reporter down.[9]

Edison's first impulse seems to have been in the direction of immediate publicity. On November 7 he cabled *Scientific American* asking whether a column in the next issue could be held open for him.[10] But he soon thought better of it, and for the rest of November he remained closed-mouthed about his success. As late as November 28 he responded a request from the *Chicago News* for firm details as follows:

All the statements telegraphed were mere guesswork. I shall however make a public exhibition within four weeks from which the public can judge of my success.[11]

On December 1, when he wrote to his London agent George Gouraud, Edison was still cagey: "Public exhibition takes place during holidays. It is an immense success. Say nothing."[12] Perhaps he was concerned with avoiding major snags; perhaps he remembered the untoward publicity that had been evoked by the premature announcement of "etheric force" and the excellent publicity that had arisen out of the delayed announcement of phonograph. Whatever the reason for it, his anticipatory silence was in marked contrast to the aggressive use of publicity in the previous year when he was seeking support for his unrealized idea.

Edison timed his announcements and displays strategically. His early demonstrations to the public were timed to impress the backers. Though he granted many interviews throughout the development period, he allowed only his backers to examine his actual results, and then only when pressed by them, particularly in relation to requests for additional funds.

A Timely Announcement

On December 4, with planning and construction moving forward, Edison told the editor of the *North American Review,* to whom he had promised an article a year ago, that "Mr Upton can prepare the article for him so it will be ready before public exhibition."[13] The article written by Upton actually appeared in the February 1880 issue of *Scribner's Magazine* (Upton 1880); an entirely different piece, written by Edison, appeared in the November 1880 *North American Review* (Edison 1880a).

In December, in the course of a series of interviews, Edison demonstrated and explained his electric light system to Edwin Marshall Fox. These interviews resulted in an article announcing the success of Edison's light in the December 21 *New York Herald.*[14] Although Edison expressed some irritation that the article was not withheld for a date even closer to the demonstration, it is hard to imagine that it could have been withheld more than a few more days. The article was clearly coordinated with the demonstration, making the demonstration equivalent to the announcement of success. The headlines present the story as one of final triumph after a long a difficult process of struggle:

Edison's Light

The Great Inventor's Triumph in Electric Illumination

A Scrap of Paper

It makes Light, Without Gas or Flame, Cheaper than Oil

Transformed in the Furnace

Complete Details of the Carbon Lamp

Fifteen Months of Toil

Story of His Tireless experiments with Lamps

Burners and Generators Success in a Cotton Thread

The Wizard's Byplay, with Bodily Pain and Gold "Tailings"

History of Electric Lighting.[15]

Recasting Edison as a man who has struggled to bring the future into being, rather than as a man of promises, the article began by announcing the upcoming demonstration:

> The near approach of the first public exhibition of Edison's long looked for electric light, announced to take place on New Year's Eve at Menlo Park, on which occasion that place will be illuminated with the new light, has revived interest in the great inventor's work, and throughout the civilized world scientists and people generally are anxiously awaiting the result.

That the reporter was likely to have seen the working lamp, and that Edison's backers and some members of the public in the Menlo Park area had witnessed the light, were never mentioned in the *Herald* article; the entire story is perched on the edge of the light's appearance. A chronological account of Edison's work is interspersed with detailed descriptions and illustrations of several of Edison's interim and final light and dynamo designs. Indeed, these illustrations and their explications are the backbone of the article. Edison is portrayed as moving from one set of designs to another before attaining success with the light and the dynamo (and with an electric motor). Although the illustrations are based on illustrations in the patent applications, here they take on the added character of completed working objects (and the experimental objects leading to the final results) that will shortly be unveiled to the public.

The Public Sees and Reads

Egisto Fabbri, aware of the importance of the public unveiling, cautioned Edison on December 26 to carry out an additional full week of testing before inviting the public,[16] but Edison confidently went forward.[17] Between Christmas and New Year's Day, many guests and curiosity seekers found their way to the Menlo Park laboratory, which every evening was illuminated inside and outside. The news stories that appeared on those days focused on the working system and the reaction of those who witnessed it.

The *Herald* of December 28 carried the first account of an actual public demonstration:

The laboratory of Mr. Edison at Menlo Park was brilliantly illuminated last night with the new electric light, the occasion being a visit of a number of the inventor's personal friends. Forty lamps in all were burning from six o'clock until after ten. The various parts of the system were explained by the inventor at length....

Twenty lamps burned with exactly the same brilliancy as did one when nineteen were disconnected. The light given was of the brilliancy of the best gas jet, perhaps a trifle more brilliant. The effect of the light on the eyes was much superior to gas in softness and excited the admiration of all who saw it.[18]

Upton wrote his father on the same day that the price of Edison stock was up to $3500 a share, and that he had received $50 for a one-week option on five of his shares at $5000 a share; furthermore,

...last night we had an exhibition and several million of capital were represented. Evening went off splendidly.... I had a number of gentlemen in my parlor among them the correspondent of the London Times[19]

The following day, the reports grew more enthusiastic as more witnessed the accomplishment. The *Herald*'s story of the December 29 begins:

'Wonderful!' When you hear this word rolled out once in a while with the proper intensity of accent, as in the grand chorus of the Messiah, it has a lifting effect on the emotions. When you hear it three or four hundred times in a day it begins to lose force and you search about for synonyms.... All this while you are under the influence of the wonderful yourself....

All came with one passion—the electric light and its maker. They are of all classes, these visitors, of different degrees of wealth and importance in the community and varying degrees of scientific ignorance. Few, indeed, are they who can approximately measure what has been done in this matter, and still fewer those who, knowing its worth, admit it. But the homage of the mass to genius which they cannot comprehend makes up in quantity at any rate for any shortcomings in quality. It is a time of rapid conversations. The little horseshoe holds its own.

Outside the building called the office, on two ordinary lamp-posts, gleamed, last evening, two little electric lamps. Afar off they seemed two large globes of fire.

It was not until you came close by that you saw the little incandescent hoop of carbon held by its delicate platinum clamps inside the small globe exhausted of air.
In the office the lights were all electric. In the library it was the same. Over in the
laboratory, upstairs and downstairs it was the same. . . .

   At last [the visitors] go in twos and threes down the hill to the railroad track,
and it is all 'wonderful!' 'marvellous!' 'wonderful, wonderful!' among them till
the train takes them away and Menlo Park is left to itself.[20]

The light has now been revealed to representatives of all stations of society.
They serve as surrogates for all the readers of the newspaper, who are able
to share vicariously not only in what they saw but also in their overwhelming response. What they have seen and what has been reported is the transformation of the world.

The Light as a Symbol of Edison's Accomplishment

Two more days of stories of visitors, what they saw, and their response followed before the official presentation of the light on the world stage. The
stories also presented Edison and his co-workers as basking in their
achievement—as proud parents of the wonder all could now witness. The
following is from the *Herald* of December 31:

All day long and until late this evening Menlo Park has been thronged with visitors
coming from all directions to see the wonderful 'electric light.' Nearly every train
that stopped brought delegations of sightseers until the depot was overrun and
the narrow plank road leading to the laboratory became alive with people. In the
laboratory the throngs practically took possession of everything in their eager
curiosity to learn all about the great invention. In vain Mr. Edison sought to get
away and do some work, but no sooner had he struggled from one crowd than he
became the center of another equally as inquisitive. The assistants likewise were
plied with questions until they were obliged to suspend labor and give themselves
over to answering questions.[21]

The light is not only present; it also stands for its own success. Even more,
it stands for its own miraculousness. The newspapers, virtual witnesses for
the rest of the world, report the light as an object of pilgrimage. Through
the amplification of telegraphy and the news, the whole world seems to
share in this emergence. *Puck* symbolized the new presence in its New
Year's illustration, titled "A New Light to the World."[22]

   In view of the suspense created by Edison's announcement 16 months
earlier and by the strategic glimpses into the secrets of Menlo Park that
had been offered, the perceived drama of the events cannot have been
entirely unanticipated. Edison had long planned the demonstration as a

public redemption for his unwarranted claims, which he had refigured into dubious promises. In addition, the demonstration kicked off the merchandising of a major industrial enterprise, drawing on the symbolic capital Edison and Menlo Park had gained through press coverage. The insistence on holding the first demonstration at Menlo Park, the rush to have it ready for the New Year, the rapid expansion of the display with new installations every day, the desire to transform all of Menlo Park inside and out, the withholding of announcements and wider public demonstrations until all was ready, the nurturing of the press, and the close cooperation with one reporter all point to Edison's desire to create the biggest possible public event as possible. The drama of the New Year's Eve demonstration, with all its symbolism of a new future, was reported by Edwin Marshall Fox in the *Herald:*

Edison's laboratory was tonight thrown open to the general public for the inspection of his electric light. Extra trains were run from east and west, and notwithstanding the stormy weather, hundreds of persons availed themselves of the privilege. The laboratory was illuminated with twenty-five lamps, the office and counting room with eight, and twenty others were distributed in the street leading to the depot and in some of the adjoining houses.[23]

Even though Edison had stage managed the event to display incandescent light to its best advantage, the public response and the reporting turned the demonstration into far more than he had hoped for. Part of what was demonstrated was great public enthusiasm and excitement. On January 2, the *Herald* summed up the events of the previous week:

During the first few days the crowds were not too large to interfere with the business of the inventor's assistants, and all went well. Every courtesy was shown and every detail of the new system of lighting explained. The crowds, however, kept increasing. The railroad company ordered extra trains to be run and carriages came streaming from near and far. Surging crowds filed into the laboratory, machine shop and private office of the scientist, and all work had to be practically suspended. Yesterday the people came in hundreds by every train. They went pellmell through the places previously kept sacredly private.[24]

The Menlo Park laboratory remained continuously illuminated well into the next year as visible and public proof of the viability of Edison's system.[25]

Technical Derision

The technical community had been cautious, even derisive, toward Edison's accomplishments. Their derision would not stop until Edison's lighting system had been examined, measured, and certified by the most respected authorities. If his system was to gain wide acceptance, Edison would have to address the deep and widespread technical skepticism.

Skepticism about Edison's project existed on both sides of the Atlantic. In Britain, William Preece, William Thomson, John Tyndall, and others had belittled Edison's attempts. The February 12, 1880 issue of *Nature* accused Edison of being unaware of the long history and difficulties of attempts:

The more we study the detailed accounts of the new inventions, the more we regret that Mr. Edison does not devote some time to learning what has been already done in this field.[26]

Moreover, on January 1 a British inventor named Joseph Swan had telegraphed Edison claiming he also had been working with carbon hoops and had overcome all the difficulties in his own design.[27]

In the French journal *La Lumière Electrique*, Theodore du Moncel recounted the well-known difficulties associated with electrical illumination, the lack of novelty in Edison's designs as reported in the press, and the impossibility of Edison's having overcome the difficulties while using the same methods everyone else had tried without success. He left no doubt about his judgment:

I think that the preceding is sufficient for the public to be on guard against the pompous announcements which come to us from the new world. (du Moncel 1880)

In the United States, Henry Morton, president of the Stevens Institute of Technology, belittled Edison's claims as commercial puffery. In a letter to the editor published in the December 22, 1880 issue of the journal *Sanitary Engineer,* he wrote:

Having a sincere respect for Mr. Edison as an enthusiastic and ingenious investigator, I am sorry to see his name used by writers who evidently are quite ignorant of the subjects about which they treat in a way that will inseparably connect it with discreditable (because false) claims, evidently made in the interest of financial speculators.

[In examining the *Herald* story,] when I examine the conclusion [made from the experiments], which everyone acquainted with the subject will recognize as a conspicuous failure, trumpeted as a wonderful success, I have only left before me the two alternative conclusions that the writer of such matter must either be very ignorant, and the victim of deceit, or a conscious accomplice in what is nothing less than a fraud upon the public. (Morton 1880)

This criticism was immediately reprinted in both the *Sun* and the *Herald,* along with a barbed response from Edison and an invitation for Morton to visit the exhibition.[28] On December 28 Morton declined the invitation, saying that more knowledge about the durability and actual economy of Edison's lamp was needed. He reminded the interviewer for the *New York Times* of Edison's earlier unfulfilled boasts and of the difficulties all investigators had encountered.[29]

A Competitor's Challenge

William Sawyer, backed by Albon Man, challenged Edison aggressively, especially after word of Edison's planned demonstration got out. On December 22, 1879, in the *Sun*, Sawyer offered $100 to Edison if he could prove a specific series of claims.[30] Two days later, Sawyer wrote to the *Herald* claiming he would make his own demonstration to the world. His criticisms of the Edison claims had moderated, but Sawyer remained bitter about how much publicity Edison had received relative to others who had worked longer in the field and who had been more forthcoming with their results. Sawyer complained that, despite the long article in the *Herald* three days before, detailing Edison's work over the past sixteen months, "all we have heard from Menlo Park is that Mr. Edison is a great and eccentric genius who divides his time between eating herrings, wearing old hats, rolling tar abstractedly in his fingers, going without his dinner, and finally founding his great achievement upon, as has always really been the case, cotton thread and paper."[31] Sawyer continued his offensive for a month in the *World,* the *Tribune,* the *Sun,* and the *Herald*,[32] maintaining throughout that Edison did not have a functioning, economically viable system.

The Validity of the Complaints

The complaints of the critics had some merit. From the first, Edison had tried to put his own stamp on all he had done, and to a large extent he had succeeded with the public. Yet in fact he had relied heavily on earlier work with which he had become acquainted through his widespread personal contacts, his copious reading, his extensive clipping files, and Francis Upton's patent searches. Although Edison did make specific advances in lamp, generator, and distribution design, his technology was built on the patents of others. Whether he invented a crucial novelty had to be adjudicated in the courts over the next ten years. (see chapter 12.) Eventually the courts ruled that much of the technology of the electric light was already common knowledge, and that Edison's ownership rested on the single innovation of high resistance in the lamps. This novelty was not emphasized in the publicity of December 1879 and early 1880.

Moreover, after his New Year's demonstration Edison continued to make unduly optimistic representations of the reliability and economy of his system, when in fact a full year of further improvement would be needed to launch the commercial central power system. In January of 1880, for example, he was still promising to send lamps to Paris soon for testing.[33]

Finally, Edison's project was deeply enmeshed with financial backers seeking quick return and with newspapers seeking circulation. Edison mobilized resources that transgressed the limits of a pure science, contrary to the advice of the journal *Nature:*

> Let public opinion insist that the inventor shall be allowed to pursue his way unhampered by the officious interference of the unprincipled speculators whom his soul abhors, or by the irrepressible scientific reporter, who is only one degree less reprehensible for the part he plays. . . . But if he succeeds ultimately, it will be in spite of the vampires of the stock exchange and the hangers on of the New York press, who dog his steps for their own selfish ends.[34]

Though Edison did not always have an easy time with the press and his financial backers, he very much needed them. He spent more than a little time enlisting them and managing their participation to meet his ends. Such unconventional and successful mobilization of social resources was bound to call attention to itself and to set heads wagging among those who were not comfortable with technology located so close to the marketplace.

Demonstrations for the Experts

Because the complaints against him had some validity, Edison needed to rebuild his credibility so as to buy additional time and support for bringing the system to commercial viability, even though he now had a functioning lamp. Scientific demonstrations were needed to back up the public spectacle. To establish technical credibility and priority, Edison enlisted scientific and technical experts as witnesses as soon as he had actual results to show. For example, the *Scientific American* of October 18, 1879, contains a photograph of an Edison dynamo being inspected by a professorial figure—perhaps George Barker of the University of Pennsylvania.[35]

Among Edison's visitors in the days leading up to the New Year's demonstration were several scientists who, according to the *New York Herald,* "minutely examined the light in all its parts" and concluded that "Edison in reality had produced the light of the future."[36] Perhaps it was this event that *Scientific American* of January 10 reported as demonstrating "a single lamp...giving a light that enabled us to read the *Scientific American* 100 feet away." "This," the report continued, "was certainly an extraordinary performance for a piece of carbon having a surface no larger than that shown in Fig. 1."[37] The article concluded as follows:

The light certainly leaves nothing to be desired so far as efficiency is concerned, and we are assured by Mr. Edison that, on the score of cheapness or economy, his system of illumination is far in advance of any other, not excepting gas at the cheapest rates. It seems that the subject of general electric lighting is now reduced to a matter of time. If Mr. Edison's lamps withstand the test of time, he has unquestionably solved the vexed question and has produced what the world has long waited for; that is, an economical and practical system of electric lighting adapted to the wants of the masses.

At the New Year's Eve demonstration, extensive experiments were displayed. "Among the visitors present," the *Herald* reported, "were several gas company officials and electricians who subjected the system to close examination, Edison ordering every facility to be given them for making all the tests they required."[38] The success of these displays was enough to satisfy the backers to provide another $52,000 to fund another year of development.

Technical Disbelief

Newspaper reports of unmonitored and unmeasured demonstrations, even when witnessed by knowledgeable specialists, could hardly satisfy the scientific community. A *Herald* article of January 23 told of the range of opinions among French scientists, with Theodore du Moncel and several others expressing disbelief, others cautiously uncertain, and others ready to believe that Edison might have made a significant advance. These scientists are quoted as wanting to make their own inspection, or at least to have a credible report.[39]

Competitors in the lighting market were quick to play on the limitations of Edison's system. The *American Gas Light Review* of February 2, for example, published a report, based on an interview with an "expert" who had visited Menlo Park in January, that denied that Edison's system presented any threat to the gas lighting industry. Although this witness also questioned whether Edison's lamps really were equivalent to 16 candles of luminosity, his main complaint was based on cost calculations; moreover, it was noted that a complicated system to deliver electricity would have to be built from scratch, while a simpler system for distributing gas was already in place. "At Menlo Park," the article concluded, "they do not hesitate to say that they are trying to make something as good as gas, and they admit that they have not yet succeeded.... Altogether, I do not think that the holders of gas stock need to fear that Mr. Edison is going to impoverish them."[40] On February 7 the *New York Times* published a report by a well-known Cleveland electrician (likely Charles Brush) criticizing Edison's claims to novelty and asserting the unworkability of what he had claimed to accomplish.[41]

Even greater skepticism came from the other side of the Atlantic. The British journal *Engineering* had reprinted the *New York Herald*'s December 21 article in its January 9 issue,[42] but the February 6 issue of the journal derided that account as "the flimsy sensations of newspaper correspondents."[43] Based on Upton's more complete account in *Scribner's Magazine,* the latter article criticized the economic viability of Edison's system on the grounds of the limits of Edison's generator and the inaccuracy of his method of measuring efficiency. The most positive comment was that his meter was clever.

In the eyes of the technical community, Edison was still on the hook, unable to match his projected claims with believable material reality. The crucial technical issue for the workability of the light, however, had shifted, implicitly granting him some grudging acknowledgment. The issue was no longer whether incandescent lighting could be made to work; thousands of people had seen it, and this had been reported internationally with the rapidity of telegraphy, the graphic presence of the new engraving technology, and the multiplicity of the steam rotary press. As early reports put it, there was no need to understand the principles or be a scientist to comprehend and witness this. At question now were the system's efficiency and longevity, measured with scientific devices, represented in numbers and judgements, and evaluated by the technically knowledgeable.

Technical Tests

Exclamations of wonder from lay audiences were no longer needed. What were needed were statements that the system was efficient, reliable, and economically viable from recognized experts in the field who would back up their claims with data collected by the approved methods and then elaborated through calculations in a manner consonant with current theory. Edison now had to stage a new kind of demonstration for select audiences, who would then report their reactions in appropriate professional forums.

Directive from TAE, February 19, 1880 (54: 368).

By Edison's own admission, the system needed more work before it was ready to have its efficiency evaluated.[44] But in the meantime competitors were using data gained from the access Edison had given them to deride the system's efficiency. Consequently, in the middle of February 1880 Edison limited detailed inspection to chosen scientists. On February 19, 1880, he issued this firm directive:

Employees will treat visitors courteously but under no circumstances will they leave their work or give information of any kind to visitors. No information will be furnished except by myself or Messrs. Batchelor, Upton, Krusie [sic] and Carman.

You will regard this as a positive Order.

Thomas A. Edison[45]

Laypeople could still come to look at Edison's lights and be struck with wonder, but the lights would have to speak for themselves to these non-scientific viewers. Any other observers who wanted to look more deeply into the matter would have to seek information through the official spokesmen for the lights. Edison designated two related teams of scientists to determine the official technical story of the efficiency of his generator and his lamps.

The first scientific team invited to Menlo Park consisted of two Princeton professors, C. F. Brackett and Charles Young, with whom

Edison had previously corresponded and traded loans of equipment. Both had also been on a summer 1878 eclipse expedition on which Edison's tasimeter had been used. Young, moreover, had been one of Upton's teachers. Brackett had been invited by Edison to measure the lamps as early as February 29, 1880, but he had not been available at the time.[46] On March 19, Brackett and Young went to Menlo Park with the specific purpose of measuring the generator's efficiency, in direct response to claims that efficiency of the magnitude that Edison claimed could not be achieved.[47] Edison acknowledged their report on March 30: "Your report received, Very Much Pleased."[48] On April 3, 1880, Brackett and Young returned to perform a second test of Edison's dynamo.[49] Both tests produced results confirming Edison's claims. Upton forwarded these, in the form of a spare technical report, to *Scientific American,* which published it in the May 15 issue. Upton, in a prefatory letter, commented pointedly on the specific import of the result for specific objections that had been raised earlier:

Professors Brackett and Young show 90.7 per cent converted and 83.9 available outside.

I hope this statement will be sufficient to end the discussion into which I was drawn some time since regarding Mr. Edison's machine. He then claim that 9/10 of the power in the current could be made available; now tests show 12/13 of the energy current is available.

It is not "childish," then, to make an armature with about one eighth of an ohm resistance, as was claimed by others at that time. (Upton et al. 1880a)[50]

Around the same time, George Barker of the University of Pennsylvania and Henry Rowland of Johns Hopkins University made similar visits to Menlo Park with the intention of measuring the efficiency of the system in terms of light produced relative to work input. From correspondence, it appears that Rowland ran tests shortly before March 20 and Barker shortly thereafter.[51] Barker almost immediately presented a positive report to the Franklin Scientific Society in Philadelphia.[52] Rowland and Barker then sent their results to both the *American Journal of Science* (which published them in its April issue) and the British journal *Electrician* (which published them in its issue dated April 24[53]). The rapid publication suggests that both Edison and the journals desired to provide definitive results for the scientific community.

As Rowland and Barker made explicit in the introduction to their article, their experiment, like that of Brackett and Young, had been specifically designed to answer questions that had been raised in the technical literature:

The great interest which is now being felt throughout the civilized world in the success of the various attempts to light houses by electricity, together with the contradictory statements made with respect to Mr. Edison's method, have induced us to attempt a brief examination of the efficiency of his light. We deemed this the more important because most of the information on the subject has not been given to the public in a trustworthy form. We have endeavored to make a brief but conclusive test of the efficiency of the light, that is, the amount of light which could be obtained from one-horse power of work given out by the steam engine. For if the light be economical, the minor points, such as making carbon strips last, can undoubtedly be put into practical shape. (Rowland and Barker 1880)[54]

Rowland and Barker described in detail the method they had used to measure efficiency and why it had been chosen over two alternatives. The results, presented both in text and in tabular form, led them to conclude that approximately 1000–1500 candlepower could be produced per horsepower—sufficient justification for the authors to return to the original concern of longevity:

Provided the lamp can be made either cheap enough or durable enough, there is no reason to doubt of the practical success of the light, but this point will evidently require much further experiment before the light can be pronounced practicable.

The prior publication of this technical article (including even the integral equations) in the *Herald* indicates how important scientific approval was to the public success of Edison's endeavor.[55] Most of the *Herald*'s readers lacked the knowledge needed to follow the argument in detail or to evaluate it. Rather, they would see the statement as an official answer resolving scientific doubt.

Hostile Interpretations

The only independent test was performed by Henry Morton, who used a single lamp that *Scientific American* had given him. Morton's results and his calculations of efficiency (Morton et al. 1880) projected that electric lighting would cost about the same as gas lighting, even though he used estimates of generator efficiency and lamp power lower than those obtained by Edison. Despite this result, Morton concluded, the added machinery, complexity, and delicacy of the electric system made it less preferable and not worth pursuing. This hostile test, with its hostile conclusions, nonetheless confirmed that Edison had a plausible, operating system.[56]

The British journal *Engineering* summarized Rowland and Barker's results and Morton's in a single article, concluding that they conclusively

14/6

THE ELECTRIC LIGHT.

RESULTS OF A SCIENTIFIC EXAMINATION INTO EDISON'S METHOD — EXHAUSTIVE EXPERIMENTS — INTERESTING DATA AND DEDUCTIONS.

[From the American Journal of Science.]

The great interest which is now being felt throughout the civilized world in the success of the various attempts to light houses by electricity, together with the contradictory statements made with respect to Mr. Edison's method, have induced us to attempt a brief examination of the efficiency of his light. We deemed this the more important because most of the information on the subject has not been given to the public in a trustworthy form. We have endeavored to make a brief but conclusive test of the efficiency of the light—that is, the amount of light which could be obtained from one horse power of work given out by the steam engine. For if the light be economical the minor points, such as making the carbon strips last, can undoubtedly be put into practical shape. Three methods of testing the efficiency presented themselves to us. The first was by means of measuring the horse power required to drive the machine, together with the number of lights which it would give. But the dynamometer was not in very good working order, and it was difficult to determine the number of lights and their photometric power, as they were scattered throughout a long distance, and so this method was abandoned. Another method was by measuring the resistance of and amount of current passing through a single lamp. But the instruments available for this purpose were very rough, and so this method was abandoned for the third one. This method consisted in putting the lamp under water and observing the total amount of heat generated in the water per minute. For this purpose a calorimeter, holding about one and a quarter kilometre of water, was made out of very thin copper; the lamp was held firmly in the centre, so that a stirrer could work around it. The temperature was noted on a delicate Baudin thermometer graduated to 0.1 C.

As the experiment was only meant to give a rough idea of the efficiency within two or three per cent no correction was made for radiation, but the error was avoided as much as possible by having the mean temperature of the calorimeter as near that of the air as possible, and the rise of temperature small. The error would then be much less than one per cent. A small portion of the light escaped through the apertures in the cover, but the amount of energy must have been very minute. In order to obtain the amount of light and eliminate all changes of the engine and machine two lamps of nearly equal power were generally used, one being in the calorimeter while the other was being measured. They were then reversed and the mean of the results taken. The apparatus for measuring the light was one of the ordinary Bunsen instruments used for determining gas lights, with a single candle at ten inches distance. The candles used were the ordinary standards, burning 120 grains per hour. They were weighed before and after each experiment, but as the amount burned did not vary more than one per cent from 120 grains per hour no correction was made.

As the strips of carbonized paper were flat very much more light was given out in a direction perpendicular to the surface than in the plane of the edge. Two observations were taken of the photometric power—one in a direction perpendicular to the paper and the other in the direction of the edge, and we are required to obtain the average light from these. If L is the photometric power perpendicular to the paper and *l* that of the edge, then the average λ will evidently be very nearly

$$\lambda = \frac{L}{\frac{1}{2}\pi}\int_0^0 \cos\alpha \sin\alpha \, d\alpha + l \int \sin^2\alpha \, d\alpha$$

$$\lambda = \frac{1}{2}L + \frac{\pi}{4}l$$

In the paper lamps we found $l = \frac{1}{2}L$ nearly; hence $\lambda = \frac{3}{4}L$, nearly.

The lamps used were as follows:—

No.	Kind of Carbon.	Size of Carbon.	Approximate Resistance When Cold.
580	Paper.	Large.	147 ohms.
201	Paper.	Large.	147 ohms.
850	Paper.	Small.	170 ohms.
809	Paper.	Small.	154 ohms.
817	Fibre.	Large.	87 ohms.

The capacity of the calorimeter was obtained by adding together the capacity of the water, the copper of the calorimeter and the glass of the lamp and thermometer. The calorimeter and cover weighed 0·103 kil. and the lamps about 0·035 kil.

First experiment, No. 201 in calorimeter and No. 580 in photometer; capacity of calorimeter, = 1·153 + ·009 + ·007 = 1·169 kil. The temperature rose from 18°·28 C. to 23°·11 C. in five minutes, or 1°·75 F. in one minute. Taking the mechanical equivalent as 775·, which is about right for the degree of this thermometer, this corresponds to an expenditure of 3,486 foot pounds per minute. The photometric power of No. 580 was 17·5 candles maximum or 13·1 mean λ.

When the lamps were reversed the result was 3,540 foot pounds for No. 580, and a power of 13·5 or 10·1 candles mean.

The mean of these two gives, therefore, a power of 3,513 foot pounds per minute for 11·6 candles, or 109·0 candles to the horse power.

To test the change of efficiency when the temperature varied we tried another experiment with the same pair of lamps, and also used some others where the radiating area was smaller, and, consequently, the temperature had to be higher to give out an equal light.

We combine the results in the following table, having calculated the number of candles per indicated horse power by taking seventy per cent of the calculated value, thus allowing about thirty per cent for the friction of the engine and the loss of energy in the magneto-electric machine, heating of wires, &c. As Mr. Edison's machine is undoubtedly the most efficient now made, it is believed that this estimate will be found practically correct. The experiment on No. 817 was made by observing the photometric power before and after the calorimeter experiment, as two equal lamps could not be found. As the fibre was round, it gave a nearly equal light in all directions as was found by experiment.

Lamps Used in		Photometric Power			Capacity of Calorimeter in Pounds	Rise of Temperature in Degrees F.	Energy Per Minute in Foot Pounds	Mean Number of Candles Per Horse Power of Electricity	Mean Number of Gas Jets Each Per Horse Power of Electricity	Mean Number of Gas Jets Per Indicated Horse Power
Calorimeter	Photometer	Measured Perpendicular to Paper.	Average, λ.							
201	580	17·5	13·1	2·57	1°·75	3486·	109·0	6·8	4·8
580	201	13·5	10·1	2·82	1°·62	3540·			
580	201	38·5	28·9	2·74	2°·44	5181·	204·3	12·8	8·9
201	580	44·6	30·5	2·76	2°·29	4808·			
850	809	19·0	14·3	2·81	1°·14	2483·	133·4	8·9	5·8
809	850	12·2	9·2	2·79	1°·54	3330·			
819		17·2	2·73	1°·28	2708·	209·6	13·1	9·2	

The increased efficiency, with rise of temperature, is clearly shown by the table, and there is no reason, provided the carbons can be made to stand, why the number of candles per horse power might not be greatly increased, seeing that the amount which can be obtained from the arc is from one thousand to one thousand five hundred candles per horse power. Provided the lamp can be made either cheap enough or durable enough, there is no reasonable doubt of the practical success of the light, but this point will evidently require much further experiment before the light can be pronounced practicable. That Mr. Edison will finally overcome the difficulty, however, no one who knows him can doubt.

In conclusion, we must thank Mr. Edison for placing his entire establishment at our disposal in order that we might form a just and unbiassed estimate of the economy of his light.

1477

A technical paper published in the **New York Herald** *on March 27, 1880 (94: 591).*

"confirm in every point the adverse criticisms" the journal had previously published.[57]

Another Kind of Public Demonstration

By the end of April, Edison had already dismantled his Christmas demonstration and had entered a new period of improvement, which would lead to a new demonstration of a commercially ready system at the end of 1880. In the interim, Edison had another opportunity to display the viability of his system: the outfitting of a passenger steamer, which would show that incandescent lighting could work far from the magical grounds of Menlo Park, even on the high seas.

In December of 1879, after witnessing the illumination of Menlo Park, Henry Villard, president of the Oregon Railway and Navigation Company, asked Edison to outfit his new ship, the *Columbia*, with the sponsorship of James Gordon Bennett, publisher of the *Herald,* who previously had backed the *Jeannette* expedition.[58] This project brought about many practical improvements to Edison's system, including safety wires (i.e., fuses) and screw-in sockets (Friedel and Israel 1986, pp. 140–141).

The *Columbia* installation consisted of 115 cardboard-filament lamps and three generators of the sort used in the 1879 demonstration. While the equipment was being installed and tested, many visitors came to the dock to see the lights.[59] In early May the ship set sail with a cargo of railroad equipment. On July 26 it arrived in Portland, the lighting system still working satisfactorily. Edison then sent new lamps with bamboo filaments. Two years later, the ship's chief engineer reported that not one lamp had failed.[60] The system remained in working order until 1895, when the *Columbia* was overhauled and modernized.

The Complete System Unveiled

Edison next went public when he was on the verge of demonstrating a system ready for commercial installation. In the October 1880 *North American Review*, Edison addressed the "seemingly unaccountable tardiness with which the work of introducing the 'carbon loop' electric lamp into general use has hitherto progressed," but he also pointed to the completion of a commercially viable system:

It is now several months since the announcement was made through the newspapers that all the obstacles in the way of utilization of the electric light as a convenient and economical substitute for gaslight had been removed. . . . But, so far as the public can see, that project has since that time has made no appreciable advance toward realization. The newspapers have reported, on the whole with a fair degree of accuracy, the results of the experiments made with this system of lighting at Menlo Park. . . . Still it must be confessed that hitherto the "weight of scientific opinion" has inclined decidedly toward declaring the system a failure, an impracticability, and based on fallacies. . . . Under the circumstances, it was very natural that the unscientific public should begin to ask whether they had not been imposed upon by the inventor himself, or hoaxed by unscrupulous newspaper reporters. . . .

The delays which have occurred to defer its general introduction are chargeable . . . to the enormous mass of details which have to be mastered before the system can go into an operation on a large scale, and on a commercial basis as a rival of the existing system of gas. (Edison 1880b)

He went on to describe improvements in the filament, the lamp, and the generator.

During 1880, the team at Menlo Park had indeed been working on all the parts of the system. The discovery that bamboo was a good material for filaments initiated a widely publicized search for the world's best bamboo, which provided an occasion for exotic news stories.[61] With far less publicity, generators and connectors were improved, melting fuses were introduced, and insulation for underground power cables was developed.

At the end of 1880 these improvements were put together at Menlo Park in a demonstration that emulated a commercially viable central power system suitable for New York and the other cities. The 425 lamps (both indoor and outdoor) were powered by eleven small generators, the current being carried by underground cables.

As the display was being built, the public was given a few previews.[62] A test of a series of street lamps on November 1 and 2, visible from the railroad, was later said to have been in honor of James Garfield's victory in the presidential election. In November and December various celebrities, including Sarah Bernhardt, visited Menlo Park to see the new display.[63] On the evening of December 20, New York's mayor and aldermen were treated to a reception at Menlo Park featuring dramatic displays of the lights, some words from Edison, an opulent banquet catered by Delmonico's, and Cuban cigars.[64]

Demonstrating the Economic Efficiency of a Working System

The late-1880 demonstration provided the full test of efficiency that Edison and his backers needed before proceeding to build a commercial installation. Once an improved steam engine and dynamo were installed, test were run overnight on January 28 and 29, 1881. A full report of these tests was prepared by Charles Clarke, a physicist in Edison's employ. (Although Edison wanted to publish the report, it was kept for internal use only at director Henry Villard's request. In 1904, however, the Association of Edison Illuminating Companies included it in a volume titled *Edisonia.*)

Clarke's report, which ran to thirteen printed pages, describes the original motivation of the installation as follows:

To demonstrate the practicability of this system a plant had been installed at Menlo Park by which all the buildings and streets were lighted. And to satisfy himself and the capitalists interested with him in the enterprise as to the commercial efficiency of the lamp, Mr. Edison had the test made upon the plant that was embodied in this report.[65]

Clarke gives details of the equipment, the parameters of the system during the test, the methods of measurement, the equations of calculation, and the specific events of the test, as well as reporting the results. The conclusions are quantitative, the bottom line giving the weight of coke necessary to light one lamp for one hour: 0.414 pound on this test, and 0.331 pound for newly available improved lamps.[66]

Even before they saw the final results of the test, Edison's backers were ready to establish a power and light company in New York. On December 17, 1880, the Edison Electric Illuminating Company of New York was incorporated, with a core group of backers from the Edison Electric Light Company, including Grosvenor Lowrey, Norvin Green, and Egisto Fabbri.[67] The original Edison Electric Light Company remained as a patent holding company, to exploit the value of the patents. The backers drew the line at funding the Edison Electric Lamp Company, which Edison had to do with his own money. The other risks were distributed in a set of interlocking companies with different financial and organizational arrangements. The publicly held Edison Company for Isolated Lighting, 51 percent of its stock remaining with the Edison Electric Light Company for its license, produced and marketed small systems for individual purchasers. The publicly held Edison Electric Illuminating Company of New York constructed and operated the first central station and provided the model for the organization of other local power companies. Bergmann & Company (a partnership of Sigmund Bergmann, Thomas Edison, and Edward Johnson) became the main supplier of sockets, switches, fuses, meters, fixtures, and other parts. The Edison Machine Works (financed largely by Edison, with some help from Charles Batchelor) manufactured dynamos and motors.

The Electric Tube Company, (funded entirely by Edison) produced supplies for the laying of cable.

These complex corporate arrangements suggest, among other things, that the demonstrations had convinced backers that they had a workable system and therefore had valuable patents, but few of them were convinced that any part of the business aside from delivering the system to New York City would immediately be profitable. Wall Street would need demonstrations beyond the protected grounds of Menlo Park before it would go ahead with a full commitment. Accordingly, demonstrations were next to be made to be made at expositions in Paris and London, and then in buildings in London and New York.

Fairs and Exhibitions: Museums of the Future

Though the incandescent light was an extraordinary public sensation, it was still not a fact of life, welcomed into home and factory. Unless it became an ordinary commodity in ordinary markets, its fate would be that of the phonograph in the 1880s. For more than ten years after its extraordinary introduction with displays in concert halls and before Congress, the phonograph was not successfully mass marketed and was not supported by a substantial recording industry.

Edison sought much more rapid commercial success for electric light and power, for several reasons. Market niches already existed, established by gas lighting. Furthermore, arc lighting and other incandescent lighting systems threatened to take over if Edison did not dominate first, and Edison saw enormous potential in providing a power infrastructure for rapidly expanding industry.

With the successful demonstration of the 1880–81 holiday season, Edison and his collaborators rapidly moved to establish power and light in cities, in order to induce people to see electric lighting as ordinary and comprehensible and as something they would want. First, there would be a series of international electrical exhibitions—one in Paris, one in London, one in Berlin, one in Vienna, and one in Philadelphia. The fairs would be coordinated with the first urban central stations, the first permanent one being in New York. The fairs would make light personally visible to great numbers of people, and the documentation, testimonials, and news stories they would produce would affirm and define the value of Edison's system.

Fairs and Exhibitions

Fairs have a long history as markets for the direct sale of goods. The mass production of novel goods to be distributed over large geographic areas, however, gave rise to a new kind of fair—the industrial exposition—that was intended to foster long-distance business alliances. Production, communication, and distribution now could be accomplished over distances, but people still needed to view products firsthand and negotiate business deals face to face.

By 1851, what had started in 1761 with a modest exhibition of industrial prize winners at London's Royal Society of the Arts became the Great Exhibition of Industry of All Nations, for which London's Crystal Palace was built. In 1855, France, which had offered industrial exhibitions since the beginning of the nineteenth century, hosted an even larger world's fair at the newly constructed Palais de l'Industrie. New York, Philadelphia, Chicago, Vienna, and other cities followed suit. Through competitions and awards, the fairs celebrated the advancement of the arts, from agriculture to engineering, and gave manufacturers opportunities to tout the excellence of their products.[1]

In the 1880s, the great exhibition halls that had been built for world's fairs became the sites of the great industrial exhibitions, such as the Paris International Exhibition of Electricity (held in the Palace de l'Industrie in 1881) and the London Electrical Exhibition (held in the Crystal Palace in 1882).

Both world's fairs and industrial exhibitions, in their display of the latest products and their celebration of the fruits of progress, served as museums of the future. A new technology that gained prominence at one of these events would establish itself in the public vision of what the future would bring.

A Place Edison Had to Be

Industrial exhibitions provided the obvious place for Edison to display his new system, for they would make the technology visible in the hearts of major cities to residents and travelers. By displaying his system at the exhibitions, Edison would stake a claim for it as part of a vision of the future and thus entitle himself to make an argument for its value in that future. That argument would be implicit in the material display—what it could do, how much it would cost, how it could be implemented. The incandescent light's presence and its successful operation would symbolize its potential and be its own argument.

Moreover, Edison's light would be placed side by side with other contenders for a place in that future, to be compared viscerally and visually as well as through the persuasive words and wiles of Edison's representatives.

The visitors would include all the relevant audiences for evaluating this future. More formal evaluation would be tied to the ceremony of the celebrations, as tests would be run, judges would examine, and awards would be given for the advances judged to be the best. In competitive displays, each technology could be seen, be evaluated, and even achieve an official imprimatur as a valid technology. And if you didn't display your technology, others would claim those places.

This was a field made for Edison to play on. In quite a literal sense Edison helped create the game, for his inventions helped to foster the excitement in all areas of electrical, telegraphic, and telephonic technology that brought about the electrical exhibitions of the early 1880s. The level of excitement can be judged by the fact that in just three years there were major electrical exhibitions in Paris, London, Munich, Vienna, and Philadelphia. By the time these exhibitions were over, major commercial light and power operations had been established across Europe and the United States.

An Opportunity That Had to Be Taken

On December 21, 1880, George Gouraud, Edison's agent in London, wrote to Edison about plans for the Paris Exhibition and transmitted the request of the exhibition's French commissioners for a complete display of Edison's electrical inventions. "Great stress is laid on this matter," he noted. "England will be there in great force."[2] Gouraud also urged Edison to pull strings to get Gouraud named as a member of the U.S. commission to the exhibition, which was accomplished.[3]

Edison's accepted Gouraud's estimation of the importance of the event. He sent his closest Menlo Park associates—Charles Batchelor, William Hammer, Edward Johnson, Francis Jehl, and Otto Moses—to Europe to oversee the work along with his European agents for electricity, Theodore Puskas and Joshua Bailey. Edison also kept a close tab on events; he received regular and lengthy correspondence from all seven. These represent the most intense communication in the Edison papers directly involving Edison except in the laboratory notebooks. Moreover, Edison devoted much time and money to the development of the exhibits. The Jumbo generator, originally planned for the New York station but diverted to the Paris exhibition, cost more than $12,000. An estimate dated June 24, 1881 put the expenses to be incurred in the United States for preparation and shipment of the exhibition at $21,150,[4] and an accounting of the expenses in Paris by Batchelor near the end of the exhibition put the final on-site expense at around $10,000. These combined expenses, adding up to more than $30,000, amounted to more than half of all the research done from September 1878 to January 1880.[5]

An Impressive Display

There is no doubt that Edison intended from the first to dominate the Paris Exhibition. To display his lighting system, Edison originally requested two rooms as well as responsibility for illuminating the grand stairway, the central gallery at the head of the stairway, twelve of the remaining 24 rooms, and the outside of the entryway.[6] Such a display, of course, would have made the Edison system the unquestioned centerpiece of the exhibition.

Although this ambitious request had originally been encouraged by anxious exhibition organizers, by the time of actual allocations interest was so high that space and illumination rights had to be shared among many parties. The main hall was to be illuminated by Maxim incandescent lamps, and the entrance hall by Jablochoff arc lights. Nonetheless, Edison was allocated two exhibition rooms and the right to illuminate the grand stairway.[7] Edison lighting also illuminated some additional exhibit rooms. Further, once the exhibit closed for the evening, power was diverted to a café on the Champs Elysee where visitors could linger after hours in the glow of Edison incandescence. Finally, that fall Edison lights contributed to the illumination of the Paris Opera house.[8]

Otto Moses, upon his arrival in Paris, immediately began planning an impressive display. In a July 10 letter to Edison he outlined plans for the grand stairway to be "a piece of practical illumination on an immense scale."[9] He described the appearance of the lights from every position and how, as one walked up the grand stairway, a strategically placed illuminated E would open onto an illuminated "Edison" at the top of the steps. In all, 700 lamps would be used to illuminate the grand stairway, and there would be 300 in the exhibit halls. Moses calculated the effect of illuminating the aging tapestries decorating the stairway and arranged for additional artwork to demonstrate the quality of Edison illumination. In the exhibit rooms, Moses arranged to have artwork and new wallpaper installed to show off Edison's light to better advantage.[10]

Additionally, Puskas and Bailey rented offices of ten rooms at 33 Avenue de l'Opera (near the site of the exhibition and a few doors down from the offices rented by the competing United States Electric Light Company, purveyor of the Maxim system.[11] These offices, decorated at some expense, served as reception and conference space for a European syndicate for the Edison Light. Though Edison's light might speak for itself in its luminous presence at the exhibit, personal relations and private talk were (as any business venture would recognize) needed to supplement the light's own statement. Moreover, Batchelor, Moses, and others made themselves available at the exhibit to explain incandescent lighting and other Edison achievements that were on display.[12]

One indication of the success of the Edison group's efforts to set the terms of the discussion through tireless explanation is found in a letter from Moses dated September 7, halfway through the exhibition: "We have disseminated the word 'system' until people talk of Swan & Maxim & Fox having no system."[13]

The USELC's Allies in the Press

Plans were set in motion to influence what would be said about the incandescent light and who would say it. As soon as they arrived in Paris, Puskas, Bailey, Moses, and Batchelor began working to win journalists and electricians over to their side, to undermine the positions of the competitors, and to create media events that would embarrass and discredit the opposition. Their letters to Edison are remarkable in their competitive candor, revealing a team poised to do whatever was necessary to present their system as the best available.

As Robert Fox points out in "Edison et la presse française à l'exposition internationale d'électricité de 1881" (1986), the French technical press was predisposed against Edison. The United States Electric Light Company had already made substantial alliances with the French press. The European representative of the USELC, one N. Kabath, was in fact copublisher (with Georges Masson) of *L'Electrician*. The editor of that journal, Edmond Hospitalier, had already published an article by J. A. Berly praising the Maxim-USELC system; moreover, Hospitalier had written an article, which would soon appear in the journal *La Nature*—a journal edited by Gaston Tissandier and published by the aforementioned Georges Masson—representing the Maxim system as the first complete system of lighting in Europe.

Theodore du Moncel, editor of *La Lumière Electrique* and chief engineer of the French Telegraphic System since 1869, had long been skeptical of the upstart American Edison, as evidenced in his 1879 book *L'Eclairage*

Electrique and in numerous journalistic articles, such as the one published in the January 1, 1880 issue of *La Lumière Electrique* (Fox 1986).

Massaging Egos

Moses, as soon as he arrived in Paris, began cultivating some of Europe's leading electricians and journalists, particularly the journalists who had aligned themselves with Hiram Maxim and against Edison: Hospitalier, Tissandier, and du Moncel.

On July 10 and 12, Moses wrote of meetings with an attorney named Armengaud and with Theodore du Moncel. Moses spent two to three hours describing the details of Edison's system to Armengaud, who was more acquainted with telephone systems than with electric light systems. "I enlightened him," Moses informed Edison, "without in the slightest wounding his 'amour propre.' He talks Maxim to me and I talk system and piracy to him."[14] Moses accepted a paper of Armengaud's for Edison to read and promised in return to obtain technical and legal documents. Armengaud was to be one of the authors of the exposition catalogue,[15] as well as to serve as an attorney in Edison's French patent litigation.[16]

With du Moncel, who had previously criticized Edison's "pompous announcements" (du Moncel 1880), Moses took an even gentler approach. After some effort to get an interview with Count du Moncel, Moses reports, he resorted to flattery. Du Moncel was much pleased by Moses's request to display a relief of him in the Edison exhibit. Moses, however, did not detail the conversation, "for it would simply be disgusting" to Edison. However, Moses did recount that du Moncel responded to hearing that Edison had requested a translation of the French electrician's writings: "This seemed to touch the man; for he thawed immediately and before I left asked the privilege of describing your lamp and offered to carry on the experimental verification in the laboratory of the Institute. He can be easily led to read a paper on the subject before the Institute....I do not think there will be any more sting in him."[17]

Buying Reporters

Two weeks later, on July 29, Puskas and Bailey reported having negotiated with du Moncel on the possibility of his severing his relationship with *La Lumière Electrique* in favor of an Edison publication.[18] On August 11, Moses reported that du Moncel and his wife visited the Edison exhibit at least twice a day, and that he recently had written an article on the Edison system for *La Lumière Electrique* which both Moses and Batchelor had been given the opportunity to revise. Moreover, the wife, who seemed to have great influence, also seemed to have become an Edison fan.[19] Bailey reported negotiations with du Moncel on the possibility of his breaking away from his journal and going into the employ of Edison for the duration of the exhibition for payments of 1000 francs a month during the exhibition and 10,000 francs when it was over.[20] In addition, arrangements would be made for du Moncel to write for the French and the German press and to lecture. Du Moncel was also promised the editorship of a new journal, to be called *The Electrical Journal,* which was to be founded at some later date. As it turned out, *The Electrical Journal* never materialized, and du Moncel stayed with *La Lumière Electrique* until his death in 1884.

Simultaneously, Moses reported negotiating with other French electricians over a pamphlet that was to be released in coordination with the exhibit. Moses arranged for Edison to send a Motograph (an Edison-improved telephone) as a gift to the hard-of-hearing Tissandier, who found Edison's invention the only instrument that allowed him to hear telephone messages. Moses also learned of the flexibility of the previously hostile reporter Hospitalier.[21] And on August 8, Puskas and Bailey reported that they had put the science writer Flammarion in their employ for a 1000 francs a month:

He is to give not only his pen but his personal devices and influence with Congress and with jury in short to work with us in all ways in his power.... His excellent social position, his extended relations with the press and with scientific men make his alliance very valuable.[22]

Elsewhere in their letter of August 8, Puskas and Bailey reported that they were negotiating with another journalist, Henri de Parville, but his terms were still too high. Although further negotiations with Parville are not documented in the correspondence, he was to praise the Edison system in the *Journal des Debats* (de Parville 1881).

Buying the Catalogue

Edison's representatives controlled publicity even more directly by buying favorable coverage in the official exhibition catalogue, which they saw not only as a reference for visitors but also as providing basic information about electricity to journalists who lacked a background in the subject—as a kind of definitive press release. At first, they planned to contract the back cover of the catalogue for 1500 francs (about $300) and four pages inside for almost another 3000 francs.[23] Within two weeks, however, they decided to add several more pages, for a total of 8000 francs.

The two general articles on the exhibits in this catalogue were to be written by Hospitalier and Tissandier. To curry Hospitalier's favor, Edison's representatives arranged further work for him.[24] And after Moses offered Tissandier—an advocate of other systems—the opportunity to prepare a pamphlet on other Edison inventions, Tissandier agreed to rewrite his catalogue essay to give Edison a more prominent place.

Puskas and Bailey negotiated with the catalogue's printer, one A. Lahure, to allow Bailey to review material relating to the Maxim light and "not to allow anything to go in to the catalogue that he objected to, without giving reasons to the Maxim people."[25]

Buying Newspaper Pages

Edison's people also paid to have stories placed in the major newspapers. On July 14, Puskas and Bailey wrote that they had paid $500 for an article in *Figaro*, and that an article in the next Sunday's *Monde Illustré* would set them back 2000 francs.[26] The journal *Papillon* offered to do a biographical story on Edison, and Bailey ordered 1000 copies for distribution. To get a lead story on Edison published in the inaugural issue of *Moniteur Officiel*, the official newspaper of the exposition, Edison's men ordered 3000 copies.[27] They also arranged for articles to be placed in late-August issues of *L'Illustration* and *Le Monde Illustré*.[28]

Even more remarkable, Puskas and Bailey arranged with the editors for all articles on incandescent lighting published in *Figaro*—including a specified string of descriptive, biographical, and editorial columns—to be favorable to Edison. This arrangement, which Puskas and Bailey report being worth 100,000 francs (about $20,000),[29] was obtained for a small cash payment and an agreement to allow the editor of that prestigious newspaper to buy shares in the syndicate being formed for European Edison Light. Similar arrangements were being made with Paris's other two major newspapers, *Tempo* and *Journal des Debats*.[30] Batchelor would give competent journalists sent from each of the papers all the technical information they needed. "These three Journals," Puskas and Bailey reported in businesslike fashion, "represent very nearly everything in the Paris Press worth serious consideration & you will remark that the arrangements are made without actual outlay either by us or by the Syndicate to any considerable extent."[31]

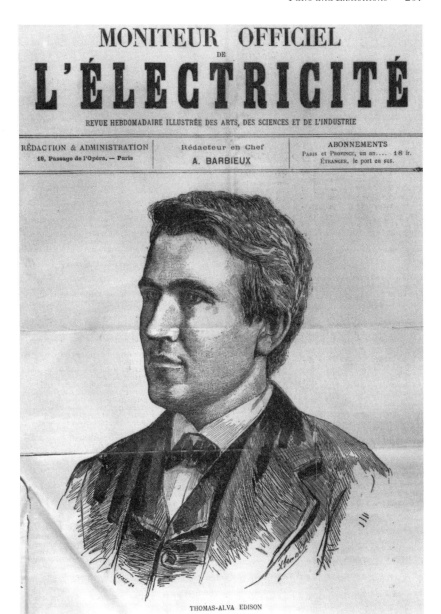

MONITEUR OFFICIEL
DE
L'ÉLECTRICITÉ

REVUE HEBDOMADAIRE ILLUSTRÉE DES ARTS, DES SCIENCES ET DE L'INDUSTRIE

RÉDACTION & ADMINISTRATION	Rédacteur en Chef	ABONNEMENTS
19, Passage de l'Opéra, — Paris	A. BARBIEUX	Paris et Province, un an.... 18 fr. Étranger, le port en sus.

THOMAS-ALVA EDISON

An issue of the official newspaper of the Paris exhibition, **Moniteur Officiel de l'Électricité** *(August 25, 1881) (89: 86).*

The Edison people seemed quite ready to enter into what they saw as the corrupt practices of the French press. Moses wrote:

If you wish anything given to the public (and there's money in it), you must pay for it. The reporters openly accost you with pen in hand (in place of pistol as brigands would do) and say they will publish such and such a notice at such and such a price. They do it as unblushingly as a bootblack would ask you for his nickle. In fact beggary is a profession here practice at times in silk or furs and sometimes in rags, but more generally the first. . . . The newspapers must be managed, however, so I hope you will arrange some way by which we can go it strong with them.[32]

Cultivating the British Press

The British electricians and the British press were not so readily bought as the French, but the Edison party cultivated them with congeniality and technical information. On August 2, Moses reported having made contact with three leading British electricians: William Preece, David Edward Hughes, and Sir Charles Tilson Bright. Although Preece, chief engineer of the British postal and telegraphic departments, had declared Edison's approach an *ignis fatuus* (Preece 1879), Moses established a positive relationship with him, and further meetings were set up. On August 9, Batchelor wrote of a long meeting with Preece at which he "seemed to be mighty glad to 'bury the hatchet'" and introduced several other leading British electricians.[33] On August 12, Batchelor wrote that he had spent four hours the previous day talking to reporters of *The Electrician*, *The Engineer*, and *The Telegraphic Journal*.[34]

Trying to Silence the Competition

While they were currying the favor of the French and British general and technical presses, Edison's men were also planning legal moves against Hiram Maxim and other competitors. On July 29, Puskas and Bailey wrote to tell Edison that they had consulted local attorneys to have Maxim's exhibit seized on the opening day of the exposition for patent violation.[35] They suggested that such a seizure would counteract the positive publicity that Maxim had been able to garner. "They should not be allowed to get possession of public opinion, and to confuse it with the idea of an even and legitimate rivalry." And on August 8, Puskas and Bailey wrote that plans to have Joseph Swan's exhibit seized were proceeding.[36]

An injunction was indeed ordered on Maxim[37]; however, the police-assisted seizure on August 12 was limited by French law to a single lamp,

Vue Generale de la Salle Edison a l'Exposition Internationale d'Électricité, La Correspondence de Paris (89: 92).

and Maxim's exhibit was not seriously disrupted.[38] Although a stir was created among the exhibitors, the event never made the papers; as Moses complained, nobody had paid for stories about it. Edison agreed to proceed with a suit; however, owing to countersuits, technicalities of the French law, and details of the French patent system, his attorneys advised against pursuing this case or the one against Swan.

A Material Accomplishment

Edison's men believed they had displayed the superiority of their lamps and their system. With some pride and glee, they wrote to tell Edison how poor the displays of the competitors looked. According to Otto Moses, Maxim kept his lights burning so low that one could barely recognize a friend across the hall and a phonograph in Maxim's room could be heard farther than it could be seen.[39] In another letter, Moses noted how uneven Swan's lamps were. "Such a mess I never did see," he wrote.[40]

Joseph Swan was impressed with Edison's system and made gracious comments to that effect, which Moses passed on to his boss: "I admit that Edison is entitled to more than I. . . . He has seen farther into this subject, vastly, than I and foreseen and provided for details that I did not comprehend until I saw his system."[41]

Stacking and Buying the Jury

The most definitive judgment as to which lamps were superior would be made by the judges. Once again, Edison's men worked to frame the conditions under which the lamps would display their merit. They intervened as actively as they could to influence under what terms the lamps would be displayed, to whom, and against what criteria. In short, they did their utmost to stack and pay off the jury, and then to define the tests in the way that would show their system at its best and the competition at its worst.

As early as July 22, Puskas and Bailey relayed intelligence about the selection of the judges to Edison,[42] and one of du Moncel's tasks for the Edison interests was to ensure a favorable jury.[43] On August 26, Moses wrote that it had finally been decided that 75 judges—half of the total number—would be from France, only ten from Britain, and only seven from the United States.[44] On September 7, Moses wrote in much greater detail about the lobbying that had taken place. Concerning George Barker, who had just finished his term as president of the American

Association for the Advancement of Science,[45] he made these only slightly veiled remarks:

… our friend to whom you gave lamps by telegraph in February, and he will make every personal effort to occupy that position. You know at what terrible expense he is here, and I am sure he will be able to do us great good if his mind were at ease. He must be perfectly independent, so telegraph me what stock can contingently be put at his disposal. Do not underestimate the importance of this. I have spoken on the subject quietly, but to the point, and this suggestion must be at once acted upon.[46]

In June, Barker had written to tell Edison that he would not need Edison to support his trip and that it would be to their mutual advantage that he be "independent in such a matter as this." However, he also referred to an Edison commitment to light his house with the new system.[47]

Barker did his work well, as Batchelor reported to Edison on October 1:

Barker is working very hard for us—the main jury have delegated 20 men to make all the experiments and Barker has worked it that there is not a single Englishman on the experimental jury in incandescent lamps. Thinking this would not stand he has proposed Crookes[48] who is very much opposed to Swan but who is favorable to Maxim.... Barker is the only man who has done anything on testing incandescent lamps so no doubt will have it all his way. He proposes to use my apparatus which I think will be accepted.[49]

Thus, the test of lamps was entirely on Edison's terms. To keep it that way Batchelor, commented delicately, "I have carefully sounded Barker and he evidently depends on something you have said to him before—anyhow he is working hard for us and ought to be taken care of as the others would gladly get him if they could."

On October 4, Batchelor wrote to suggest that Barker, though he favored Edison's system, was inclined to set test conditions that would also display competing lamps favorably. It was important, Batchelor felt, to be able to determine at what powers the lamps would be tested, so as to demonstrate the weakness of the competition. To ensure Barker's loyalty, he arranged for additional work for Barker.[50] Immediately after the awards were announced, Grosvenor Lowrey arranged to put Barker more directly in Edison's employ.[51]

The Rewards of Careful Work

Although little came of the legal maneuvering against competitors, the Edison Electric Light Company's presentation of a complete working system impressed the journalists, the judges, and the other technical observers. The Edison system was now perceived as the one to be contended with.

The official exhibition jury gave Edison's system technical credibility. Edison received five gold medals for his accomplishments in a variety of electrical fields. Further, for all his accomplishments in electricity, Edison was awarded a Diploma of Honor, the highest award for individuals. Although nearly twenty Diplomas of Honor were conferred, Edison's was the only one associated with work on lighting.[52] The Edison interests interpreted this to mean that Edison's diploma established his pre-eminence in light.

News of the awards spread rapidly by telegraph and was reported widely in the United States, in Britain, in France, and elsewhere. The Diploma of Honor became a public symbol of Edison's accomplishment.[53] An unattributed cable to Edison (which turns out to have been sent by Edison's attorney Grosvenor Lowrey) conveyed the following: "This is complete success, the Congress has nothing further to give." There were also congratulatory cables from George Barker and (even more significant) from Joseph Swan, Edison's most serious competitor.[54]

The following June, the report of the Experimental Committee of the Paris Exhibition finally appeared. The committee found the Edison lamp more efficient than any of the other incandescent lamps exhibit at both high and low incandescence, and it found Edison's high-resistance system more efficient than low-resistance lamps.[55]

The Triumph in the French Press

The French press coverage of Edison's accomplishment was widespread and effusive, and it was often accompanied by hagiographic portraits of the sort that had appeared earlier in the American press. The most important and immediate representation was that published by Theodore du Moncel in the October 1 issue of *La Lumière Electrique,* in which the carefully cultivated and amply rewarded du Moncel reversed his previous opinion and announced that the Edison system was "complete and ready to go."[56] Though he also described all the other incandescent systems, du Moncel was particularly enthusiastic about Edison's.

Versions and translations of the du Moncel article were published by Edison interests in France, Germany, and the United States.[57] The French and German versions included appendixes with extensive reactions to Edison's display and other testimonial documents. The article was also translated, reprinted, and summarized by periodicals around the world, and its illustrations were widely reproduced.[58] This article, for which du Moncel was so carefully primed, provided the standard representation of the value and viability of the Edison system and served as evidence that Edison's triumph at the Paris exhibition had won over his chief French critic and the chief journal of opposition.

The other journalists and publishers who had been courted by Edison's representatives also came through on their commitments. Major articles appeared in *Journal des Debats, Le Papillon, Du Temps, Moniteur Officiel, Le Figaro, L'Illustration, Republique Français, Le Voltaire, La Science, La Correspondence de Paris,* and *L'Écho Industriel.*[59]

The British Technical Press Takes Edison Seriously

The British journal *Engineering,* which had been consistently opposed to Edison before the exhibition, began to take a different view as soon as the exhibition was opened. A September 2 article about the opening announced that Edison's lights "seem to burn with great beauty and steadiness" and that "the form is almost identical with that of Mr. Swan's lamp."[60] By equating Edison with Swan, *Engineering* accepted him as a serious competitor. Maxim received less favorable comment, and Lane-Fox, although British, only a brief descriptive notice.

The Edison system also received favorable comment, though not quite the enthusiasm granted Swan, in *Engineering*'s description of the lighting of the Paris Opera in the issue dated October 21, 1881.[61] In the same issue, Edison's Jumbo generator was described at some length as "undoubtedly the largest electro-dynamo-electrical machine that has ever been constructed."[62] Although some opinions were hazarded on the possible benefits and inefficiencies of the size, judgment was cautiously withheld: "We have no doubt that [at the exhibition] some very valuable measurements will be made of this unique and interesting apparatus of which we will await with much interest."

Now open to an objective evaluation of the Edison system, on December 23 *Engineering* published a study of the efficiency of Edison's lamp (Howell 1881). The journal was also well disposed toward Edison's plan for a demonstration system at the Holborn Viaduct, and it published an extensive description of that project.[63]

Whereas *Engineering* was merely willing to accept the Edison system as real and workable, *The Electrician,* though still granting respect to the other contenders, was willing to see the Edison system as the clear leader. In the issue dated October 22, 1881, a reporter for the journal freely admitted his fascination with the Edison exhibit and its general popularity. He also admitted that Edison's representatives had been influential:

We, like thousands of others, visited the exhibits of Mr. Edison over and over again. We had many opportunities of conversing with Mr. C. Batchelor and Dr. Moses on the subject of the exhibits, and made a few tests of the incandescent lamps in order to obtain a personal and not merely a hearsay knowledge of their capabilities. It is due to Mr. Edison's representatives to say that a full and free permission was granted for investigating.[64]

The Electrician judged Edison's light as meeting the commercial criteria of price and longevity. By citing success in the market as a criterion, the journal conceded that Edison seemed ready to test the market. Although other systems were mentioned, Edison was featured throughout. The following week's article on the Swan system began by pointing out that Edison's light was part of a system and then commented, dryly, that Swan's lamps performed well enough for lamps that were not part of a system.

Perhaps the most substantial victory for the Edison system in British science was announced in a lecture by the British electrical engineer William Henry Preece on December 14 to the Society of the Arts and published in the December 17 and December 24 issues of *The Electrician* (Preece 1881). According to Preece, the Paris exhibition marked the arrival of the epoch of electricity, and no future exhibition could match the brilliant and vivid impression made by this exhibition's electric light section. Preece concluded by identifying the Edison system as the supreme achievement of the exhibition:

The completeness of Mr. Edison's exhibit was certainly the most noteworthy object in the exhibition. Nothing seems to have been forgotten, no detail missed. There we saw not only the boilers, engines, and dynamo machine, but the pipes to contain the conductors. (ibid., p. 92)

Two lengthy paragraphs list all the admirable details, concluding as follows:

Mr. Edison's system has been worked out in detail, with a thoroughness and mastery of the subject that can extract nothing but eulogy from his bitterest opponents. Many unkind things have been said of Mr. Edison and his promises; perhaps no one has been severer in this direction than myself. It is some gratification for me to be able to announce my belief that he has at last solved the problem he has set himself to solve.

Preece said that he looked forward to more detailed results of tests from the Paris exposition and from the Crystal Palace exhibit, which was soon to follow.

Edison on Top

The game seemed to be over. The overwhelming opinion of the technical community was that Edison's system worked and was the star of the first public gathering of electrical and light technology. The carefully staged events in the meaning-laden symbolic space of the Paris exhibition and its press representations had created new value for Edison's light. The light had shone strongly within the space designated for its display, and the story had spread throughout the technical world and the popular press. The judgment was then made official with tests, results, and awards. This news was brought home to the United States through *Scientific American*,[65] *The Operator*,[66] and the popular press.

Incandescent Electric Lights, a book published by D. Van Nostrand, collected a translation of the du Moncel article, reprints of the Preece lecture, reports of Howell's tests, and an article by Charles Siemens, Germany's leading electrician (du Moncel and Preece 1882). Prefaced by an illustration of the Jumbo generator and a reproduction of the Diploma of Honor, this book proclaimed that international technical opinion supported the Edison system.

Bottom-Line Results

Perhaps most important, the Paris Exhibition set the terms on which electricity was to be established in Europe. Negotiations just before the exhibition's opening established an Edison Europe consortium, which, using the Jumbo generator brought over for the exhibition, began operations in 1882 at Ivry-sur-Seine. Moreover, negotiations began between Edison's and Swan's interests, and these ultimately led to an Edison-Swan merger in Britain in 1883.[67]

The Crystal Palace Exhibit

Edison's light was soon displayed again (although without the Jumbo generator) at the London Electrical Exhibition, held at the Crystal Palace, which opened four months after the closing of the Paris Exhibition. The installation of a working central power plant at the Holborn Viaduct was coordinated with the Crystal Palace event. The exhibit brought incandescent light to Britons who had not made the trip to Paris. The London display was larger than the one in Paris had been, and the exhibits of systems were in separate rooms, so the value of each system could be judged more definitively. The result was to make the arrival of successful incandescent light even more certain. *The Electrician* on March 4 commented:

The capabilities of the incandescent light must now force themselves upon the visitors—except such visitors as are endowed with the feelings and ideas expressed by our eloquent gas contemporaries. But who so blind as those who will not see?[68]

Elsewhere in this and other issues, *The Electrician* described the various lights and systems, taking care not to prejudge whether Edison, Swan, or Maxim would prevail; nonetheless, the Edison System evoked special enthusiasm:

From the evening when the concert room was first lighted up before a select audience till the present moment the attraction of the Edison light has been one of the most prominent in the Exhibition. . . . We are quite sure that many of our English firms might with advantage copy a little of the energy that has been thrown into this exhibit. . . . Edison and his colleagues—for such they are—have steadily and determinedly faced their difficulties, have worked hard, and so far have been awarded a fair meed of success in popular estimation. Their motto is evidently *aut Caesar aut nullus.*[69]

In October of 1882, *The Electrician* summed up the Paris and London exhibitions, demonstrations, and tests:

At the Commencement of the autumn of 1881 we often heard the exclamation that during the next six months "electricity would be on its trial," and now at the commencement of the autumn of 1882 the same phrase is heard on every hand. There is just as much truth in the saying as in ninety-nine of the ever recurring popular watchwords. To a certain extent electricity is on its trial, in so far as its application to electric lighting is concerned, but even in this direction we may safely say that the critical period is past, and neither local boards nor national parliaments can stop its progress. Its advancement is as certain as that of the steam engine in the time of Stephenson. . . . France, or rather Paris, laid the foundation stone, she really brought conviction to many minds as to the possibilities of the electrical age, and every large town seems inclined to build upon the foundation. Exhibitions electrical are the order of the day, and every succeeding exhibition proves to be of greater importance than appeared at first sight.[70]

Other Exhibitions

Further exhibitions allowed finer comparisons and judgments, and displayed how incandescent lighting could be brought into everyday life. An exhibition at Munich in the autumn of 1882[71] and another in the autumn of 1883 emphasized the aesthetics of lighting fixtures and the use of lighting to create pleasant effects and to enhance architecture and the exhibition of artwork.[72] Isolated installations lit opera houses in Paris and Brunn, museums in London and Munich, and department stores in Paris, New York, Chicago, and Boston. The first permanent central power station went into operation on Pearl Street in New York in September of 1882, and soon there were others around the world.

By the autumn of 1883, when the Vienna exhibition opened, there seemed to be little new to say about the technology, which now was becoming familiar at least as a display object; furthermore, Edison's system had achieved prominence among its competitors.[73] Other displays— one at the Southern Exposition at Louisville (summer 1883), one at the Philadelphia International Electric Exposition (autumn 1884), one at Turin (winter 1884–85), and one at the Cotton States Exposition in New Orleans (1885) made the technology visible and real to curiosity seekers and industrial developers in the United States and Europe. The Great White Way aesthetic of electric illumination, which flowered at the Chicago World's Fair of 1892 (Rydell 1984; Marvin 1988; Nye 1994), was developing as a celebration of the accomplishment, but that accomplishment was the primary message. Incandescent lighting was now a notable feature of the landscape.

11

Lighting New York: Urban Politics and Pedestrian Appearances

One crucial display remained to be accomplished: the lighting of New York City on a regular basis, which would establish the reality of Edison's system. Although the system had been demonstrated and tested in extraordinary venues, it had not yet faced the rigors of city life and of the market—keeping the system up and running, gaining and keeping customers, servicing installations, and sustaining an economically viable corporation. Edison's promise of creating a viable system of electric lighting could not be said to be fulfilled until the Pearl Street station was in regular operation.

Yet before the Pearl Street station opened, the news and the hoopla seemed over. The only ones who seemed to be anxiously waiting were the prinicipals and stockholders of the Edison companies, uncertain about the fate of their investment of time and money. To the public and the press, electric lighting was already a part of everyday reality.

Political Interests in Incandescent Lighting

Before work could begin on the Pearl Street station, the politicians and governmental officials of New York had to be convinced that light and power promised benefits to them. Their approval and cooperation were needed for constructing the plant, laying underground cable, and maintaining a favorable regulatory, tax, and fee structure.

There were many reasons for local governments to be interested in electricity. It could have been construed as a sign of progressive leadership and economic dynamism by political leaders, as it was in San Francisco and Louisville.[1] It could have meant economic development, as in Harrisburg.[2] Partaking of the technological sublime, electric lighting asserted metropolitan power (Marvin 1988; Nye 1994). Providing for safety and amenities and fostering and regulating electric lighting were responsibilities of good government. Electric light could also have provided tax revenues, new jobs, and even payoffs to officials.

Yet competing industries made some local governments sites of contention to restrict the growth of electric light and power. In 1886, for example, the president of Cincinnati's local gas company, Major A. Hickenlooper, attacked the unreliability of Edison's incandescent lighting before that city's Municipal Council in an attempt to maintain his company's hold on street lighting contracts (Hickenlooper 1886).

Harold Platt tells in the early chapters of his 1991 book *The Electric City* how many of the above-mentioned issues came together to make electric power a major political issue in Chicago, where it was deeply tied to the struggle over corruption that pervaded city government. Yet even in this detailed story of the politics of electric power, we only see the general position and movements of the actors. We see little of the discourses by which electricity was represented in the halls of power. We witness only the political postures, and not the political action.

Covered Political Tracks

The problem in understanding politics as a discursive activity, despite the long tradition of rhetoric as political art, is that crucial arguments and negotiations usually do not happen on the public podium. Politics is always a matter of what is on the record and what is off. On the record we get the publicly legitimating and legitimated accounts, the attempts to gain and cement the backing of large publics, the announcements of alliances that aggregate power, the campaign speeches, the city council minutes, and the official handshakes. But off the record is where the political players come to terms, some of which are then displayed in public for the record and some which aren't.

This was especially so late in the nineteenth century, when American cities typically were run by small groups of business leaders or by political clubs, when small towns were turning into modern urban centers, and when old personal ways of transacting city business were just starting to meet modern record keeping and investigative journalism.

We have only hints of the kinds of political negotiation that must have gone on to bring Edison's light and power from a converted farm in Menlo Park to the center of New York's financial and newspaper district. Nonetheless, it is clear that Edison and his colleagues had to deal with politicians who expected special deals.

New York City in 1880

New York, which had been the largest city in the United States since the 1820s (when it eclipsed Philadelphia), was still expanding in 1880. Manhattan's population had grown from 30,000 in 1789 to more than 1.2 million, and in another 20 years it would top 2 million. Across the East River, and not to be unified with Manhattan until 1898, was Brooklyn; with a population of 600,000, it was the nation's third-largest city. After the Civil War, when building and public projects boomed, city improvement was the major civic theme, pervading the municipal elections in the 1870s and the 1880s. The Brooklyn Bridge, the era's most prominent symbol of urban development, was begun in 1870 and officially opened on May 24, 1883, eight months after the Pearl Street station.

Moreover, the jobs and profits generated by extensive public improvements became the foundation of political power in the city during this period, as the conviction of William Marcy Tweed in 1873 had made more direct profiteering from city coffers dangerous and unpopular. "Boss" Tweed's successor at Tammany Hall,[3] "Honest John" Kelly, had mastered the art of staying legal[4] while gaining personal wealth and maintaining a political machine that ran on profits and patronage (Allen 1993, chapter 4; Mandelbaum 1965, chapter 15; Myers 1968, chapter 27; Werner 1968, chapter 5). After 1900, George Washington Plunkitt, one of Kelly's lieutenants, explained the philosophy of "honest graft" practiced by the reformed Tammany in his "plain talks on very practical politics." Tammany, according to Plunkitt, "looked after their friends, within the law, and gave them what opportunities they could to make honest graft.... Every good man looks after his friends, and any man who doesn't isn't likely to be popular." (Riordan 1948, p. 7) Such opportunities included using inside information and influence to buy land that would soon be more valuable, to buy surplus at bargain rates, and to raise the salaries of loyal workers.

After Tweed's fall, other political forces counterbalanced Tammany. One was reformist Democrats organized in two other clubs: Irving Hall (founded in 1870) and County (founded in 1880). More powerful, and more relevant to Edison, was a group of loosely organized but tightly networked prosperous business leaders who became known as the Swallowtail Democrats because of their frock coats. From the mid 1870s through 1888, all of New York's mayors came from this group, backed by one or more of the organized Democratic clubs and at times also by the Republicans.

Tammany Hall was a leading partner in all these mayoral coalitions, except during Edward Cooper's 1879–1880 term and William Grace's second term (1885–86) (Hammack 1982, chapter 4). Although Grace was a Tammany candidate for his first term (1881–82), he lost Kelly's support for not providing adequate patronage to Tammany; after a hiatus, he won his second term, backed by several reformist groups.

This alliance of business and urban machine politics drew together real estate interests, industrialists, financiers, and immigrant groups through personal relations, obligations, and mutual support. Both the business elites and the Democratic clubs carried out their business through personal relations. Though New York was becoming a financial and population giant, requiring enormous administrative support, it was still run informally, as though it were a small town.

Off the Record

In the early 1880s, records were still written by hand and filed in pigeonhole desks. Cabinets for horizontal files were introduced around 1868, but the modern vertical file was not generally available until 1893. The typewriter, perfected in the 1870s, did not gain commercial success until the 1880s (Yates 1988).[5] Printing was still the primary method of making multiple copies. Press books and the new carbon paper had limited use. Circulating documents among members of even small committees was expensive and inconvenient, requiring copyists. Mimeograph, stencil, and photocopying were still in the future, Edison's electric pen (which produced low-quality results) being the only available precursor technology. Thus, government documents were either private or very public, either making it into the print record or vanishing from the pigeonhole desk into the trash. Most business was undocumented, and only what people intended to be recorded was recorded.

The Municipal Archives of the City of New York contain only a few boxes from each of the mayoral administrations in this period, so we have few records of how Edison and his companies represented themselves as meaningful and valuable to the city. However, the outlines of the interests and the relations are clear. Edison was well connected in the business and financial community, especially in the telegraphic, railroad, and banking industries. Moreover, his project aligned precisely with the interests of business leaders in economic development. He also "played ball," doing what was necessary to gain the support of Tammany.

Clues to the Unknown Story

Edison later recounted how local Tammany bosses had to be taken care of. For example, in order to gain permission to store items from his over-crowded Goerck Street Works on the sidewalk, Edison had to agree to hire any man who came by with a note from the Tammany boss. Similarly, in 1882, when cable was being laid for the Pearl Street station, the Commissioner of Public Works informed Edison that he had to hire five inspectors at $5 a day. Afraid that these inspectors would slow the around-the-clock work, Edison was relieved to discover they only turned up on Saturdays to pick up their pay (Dyer and Martin 1910, pp. 380–381, 393).[6]

Edison also found ways to deal with politicians on bigger issues, such as gaining the City Council's approval to lay underground cable. In November of 1880, Tracy Edson, president of the American Bank Note Company and one of the directors of the Edison Electric Light Company, approached Commissioner of Public Works Allan Campbell on Edison's behalf to invite him to Menlo Park for a private demonstration. Campbell and Mayor Edward Cooper were anti-Tammany reformists who believed in public improvements. Both were also lame ducks, as Tammany-backed William Grace had just won the mayoral election.[7] Nonetheless, Campbell and the Superintendent of Gas and Lamps (a man named McCormick) were invited down to Menlo Park for a "private and quiet" demonstration,[8] which was to take place on Wednesday, December 1. Edson, writing to Edison, spoke guardedly but emphatically about the need to gain their personal cooperation:

Now as I think it would be a great assistance to us in getting the rights we desire if these Gentlemen should be favorably impressed in regard to our Light, I would like it very much if you would make arrangements to exhibit it to them in the manner above indicated, as soon as you conveniently can.[9]

(The phrase "the manner above indicated" refers to the phrase "in a private and quiet way" on the previous page of the letter.)

Edison and the New York Aldermen

Whatever was hoped for and whatever transpired shortly thereafter, the president of the New York City Board of Alderman, Joseph Morris, seems to have objected to granting Edison a license on the legalistic grounds that Edison had not carried out his experiments and demonstrations within New York City, as a resolution passed two years earlier had required. In an apologetic letter to Morris, Edison explained that his work could not easily be moved from Menlo Park or easily accomplished "in the streets of New York," in view of the many experiments and changes that were a necessary part of the development of the system. Edison requested a personal meeting with Morris.[10]

Two days after posting his response to Morris, apparently as a result of a further exchange, Edison invited New York's entire city government to Menlo Park for a private demonstration, to be held on December 20.[11] Seven members of the City Council, the Parks Commissioner, the Superintendent of Lamps, and some other guests attended, the party numbering about forty.[12] Representing the Edison interests, in addition to Thomas Edison himself, were some of his politically best connected directors, including Tracy Edson and Grosvenor Lowrey. This event, publicized in the press[13] on December 21 and later reported in histories, has been taken as the turning point in Edison's relations with the city. A lavish dinner was prepared by Delmonico's Restaurant, and the best cigars and wines were served. Lowrey gave a flowery speech flattering the city fathers, who reciprocated with effusive toasts. Francis Jehl reported that the event created some rapport with the aldermen, one of whom commented that Edison handled his cigar as well as any of the boys in Tammany (Jehl 1937, volume 2, p. 780). However, Tax Commissioner Mitchell told an anecdote about a sweeping machine that was never used because it would put the sweepers out of work and because a machine "can't cast a single vote," thereby raising the issue of jobs and patronage (ibid., p. 781). Some serious talk, unreported, occurred after dinner.

The aldermen seemed to be won over. However, Jehl comments, "things did not go as fast as we expected. Plenty of wrangling and wire pulling followed before the company received the franchise that permitted them to tear up the streets and place underground wires in position." (ibid., p. 785)

Both sides took tough public positions. At the turn of the year, Edison complained to the press that the aldermen were demanding a fee of 10 cents per foot for the laying of cable (which would amount to $1000 a mile, since cables would have to be laid on both sides of the street); moreover, after three years a 3 percent tax on gross receipts would go into effect. Further, street lighting would have to be provided to the city at a reduced rate. The aldermen also demanded that, upon their request, Edison would have to tear up his wires and restore the streets within 30 days.[14] Edison, apparently perceiving that these were unrealistic conditions designed to keep him from doing business, commented to a reporter that

Y, DECEMBER 21, 1880.

ALDERMEN AT MENLO PARK.

EDISON GIVES A SUCCESSFUL EXHIBITION OF HIS ELECTRIC LIGHT.

The City Fathers Partake of a Collation, Swallow Innumerable Bumpers and Make the Most Scintillating Speeches.

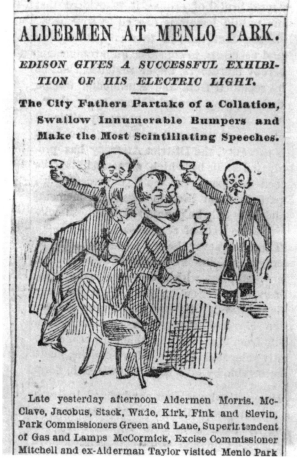

Late yesterday afternoon Aldermen Morris, McClave, Jacobus, Stack, Wade, Kirk, Fink and Slevin, Park Commissioners Green and Lane, Superintendent of Gas and Lamps McCormick, Excise Commissioner Mitchell and ex-Alderman Taylor visited Menlo Park

"Aldermen at Menlo Park," **New York Truth,** *December 21, 1880 (94: 623).*

the outgoing board's actions "savor of a body of gasmen rather than public servants."[15]

Edison threatened to work with a more cooperative city, such as Newark, Brooklyn, or Philadelphia.[16] On February 8, 1881, he arranged a Menlo Park demonstration for the leading government and economic figures of Newark; part of the pitch made by Edison's spokesman Major Eaton was that the electric light was "a Jersey invention, and belongs to Jersey alone."[17]

Immediately after New York's new administration was installed, Edison expressed his frustration with the former mayor, Edward Cooper, and his hopes for cooperation with new mayor, William Grace.[18] On March 22, the Board of Aldermen, by a vote of 18 to 1, passed a resolution to allow the Edison Electric Illuminating Company to lay its cables at a one-time permit fee of 1 cent per lineal foot.[19] Four of the aldermen who voted positively had been dinner guests at Menlo Park the previous December 20; the one alderman who voted negatively had not been at the dinner.

However, on April 5, Mayor Grace vetoed the resolution, asserting that the city was not being adequately compensated:

This resolution proposes to grant in perpetuity a franchise, the value of which cannot fail to be very great, without any appreciable revenue to the city treasury.[20]

Grace requested a more general resolution that would guarantee the regular payment of revenue to the city.[21] On April 12, in defiance, the aldermen overrode the mayor's veto of the Edison resolution by a vote of 19 to 2. By two votes of 18-0, they also passed resolutions granting the United States Illuminating Company and Brush Company the privilege of laying cable for the same fee: a penny a foot.[22] On April 19, Mayor Grace vetoed the USIC and Brush resolutions.[23] On May 3, the aldermen overrode the Brush and U.S. Illuminating vetoes, again by votes of 19 to 2.

The Complex of Politics

What sense can we make of this? Mayor William Grace, founder of the Grace Shipping lines and a believer in modern improvements, was committed to economic and technological development of the city. Further, Grace had purchased a shipping line recently from Egisto Fabbri, one of Edison's chief backers. Grace, one of the Swallowtail Democrats who lent respectability and civic development to machine politics, was elected with the support of a Tammany-led coalition. Yet here and in other situations he acted independently of Tammany and its patronage demands, in what he perceived to be the best interests of the city (Breen 1899, pp. 685–687). Indeed, like the reformist Cooper, he regularly vetoed municipal franchises that the Tammany-dominated Board of Aldermen wished to grant freely or cheaply, serving their own interests rather than that of the city; the aldermen, in turn, regularly overrode the mayor's vetoes, thus keeping the patronage flowing. Grace was not renominated in 1882, but in 1884 he ran on a reformist ticket, backed by a coalition of Irving Hall and County Democrats.

Edison clearly had the backing of Tammany and of some non-Tammany Democratic aldermen, but what it took to get that backing is not exactly clear. Tammany's 1882 mayoral candidate was Franklin Edson, president of the Produce Exchange.[24] At his inaugural speech to the aldermen, Edson called for a large public works program to modernize the city and prepare the way for economic growth. He talked of improving the water system, the police department, the hospitals, the wharves, and the piers, and of the laying of underground wires. Although Edson was criticized for pandering to interests of the aldermen in his appointments, he was instrumental in preparing the way for civil service system reforms and for the creation of a board of examiners (already mandated by the state legislature). The state legislature had also mandated that all telegraph wires be laid underground. The Board of Aldermen, under suspect conditions, granted a franchise for New York Electric Lines to lay lines which would consolidate service and be leased to various users. The reformist Gustavus Myers later wrote:

Rumors of the promise of money and stock to the aldermen were general, but there was no specific proof of bribery. The methods used in attempting to pass the ordinance were such as to make the charges easily believed. The report on the committee on ferries and franchises was read so hastily that nobody except those in the secret knew what it contained. Not even the names of the men back of the company were known. (Myers 1974, p. 185)[25]

The mayor vetoed this measure, which would have affected Edison's control over the lines he had already laid. The Board of Aldermen passed the measure over his veto, but the New York Electric Lines Company never began construction, as the politics of municipal utilities was a world of shifting advantages and alliances (ibid., pp. 185–186). The control of underground cables went through several more years of political and legal wrangling. In 1887 a plan that seemed acceptable to the Edison companies was enacted.

This same Board of Aldermen, moreover, was soon to be investigated in relation to a scandal concerning a franchise—also passed over Edson's veto—for the Broadway Transit Company. As a result of the exposure of the proven bipartisan bribery (thirteen aldermen had agreed to vote in concert except on political issues), four aldermen were convicted, another three turned state's evidence, six fled to Canada, and another ten were indicted. The indictments were later dropped.[26]

Thus, during the crucial period of getting central power started in New York, Edison was successful in gaining the favor of those political forces who could press the advantage of their friends without being exposed. At times this meant the backing of the aldermen; at other times it meant the backing of the mayor. Although we cannot sort out all the details of the political alliances Edison had to forge, and although we know little of the specific communications by which Edison learned to speak the language of politics, we do know he had contacts through which that language could be spoken, and that he and his associates did speak that language to a number of the players. We also know that the language involved patronage, jobs, political support, factional infighting, and perhaps payoffs. During the first few years of the 1880s, in a complex and corrupt political environment, Edison satisfied the relevant politicians well enough to get the low fees and the cooperative regulation he needed.[27]

The Reality of Isolated Installations

Politics could pave the way for central installations, but only the light itself could fully announce its own existence. Making incandescent lighting an everyday reality was a matter of illuminating many sites, making them appear to be operating in a routine and trouble-free way, and then making many people aware of their operation by firsthand observation and through the press.

The work of illuminating ordinary workplaces, shops, and homes by means of isolated systems had been underway for some time before the first central station opened. At the Paris Exhibition, an after-hours café, a museum, and an opera house were illuminated. A small station established in London in conjunction with the Crystal Palace exhibition illuminated several streets, several hotels and restaurants, some commercial offices, and the General Post Office with 950 lamps. A café in Havana was illuminated in 1882. An Edison plant built to light the Brazil National Exhibition in Rio de Janeiro was later used to light a resort hotel outside the city. To Americans, however, these illuminations were reported as exotica. Somewhat closer to home were the steamships that traveled up and down the East Coast. In 1882 the success of the installation on the *Columbia* (both as a working system and as a publicity vehicle) led Henry Villard—an Edison backer as well as the president of the Oregon Railway & Navigation Company—to have Edison outfit a second steamer, the *Queen of the Pacific*. James Gordon Bennett, the editor of the *New York Herald,* who had done so much to publicize Edison's earlier work, outfitted his new yacht, the *Namouna,* with an Edison isolated plant. By the time the Pearl Street station opened, Edison had received orders for the outfitting of a number of steamships of various lines.[28]

Working Edison isolated systems soon began appearing on land. The first of these illuminated Edison's Manhattan office and display room at 65 Fifth Avenue early in 1881. The Manhattan wholesale grocers Thurber and Company had one location illuminated early in 1882, a second by the middle of that year.

Among the early purchasers of Edison isolated plants were flour and textile mills. The lighting of a bleachery mill in Fall River, Massachusetts in February of 1882 inspired such local enthusiasm that by October the King Phillip Mills, the Bourne Mills, the Laurel Lake Mills, and the Conanicut Mills had all purchased Edison isolated plants. In addition, a consortium of downtown offices and merchants set up a small central plant. Fall River's enthusiasm for the Edison System led to the establishment of a central station in 1884.

Banking and financial institutions with direct connections to Edison were early to illuminate. Even better for publicity's sake was the fact that many newspapers found incandescent lighting useful in their nocturnal business. Among the papers that purchased Edison systems in 1882 were the *New York Herald,* the *Ohio State Journal,* the *Philadelphia Ledger,* the *Philadelphia Public Record,* the *Davenport Gazette,* the *Baltimore Sun,* and the *Boston Herald.* And hotels provided publicity and a chance for the public to see the light in a simulated domestic setting. Among the first hotels to acquire Edison systems were the Blue Mountain Lake Hotel in the Adirondacks, the Palmer House in Chicago, the Hotel Everett and Everett's Hotel in New York City, and the Vendome Hotel in Boston. By June 27, 1882, Edison companies boasted 67 installations and more than 10,000 lamps. By the middle of October the figures had nearly doubled; there were 123 installations and almost 22,000 lamps.

A bulletin distributed once or twice a month to Edison agents and stockholders during this period described these installations, provided reports of their success, and printed testimonial letters from customers. These bulletins also reported the longevity of the Edison lamps, announcements of Edison's patent protections and patent actions against competitors, and the travails of gas lighting (fires, explosions, and disruptions of service).[29] The lists of customers and the letters of testimonials were also used in a variety of other descriptive and advertising pamphlets. The length of the lists, the variety of the sites, and the prestige of the customers attested to the reliability of incandescent lights and to the sound business judgment of those who had chosen the Edison system.

A Location for Visibility

For Edison, however, these isolated plants were only an interim measure. The first permanent central installation, long planned and truly representing the beginning of the American lighting industry, was New York's Pearl Street station. Because the technical and business details of its construction have been given many times in biographies of Edison and in histories of electrical technology, I will focus on a few points about the how the station was designed to be perceived and how it came to be perceived.

Edison chose the site for the first plant with some care. In the spring of 1880 he studied maps of Manhattan and settled on a district between City Hall and the East River. Toward the end of 1880 he had a detailed survey made of the lighting and power needs of the buildings in the district. This detailed market survey inquired about the nature of fixtures and shades on existing lighting as well as about complaints against gas lighting utilities, gas costs, and insurance rates.[30] From these he had Francis Upton and the engineer Hermann Claudius calculate the district's power needs.[31]

The district as Edison originally anticipated it would have covered all the financial markets, City Hall, and the offices of newspapers and other publications. In short, Edison would have lighted all the institutions that were crucial to the success of his project. Even when the size of the first station was reduced, its range still included most of the financial district and bordered on both the newspaper and government districts, so that the incandescent light and its effects would be visible there. The early customers of the Pearl Street station included Kidder, Peabody & Company, Drexel Morgan & Company, the Merchant's Bank of Canada, the Continental Bank, the Third National Bank, Great Western Insurance, New York Insurance, Guardian Insurance, Continental Insurance, Home Insurance, Knickerbocker Insurance, National Fire Insurance, the New York Stock Exchange, the *New York Times,* the *New York Truth,* the *Mail and Express,* the *Sun,* and the *Commercial Advertiser.*[32] The *Herald* and the *Telegram,* just outside the district, had their own Edison isolated plants.

Laying the Electrical Tubes, **Harper's Weekly,** *June 24, 1882 (24: 254).*

The Construction of the Ordinary

The simple act of beginning construction of the New York system brought the electrical incandescent light into the realm of the familiar. As the Paris exhibit was ending, *Scientific American* reported that trenches were being opened in the New York streets for the installation of Edison's system. The story paired the Edison excavations with a project intended to distribute steam heat centrally. Both were presented as the next stage in the centralization and organization of all services, which had been accelerating over the past few decades with rail and horsecars, railroads, the telegraph, the telephone, and water and gas service. "And the next steps of social and domestic organization," *Scientific American* noted, "promise to be the distribution of motive power with our illuminant, and the displacement of our heaters and cooking stoves by steam conveyed through the streets in pipes."[33] *Scientific American*'s year-end summary listed the work of putting down steam mains as a part of the industrial advance of the United States, along with the opening of new mines and opening new areas for cultivation.[34]

Throughout 1882 the *Bulletin of the Edison Electric Companies* turned the construction of the Pearl Street station into a gradually proceeding reality for the agents and stockholders of the Edison companies. *Bulletin* number 3, dated February 24, 1882, announced that six miles of mains had already been laid and that work was going forward on the six Jumbo dynamos for the Pearl Street station. From then on, every issue of the *Bulletin* provided at least a short report about the number of feet of conductors laid, the development of meters, or the testing of the dynamos. In the seventh *Bulletin*, Pearl Street was the lead story, and it held that status almost until the publication vanished in 1883, no longer necessary once the Edison light was a well established reality.

The *Bulletin* reported the station's opening with an overall sense of normality:

This plant was started and the district was lighted up for the first time at 3 p.m. September 4th. Since then the station has been running day and night without stopping. The statement was made in the last Bulletin that up to that time no serious obstacle had been met with. The same is true now. Indeed, we can go further and say, that as regards Mr Edison's part of the work, namely the electric apparatus and every thing appertaining thereto, the result has exceeded our anticipations, the only delay that we have had having been caused by purely mechanical matters, such as the regulation of engines, and other usual engineering annoyances incidental to starting for the first time a number of high-speed engines.[35]

The rhetoric of normality was maintained despite the technical difficulties later reported in the heroic stories of Edison's last-minute repairs to the system and despite the fact that customers were not charged for the first five months of power because the delivery was not yet stable or regular.[36]

Reports in the *Bulletin* in the year following the opening continued to present a picture of reliability and business as usual, detailing the numbers of customers, the hours of operations, and other signs of normal success.

Lighting Up the City

In 1882 the press and the public also started to see incandescent lighting as a normal part of life, worthy of only a little special notice. When the Pearl Street station was finally opened, on the evening of September 4, the *New York World* gave it only two paragraphs in the next day's edition.[37] The *New York Herald* provided half a column of details, but only in the back pages, in a story only slightly longer than a neighboring one about a training encampment of the Connecticut State Guard. The *Herald* noted that the event had gone by almost without notice or ceremony.[38] The *Sun* ran a slightly shorter article on its front page September 5, quoting a pleased Edison as saying "I have accomplished all that I promised."[39] But this accomplishment was received as so ordinary a fact of life, so without ceremony, that the third paragraph was devoted to Edison's showing off the red and blue lamps that served to warn a young employee to regulate the power. The fourth paragraph has Edison winking his eyes to show how pupils dilate, so as to convince the reporter of the benefits of a steady light. Though Edison obviously was doing his best to gain what marketing mileage he could out of the event, it was hardly a spectacular public occasion.

The *Tribune* showed more enthusiasm and provided a somewhat grander account, describing how workers thronging out onto Fulton Street "had their attention attracted yesterday to the lights in several of the stores on both sides of the street."[40] After a paragraph of descriptive facts, most of the enthusiasm of the article is embedded in long quotations from Edison and his engineer Edward Johnson, who are described as "in a high state of glee."

The *New York Times,* its own building served by the Pearl Street station, provided the most extensive and enthusiastic coverage, opening as follows: "Edison's central station was one of the busiest places down town, and Mr. Edison was by far the busiest man in the station." However, after the opening paragraph, the rest of the story was about the *Times*:

Yesterday for the first time the Times building was illuminated with electricity. Mr. Edison had at last perfected his incandescent light, put his machinery in order, and last evening his company lighted up about one-third of the lower City District in which the Times Building stands.[41]

This article appeared on page 8.

Multiple Promises, Delivery Dates, and Goods

Despite the enormous pressures Edison had been under to make good on his original claims, by the time the promise was fulfilled few people were still anxiously awaiting the fulfillment of his word. The public and the press, as onlookers of spectacle, had had their appetites satisfied by the Menlo Park exhibits. The technical community had been satisfied by the displays, tests, and awards associated with the Paris and London exhibits.

Most of the public had yet to be enlisted as customers. New York City's politicians seemed only anxious to make sure they got their cuts and patronage. Only Edison's backers had anything seriously riding on the opening, for they had literally invested in Edison's word against this day. But they could not really light their cigars until the company was profitable. Thus, although Edison could say that he had delivered on all he had promised, the promises were many and were diffused through many communities. Few saw the local event of the opening of the Pearl Street station as a moment of resolution.

On the other hand, the ordinariness of the event marked another kind of success. Incandescent lighting now had an unremarkable place in the daily life world—an actual competitive presence that, over time, would have to prove itself on its ordinary merits.

Searle (1969, p. 59) notes that a promise is associated with an action that would not be carried out in the ordinary course of events. Now, with incandescent lighting part of the ordinary course of events, Edison's material production of light was no longer a major promise. The physical reality of the light was now just the physical reality of the light; it was no longer an awaited sign symbolizing Edison's fulfillment of his promises.

IV

Establishing Enduring Values

Patent Realities: Legal Stabilization of Indeterminate Texts

The story of the book to this point has been of Edison establishing presences within systems of meaning, which the delivery of the actual technology would then realize in material value. Edison announced in the press that he would produce incandescent lighting, and the press eventually reported light produced and witnessed in Menlo Park, Paris, and downtown Manhattan. Edison claimed patents for incandescent lighting, and he turned the ownership of a potential into the ownership of a working system. Edison promised financiers his best efforts to create a working system, and he delivered. However, once the technology was a material accomplishment it had to take on new, long-term meanings if it were to endure. Ownership of the technology had to be stabilized despite contested patents. The informal work arrangements of Menlo Park had to be stabilized into corporate communicative structures. Edison companies had to become valuable long-term investments on the financial markets. And incandescent lighting had to take on meanings that were compatible with people's perceptions of domestic life.

As we have seen, a successful patent establishes within a legal framework an invention, an inventor, and a period of protection. The designated inventor (along with partners, backers, or assignees) now owns a property to trade on and with. The patent stabilizes the fluidity of development, defining specific achievements and allocating the rewards. But such stabilization is not in the interests of all concerned. First, the Patent Office does not want to give things away too freely; promiscuous approval might lead to excessive litigation and constraint of trade with no counterbalancing benefit of substantial new invention. Second, competitors are restrained by the same patents that reward their owners.

Therefore, obtaining a patent is not just a technical matter of fulfilling paper requirements. Individuals with contending interests may attempt to undo the patent grant or limit its applicability. They may attempt to place

A lamp package listing patent numbers (48: 754).

obstacles to the speech act of a request for a patent before it is realized, or they may attempt to undo the act once it is seemingly done and stabilized. An attack may be carried out on the formal perfection of the speech act or on the link between the speech act and the events, objects, and arrangements it purports to display.

Challenges to the validity of a speech act can tell us much about how the standard forms of speech acts sit within contexts of taken-for-granted assumptions, other texts, and events. But if the textual performance of the speech act stands, it reaffirms its relationship with the contexts that are bound to the performance and thereby strengthens the taken-for-granted robustness of those contexts. That is, people may question a patent by doubting its relation to other patents, to the work of other inventors, to laboratory events, to the claimed inventor, to the law, to contractual arrangements, or to prior knowledge. By breaking these bonds between text and context, the challengers try to make the patent text lose meaning. Insofar as the patent stands, it reaffirms the accounts it embeds of laboratory events, legal understandings, relationship to the work of others, the state of knowledge, contractual arrangements, and interpretations of the patent record.

Two Stories of Indeterminacy and Determination

In this chapter we will examine two related stories of challenges to would-be patents and already-accomplished patents. These stories will lead us to see how patents sit on top of complex contexts, giving order, meaning, and certainty to arrangements of benefit to particular parties. The patents, nonetheless, are always open to challenge that can potentially establish different orders, regularized through alternative texts to stabilize alternate realities.

The first story is of a patent application that went through many rejections, revisions, and appeals, only to vanish when the passing of time made the issues moot. At one level, this is a story about the Patent Office's cautiousness about recognizing novelty so as not to grant too much on too little basis or to give away what already belongs to others. At another level, it is a story about how patent claims serve changing needs, stakes, and functions.

The second story, which intertwines with the first, is of a conflict between Edison's patents and those of William Sawyer and Albon Man. This conflict, involving the fundamental issues of who invented incandescent lighting and who deserved the reward from it, was first fought out within the patent system as an interference and then as an appeal; consequently, it moved to the courts, where it produced eight volumes of documents totaling more than 5000 pages. Within some limitations, this case established the Edison companies and their successor, General Electric, as the primary owner of light technology as the industry expanded. The case also stabilized a historical account of Edison's invention and priority.

Together the two cases show that, although legal action is almost entirely a matter of internal discourse within government offices and courts, the legal system does not carry out a purely closed set of symbolic operations. To enter into the legal discourse of the government and courts, all considerations must be translated into the symbolic system by which rights and obligations are sorted out; however, these symbolic activities exist in continuing relation with the realities that people want adjudicated, that provide the sources of evidence, that offer constraints and pressures influencing the legal process, and that are the sources of interests that keep cases in motion. When the legal system does achieve stable symbolic resolutions in its own terms, it then binds people to sort out their affairs in relation to the legal mandate, though this process may not be simple or complete.

When patents are contested, it becomes unclear who owns what, who invented what, what is original, who should receive what proceeds, and what is technologically significant. The legal system provides a method of stabilizing answers of what constitutes the property and who owns it, but not until these issues have been translated into legal terms within legal institutions. Even those translation processes that create legal meanings and values, however, can be challenged by opponents who wish to diminish the standing of the first party's case. Once issues and evidence have been successfully brought into court or other legal forums, they are adjudicated through processes of legal discourse consisting of series of discursive moves by the various participants. The resulting judgments are then brought out of court with consequences for other discursive realms, but how those judgments apply to specific situations can again be litigated.

A Failed Patent Application

The first case concerns a patent that never came into being but which Edison pursued for more than 15 years. The process was instigated by an application, dated February 5, 1880, for "An Improvement in Electric Lighting & System of Electric Lighting." At least 42 pieces of correspondence traveled between Edison's attorneys and the U.S. Patent Office before the file was officially declared dead on April 27, 1895.[1]

The Edison papers contain only copies of the figures from the original patent application, but from surrounding documents it appears that the application claimed a method for increasing resistance of the filament while maintaining the light-giving surface.[2] This application was filed on February 5, 1880. Patent 223,898, granted just previously (January 27, 1880), was later judged to be crucial patent covering high resistance, although high resistance was only a subsidiary part of that patent. Since Sawyer and Man had applied for a patent for their lamp on January 9, 1880, the purpose of Edison's application may have been to make his claim to the principle of high resistance more explicit. This interpretation is supported by a notebook entry—dated November 28, 1878—in which Edison's assistant Charles Batchelor comments that the weakness of the Sawyer-Man lamp is its low-resistance bulb, which would require very large conductors to supply sufficient electricity for multiple lights.[3] Furthermore, Edison's attorneys cited this patent application as being at the center of Edison's case in the 1885 action against the U.S. Electric Light Company.[4]

Ambiguity, Visibility, and Accountability

In a communication dated March 30, 1880, an examiner rejected the February 5 application as "ambiguous" on the ground that he could not understand from the specification how the surface area would not be increased in proportion to the increased resistance. The examiner, in pursuing his institutional role, carried out a mandate that a patent cannot be granted for any undefined item. Even though there is no test for a working item, the patented item must plausibly specify intelligible mechanisms by which the object would work. Moreover, the object must be described in sufficient detail to show how it would operate according to those principles. The examiner concluded: "The claims in view of the indefinite description are rejected as being vague."[5]

In response, Edison's lawyers explained the principles involved and offered amended language to clarify how the improvement worked.[6] Thus, under the pressure of the examiner's power to accept or deny the application, Edison's attorneys, in order to pursue the patent application, needed to provide a more detailed and intelligible account of the purported invention, which they then did.

Once a claimed invention is made intelligibly visible, the would-be patent is held accountable for being a significant advance against what is known. Since it is only the novelty, or invention, that can be owned, if there is no significant improvement there is nothing to be owned. In a communication dated September 20, 1880, the examiner rejected the application for representing only a deduction of well-known scientific principles, not the discovery of a new law.[7] After an exchange of letters with Edison's lawyers pointing out that patents concern not the discovery of scientific laws but the invention of practical objects,[8] the commissioner of patents reiterated that the application in question was only an obvious extension of well-known principles and thus was not an invention.[9] That is, it was not sufficiently novel.[10] To contest this rejection, the applicant and his agents must provide a persuasive alternative account of what is known in the technical area and how this application extends existing art. Edison and his agents let this challenge slide until they had compelling reason to pursue the issue. That is, they did not pursue the task of making an acceptable account of originality until they perceived that the advantage to be gained outweighed the costs and trouble of making the case.

Changing Conditions, Changing Needs, and Changing Claims

After being rejected on December 15, 1880, the application lay dormant until July 26, 1882 when Edison's attorneys proposed extensive amendments. By this time the Pearl Street station was preparing to start operations, and contention over a patent interference filed by Sawyer and Man had heated up. The emergence of a workable system in the marketplace raised the stakes in ownership of what once had been only a hopeful idea.

The amendments to the original patent application, offered by Edison's attorneys in a nine-page document,[11] amount to a complete rewrite. Only the original diagram and the first eleven lines remain as in the original. In addition, there is a new figure. The continued use of the original diagram implicitly argues that it is still the same invention for which a patent is sought, with revisions only to clarify the nature of the invention. The new text emphasizes the value of increased resistance, suggesting that as the significant advance being claimed as invention. There are several phrases describing as benefits the decreased size of the conductors required, the decreased cost of metal, the practicability of delivering electricity to sparsely settled places because of the decreased metal in the conductors, and so on. The first summative claim reiterates these themes:

First: In a system of generation, distribution, and translation of electricity for the purposes of light, the method of diminishing the amount of metal required in a given length of main conductors by increasing the resistance of the lamp, substantially as described.[12]

The invention now is more than a means of resistance; it is a means of transporting electricity more economically.

On September 13, 1882, the examiner found some difficulties with the phrasing of parts of the descriptions and claims, and discounted one of the four claims altogether as being covered by a previous patent. The first phrasing difficulty was a matter of accurate correspondence between illustration and text, so that they would clearly appear to refer to the same object or idea without contradiction. The second phrasing objection pointed to the inappropriateness of a claim:

The 1 claim is informal [not formally correct], it being for the construction of the device or rather for the conception of the manner of constructing the device or system instead of being for the system itself. Furthermore a method must form a part of an art & not of a system.[13]

The claim must fit within the standards of the kind of thing that may be claimed. Since this patent is of a device rather than of a system, claims concerning the system as a whole are not appropriate. This issue of form has to do with the relationship between the claim and the regulations governing the claim. Patents occur only within the context of the law and regulations, which must constantly be observed.

The third question of phrasing involved how the idea being claimed related to the specifics of the object described. In particular, the claim referred to "a uniform radiating surface," which the examiner did not find precisely correlated with anything in the description.

Edison replied on September 13, 1884 by making further amendments, primarily substituting four new claims for the previous ones.[14] These new claims respond directly to the examiner's desire to see a closer correlation between object and description. The claims specify the exact configuration of elements in the bulb design rather than general properties of high resistance and series connections. The issues of cost and high resistance had now vanished from the application.

Why did Edison revive the application after two years, in a narrowed version that eliminated the general issues of resistance and cost and pursued only a particular configuration of elements in the bulb design? We have no evidence as to what Edison and his colleagues thought about this patent between the 1882 rejection and the 1884 resubmission. However, in 1883 Sawyer and Man had gained a major advantage by winning an interference that limited Edison's rights and increased theirs, and had begun producing lamps under contract to Thomson-Houston. It appears that Edison, no longer sure that his general claims would be protected, sought to protect his specific current designs by reinterpreting the rejected patent application on narrower grounds. The revision of the claims allowed the application to became a variable tactical resource to fit the changing circumstances.

Edison's 1884 amendment was rejected on the ground that the first two claims were covered by a separate British patent.[15] After another two-year delay, Edison proposed further amendments to delete claims which he felt were disallowed by the competing patent but to retain his final two claims (one concerning the construction of the filament and one concerning the arrangement of lamps in series).[16] The rejection this time specified further that there was no novelty here in the shape of the filament, which was covered by other British patents.[17]

The Stabilizing Context

In February of 1888, Edison provided extensive new amendments. In essence, he rewrote the whole patent again, returning to the general principals of resistance and explaining how high resistance permitted lamps to be arrayed in series (but now asserting that the latter idea was established in patent 223,398, which was just about to be judged as the crucial patent in the a trial involving the McKeesport Light Company). Furthermore, Edison cast this as an application for a "specific lamp and system of this character."[18] That is, now the application was for a particular realization of the general system, which was protected under the previous patent. The claims returned to a higher degree of generality than those of 1884 and 1886, but they were more detailed and more specific than those of 1882. Some new figures accompanied this application, including one illustrating the connection of the lamps to the circuit (something not discussed or illustrated in previous versions of the application). This application seems to have been a vehicle for protecting previously unnoted or unprotected aspects of what eventually turned out to be the produced technology. When the original patent was first proposed, in 1880, it was not clear what the system would ultimately be, so the patent was cast in more general terms. Now, however, the application could be reinterpreted as a specific claim in relation to what had since evolved as the economically valuable system. In the 1888 version, an argument several paragraphs in length explained how this particular advance lowered the costs of the system and its maintenance. That Edison was completing a web of protection around the existing and now stabilized product is further supported by an accompanying deposition in which he claims that this improvement is not covered by any of his existing patents, but that he enjoys patent protection in other countries (including Australia, India, Sweden, and Portugal) through seventeen foreign patents.

In response, the examiner pointed out additions that were not implicit in the original concerning the arrangement of the bulbs in circuits and commented that this suggested that without these additions the improvement was unworkable. He therefore judged that this was all new material and that it should be disallowed as an amendment.[19] Edison's attorney responded that this arrangement was evident in other patents and applications of Edison's dating from the original submission and would have been obvious to anyone versed in the art.[20] The examiner again rejected the change as not merely making explicit what was implicit but changing the fundamental character of the claim.[21] A series of letters, depositions, judgments, and hearings followed, arguing over what was and was not new. In April of 1890 the examiner identified specific new material and material redundant with other applications that was to be removed if the application was to be considered. Such deletions, the examiner ruled, would leave only material already covered by Edison's other patents.[22]

In 1892, Edison's attorney came back with further amendments, which the examiner rejected on grounds "fully and explicitly stated in the previous action." The 1892 communication ends with the examiner's testy comment about this being the end of the matter, which had dragged on for twelve years.[23]

However, correspondence continued, and in May of 1894 there was an appeal before the board of examiners.[24] Edison's attorney, Richard Dyer, argued that the claims were not covered by previous patents and that these improvements had current value in the now-competitive industry.[25] This new economic rationale highlighted the coiling of the burner, another innovative feature of the design expressed in the original application. Only recently had the value of this innovation become evident. In further correspondence, arguments as to whether this innovation was covered by previous patents ensued. On April 27, 1895, after several more appeals, amendments, and rulings, just when the case was about to be reconsidered once more, the file was declared dead because of the expiration of foreign patents.[26]

Some Lessons

This involuted story has several lessons.

First, a single patent application can be a variable resource, attempting to cover different aspects of the emerging technology in relation to changing contexts of actually produced designs, economic values, legal rulings, competitor's positions, and the accumulated file of legal documents. While the specific object portrayed in the patent application remained fairly constant (except for additional diagrams making visible new aspects, which may or may not have been there at the original time), the meaning of the object varied, as was made explicit in the claims.

Second, a patent gains meaning as part of a general strategy of protection in relation to a particular technology within a particular configuration of production, market, financial, and competitive circumstances. The choices made for the tactical use of this application are interpretable only in relation to other patents owned by Edison and his competitors, the changing technology actually in use, and the changing business situation. Edison used the application to shore up whatever he saw at the moment as the weak point in his interlocking networks of patents in an attempt to maintain stable ownership of a technology competitive on the market.

Third, it is the examiner's task to try to create regulative order out of the competing perceptions of various parties and the protean situation. To create a stable determination, the examiner attempts to hold the patent application accountable to precision of ideas, visibility of novelty, originality of events, and the archive of prior ideas as they unfold historically and as they are incorporated in the historical record in prior documents. That is, while the inventor wants to stabilize ownership of technology in the broadest and most comprehensive terms, the examiner tries to make those claims accountable to material, historical, technical, and linguistic criteria by which the patent is to be stabilized.

Competitors and the Attempt to Destabilize Acts

Competitors whose interests would be hurt by a patent may attempt to obstruct or undo the patent grant. However, there are only limited, focused ways of intervening in the patent-assignment process. Interventions can be pursued only by parties who have established their right to carry out actions, and they may be pursued only through particular forms, which must be carried out persuasively. Only certain kinds of claimants and certain kinds of claims can be heard. And once argumentative claims have been heard, only a subcategory will be deemed successful, resulting in reshaping of the owned property or reassignment of ownership.

In Edison's time, the first intervention available was an interference, which asserted that a patent application being considered overlapped or infringed on a patent that the person pursuing the interference already owned or had applied for. Interference could be asserted in the system of patent appeals or in the courts.

An inventor, on the other hand, once granted a patent, could sue competitors for infringement, asserting the patent right against imputed violations. The ideal situation for any inventor is that the patent granted sets such clear and firm boundaries around a technology that no one else will come near to producing anything like it for fear of liability in the courts. Then the costs and trouble of litigation and the likelihood of loss force the competitors to respect the inventor's ownership.

Though clearly defined patent rights tend to discourage litigation, if a technology turns out to be quite profitable the stakes in grabbing a "piece of the action" may make even a long-shot interference case worthwhile to competitors. The original patent owners may then become vigilant to quash any hint of an infringement in order to maintain the maximum integrity of their valuable property. If there is any uncertainty in the patent claims, and if the technology is valuable, the legal situation can become volatile and contested. Many perceived intrusions may occur, because each side is attending to its own self-interested interpretation of the maps of the intellectual property.

Since electric incandescent lighting was an area in which a number of people had been working for a number of years, and since arc lighting was already a marketable technology, questions of overlap were substantial. By one count (Conot 1979, p. 214), in the 15 years before Edison was granted his first patent for incandescent light, 31 patents had already been granted in that area, and in the next 4 years at least 321 new patents by 8 inventors (including 147 by Edison) were issued. Edison's claims had to assert their place in a complex and crowded terrain.

Industrial Success and Increasing Legal Stakes

As the Edison light proved successful, competitors pressed the question of ownership of the technology. In Europe, Edison quickly cleared up such questions by a combination of agreements with competitors (Swan in Britain and Siemens in Germany) and favorable court rulings. But in the United States the issue of ownership of the produced technology remained murky. Several alternative patents (the most important being those of Hiram Maxim, William Sawyer backed by Albon Man, William Stanley, and Moses Farmer) provided justification for other manufacturers to enter into competition with Edison. First, Consolidated Electric produced the Sawyer-Man lamp. The United States Electric Lighting Company, which had been using the Maxim patents, then absorbed Consolidated and produced lamps based on both sets of patents. Westinghouse began production based on the Stanley and Farmer patents in 1884, and in 1888 it took over USELC, acquiring all the serious competing patents. Edison, nonetheless, still dominated the market, with around 50 percent of the sales in 1892; Westinghouse and the smaller manufacturers divided up the rest.[27]

In 1892, court decisions granted General Electric full control over the technology. But the process turned out to be slow and complex, and the judgments could never be fully enforced in an unruly marketplace.

Litigation against Sawyer and Man

The Edison companies, in their communications with stockholders, emphasized the importance of patents to the financial value of the company quite early in the game. However, the Edison companies were not vigorous in enforcing their patent rights until 1885, by which time the legal successes of Sawyer and Man, Westinghouse's substantial entry into the incandescent lamp market, the growing complaints of the Edison licensees (who had paid handsomely for exclusive rights), and the need to protect the value of Edison stock all suggested a more aggressive assertion of patent rights. The Edison Company for Isolated Lighting began publishing a bulletin that paid considerable attention to patent issues,[28] and that company's pamphlets now regularly included lists of Edison patents.[29] The Edison Electric Light Company issued a letter warning of its intent to prosecute patent cases[30] and circulated advertisements[31] and pamphlets arguing in favor of Edison's patent rights. Edison's lawyers initiated suits, and his boards of trustees informed stockholders of litigation.[32] Such initiatives were to continue until the court cases were resolved.

The roots of the case went back to the time of Edison's first success with incandescent lighting. On January 9, 1880, two months after Edison filed a patent for an electric lamp using a carbon filament, William Sawyer, backed by the financier Albon Man, filed a similar patent application for a carbonized paper filament, claiming it as an improvement to a patent granted to Sawyer and Man on June 18, 1878.[33] Sawyer then filed an interference, asserting that his claim had priority over Edison's. Between June 10 and July 20, 1881, a hearing was held in New York to determine the date and the scope of each inventor's accomplishments. Whereas Edison's original application (November 1879) was limited in scope to the carbon filament itself, Sawyer's January 1880 included as essential the placement of the filament in a near vacuum. Edison's lawyer explained Edison's failure to claim the vacuum technology by a desire to perfect the vacuum technology before claiming it.[34] Throughout the proceedings, the lawyers for Sawyer and Man objected to any testimony directed toward establishing Edison's working with carbon in a vacuum or establishing a date for Edison's working with a carbon filament before the filing date. Edison attempted to have his original application amended to reflect what he claimed was the history. In 1883, however, the patent commissioner ruled in Sawyer's favor, saying the history could only be considered as originally represented in the filed applications. That is, within the highly specialized discourse of legal deliberations, only information that had been translated into the legal system properly and in a timely fashion could count.

Competing Patents and Control of the Industry

Edison's lawyers delayed the issuance of the patent until 1885, but in effect Sawyer and Man immediately had the right to manufacture. They sold this right to Consolidated Electric, which then came under the control of USELC. The Sawyer-Man patent left the Edison companies under a cloud of insecurity. Edison, his backers, and the Edison companies had invested heavily in an interlocking system of inventions and had established an extensive corporate enterprise that served many customers. All of it could be lost with the loss of patent rights.

To reassert control, the Edison Electric Light Company filed an infringement case against the United States Electric Lighting Company in 1885 in the United States Circuit Court of the Southern District of New York. On December 8, 1887, while this case was pending, Westinghouse— now in control of the Sawyer-Man patents—filed another suit in the Western District of Pennsylvania, in the name of its subsidiary Consolidated Light Company, against the McKeesport Light Company (an Edison subsidiary). The Pennsylvania case was heard before the New York case, and it provided the crucial decision, handed down in an opinion of

Justice Bradley in May 1888. Bradley upheld the Edison claims and invalidated the Sawyer-Man claims. This precedent led to a similar decision upholding the Edison claims in the New York case in the following year (although appeals delayed the final resolution until 1892, during all of which time Westinghouse and other competitors continued to manufacture lamps).

Let us now examine the progress of the McKeesport trial to see how claims were reconstructed in contention and stabilized by court decisions, how the court's decision affected the historical understanding, and what difficulties arose in making the corporate world correspond to the dictates of the court.

Creating a Historical Record

The testimony and evidence presented in the McKeesport trial bore on all aspects of the history of the invention and manufacture of incandescent lighting and the relations between Edison and Sawyer.[35] The substantive testimony, however, was preceded by technical legal issues concerning the status of international patents. Rulings on the applicability and expiration dates of foreign patents had the potential to invalidate Edison's U.S. patents, essentially erasing the texts on which Edison's claims stood. Thus, Sawyer's strategy was to undermine the validity of Edison's patents as speech acts, either through expiration or other technical means or by establishing that the representations made in the patents were inaccurate (in their reference to events) or otherwise faulty (particularly in terms of novelty).

The Edison lawyers, similarly, tried to undermine the word and thus the speech acts of Sawyer (who had died in the interim). To discredit Sawyer's character and word, and thereby the general credibility of Sawyer's patent claims, Edison's lawyers presented evidence of Sawyer's general truculence, his alcoholism, his public hostility to Edison, his refusal to believe Edison's accomplishments, and his deception in the public displays of his own lights. Through such tactics, Edison's lawyers attempted to depict Sawyer as having little integrity and his patents as less than honest representations.

Edison's lawyers also elicited testimony from Edison colleagues and people in the industry concerning Edison's dominant role in the development, production, and distribution of incandescent lighting. This evidence suggested that Edison had in fact established the industry, and that social and economic disruption would take place if his de facto role were not recognized. This testimony implicitly made Sawyer look even more like a lightweight.

Thus, each side attempted to discredit the opposition's speech acts and to reassert the validity of its own patents as documents that should rule over the technology. The judge, however, resolved the case by deciding which speech acts were significant. He ruled that the inventive issues over which the lawyers had argued—priority for the carbon filament, the arched shape of it, and the vacuum—were aspects of preexisting technology. Although Judge Bradley tended to believe Sawyer's words, he found Sawyer's honesty moot because the crucial advance in the technology was outside Sawyer's claims.

The judge ruled that, since all the other aspects of the technology had been common knowledge but had not in themselves produced a working technology, the one element that made a difference was Edison's introduction of high-resistance filaments, as first declared in patent 227,229 and as established in patent 223,898. There was no question that high resistance was solely a novelty of Edison, or that Edison had clear patent control over it.[36]

Certainly Edison had been aware of high resistance as a significant advance of great economic advantage in decreasing the need for copper, as we have seen in the story of the failed application that he kept pressing. However, until this point high resistance had never been considered the sole or primary advance or the key point to be argued. Of the testimony assembled, only a fraction referred to resistance. The two moments that had been previously presented to the public as breakthroughs rested on entirely different ideas: temperature regulation (a false lead) and the use of carbonized thread as a filament.

The judge's rulings rewrote history, not by changing facts, but by putting facts into a new relation by declaring one development crucial and the others legally irrelevant. This new history of the invention served the needs of the law to stabilize its judgments and regulate the surrounding world. In the subsequent suits, resistance became the key point. The pending New York case *(Edison Electric Light vs. U.S. Electric Light)* was argued on the basis of patent 223,898. More remarkable, for nearly 100 years accounts of the development of incandescent light tended to take the judge's word that high resistance was the crucial invention. Only the detailed examination of the discovery records of Friedel and Israel (1986) put the resistance story back in the context of multiple advances.

T. A. EDISON.
Electric-Lamp.

No. 223,898.   Patented Jan. 27, 1880.

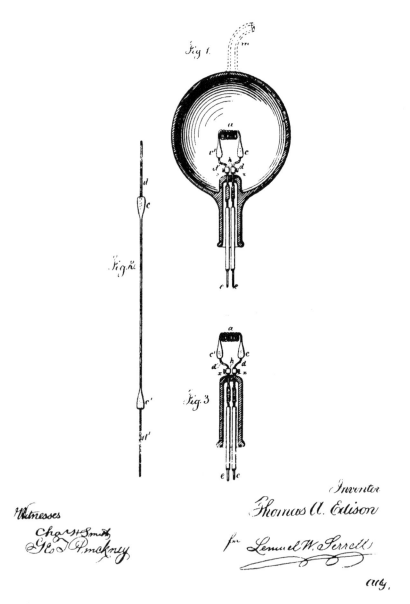

A drawing from the patent determined to be crucial: patent 223,898.

The Legal Need for a Clean and Plausible Story

Faced with a murky case that required clear resolution if the technology and the industry were to develop on a stable legal basis, Judge Bradley did what judges tend to do: find the cleanest, narrowest grounds for making a definitive ruling. High resistance provided a way to cut through the morass of questionable credibility and alternative accounts of events, and through the tangle of overlapping patent claims, and to establish a stable under-standing of the legal rights. Moreover, high resistance as the crucial novelty was a technologically plausible explanation. The judge determined the novelty, and that determination became legal history, which then was taken to be actual history. Edison, having been given a history that served his interests, readily accepted that history. Historians, needing to sort through the morass of claims and counterclaims and (even more) through the morass of a complicated process of invention and development, also took guidance from the judge's ruling.

Translating Clean Law into the Messy Marketplace

The legal judgment in the McKeesport Electric case gave the Edison Electric Light Company legal warrant to make the marketplace conform to it. However, just as the history was murky and indeterminate before it was translated into legal discourse, so the corporate marketplace into which the judgment had again to be translated was murky, complex, and indeterminate. Because the process of making the world conform to the legal ruling was complicated and incomplete, the EELC, while reaping much of the benefit of patent ownership, could not enforce its monopoly, and other manufacturers managed to gain some profit that was, legally speaking, not rightfully theirs.

    A number of factors made the enforcement of the patent rights diffi-cult. First, the easiest course—imposing a moderate royalty fee—violated the EELC's agreement with the power station licensees, who had been guaranteed regional monopolies. Second, the EELC and its successor companies (Edison General and General Electric) had no desire to alien-ate users of non-Edison lamps or independent power companies who had been supplying lamps to their consumers, since they constituted his potential expanded market; thus, he did not seek enforcement against them. In any event, a lamp burned out after 600–1000 hours, and Edison could anticipate getting the replacement business. Third, by the time the court cases were fully resolved, only a few years remained on the patents

United States Patent Office.

THOMAS A. EDISON, OF MENLO PARK, NEW JERSEY

ELECTRIC LAMP.

SPECIFICATION forming part of Letters Patent No. 223,898, dated January 27, 1880.

Application filed November 4, 1879.

To all whom it may concern:

Be it known that I, THOMAS ALVA EDISON, of Menlo Park, in the State of New Jersey, United States of America, have invented an Improvement in Electric Lamps, and in the method of manufacturing the same, (Case No. 186,) of which the following is a specification.

The object of this invention is to produce electric lamps giving light by incandescence, which lamps shall have high resistance, so as to allow of the practical subdivision of the electric light.

The invention consists in a light-giving body of carbon wire or sheets coiled or arranged in such a manner as to offer great resistance to the passage of the electric current, and at the same time present but a slight surface from which radiation can take place.

The invention further consists in placing such burner of great resistance in a nearly-perfect vacuum, to prevent oxidation and injury to the conductor by the atmosphere. The current is conducted into the vacuum-bulb through platina wires sealed into the glass.

The invention further consists in the method of manufacturing carbon conductors of high resistance, so as to be suitable for giving light by incandescence, and in the manner of securing perfect contact between the metallic conductors or leading-wires and the carbon conductor.

Heretofore light by incandescence has been obtained from rods of carbon of one to four ohms resistance, placed in closed vessels, in which the atmospheric air has been replaced by gases that do not combine chemically with the carbon. The vessel holding the burner has been composed of glass cemented to a metallic base. The connection between the leading wires and the carbon has been obtained by clamping the carbon to the metal. The leading-wires have always been large, so that their resistance shall be many times less than the burner, and, in general, the attempts of previous persons have been to reduce the resistance of the carbon rod. The disadvantages of following this practice are, that a lamp having but one to four ohms resistance cannot be worked in great numbers in multiple arc without the employment of main conductors of enormous dimensions; that, owing to the low resistance of the lamp, the leading-wires must be of large

dimensions and good conductors, and a glass globe cannot be kept tight at the place where the wires pass in and are cemented; hence the carbon is consumed, because there must be almost a perfect vacuum to render the carbon stable, especially when such carbon is small in mass and high in electrical resistance.

The use of a gas in the receiver at the atmospheric pressure, although not attacking the carbon, serves to destroy it in time by "air-washing," or the attrition produced by the rapid passage of the air over the slightly-coherent highly-heated surface of the carbon. I have reversed this practice. I have discovered that even a cotton thread properly carbonized and placed in a sealed glass bulb exhausted to one-millionth of an atmosphere offers from one hundred to five hundred ohms resistance to the passage of the current, and that it is absolutely stable at very high temperatures; that if the thread be coiled as a spiral and carbonized, or if any fibrous vegetable substance which will leave a carbon residue after heating in a closed chamber be so coiled, as much as two thousand ohms resistance may be obtained without presenting a radiating-surface greater than three-sixteenths of an inch; that if such fibrous material be rubbed with a plastic composed of lamp-black and tar, its resistance may be made high or low, according to the amount of lamp-black placed upon it; that carbon filaments may be made by a combination of tar and lamp-black, the latter being previously ignited in a closed crucible for several hours and afterward moistened and kneaded until it assumes the consistency of thick putty. Small pieces of this material may be rolled out in the form of wire as small as seven one-thousandths of a inch in diameter and over a foot in length, and the same may be coated with a non-conducting non-carbonizing substance and wound on a bobbin, or as a spiral, and the tar carbonized in a closed chamber by subjecting it to high heat, the spiral after carbonization retaining its form.

All these forms are fragile and cannot be clamped to the leading wires with sufficient force to insure good contact and prevent heating. I have discovered that if platinum wires are used and the plastic lamp black and tar

The text of patent 223,898.

material be molded around it in the act of carbonization there is an intimate union by combination and by pressure between the carbon and platina, and nearly perfect contact is obtained without the necessity of clamps; hence the burner and the leading-wires are connected to the carbon ready to be placed in the vacuum-bulb.

When fibrous material is used the plastic lamp-black and tar are used to secure it to the platina before carbonizing.

By using the carbon wire of such high resistance I am enabled to use fine platinum wires for leading-wires, as they will have a small resistance compared to the burner, and hence will not heat and crack the sealed vacuum-bulb. Platina can only be used, as its expansion is nearly the same as that of glass.

By using a considerable length of carbon wire and coiling it the exterior, which is only a small portion of its entire surface, will form the principal radiating-surface; hence I am able to raise the specific heat of the whole of the carbon, and thus prevent the rapid reception and disappearance of the light, which on a plain wire is prejudicial, as it shows the least unsteadiness of the current by the flickering of the light; but if the current is steady the defect does not show.

I have carbonized and used cotton and linen thread, wood splints, papers coiled in various ways, also lamp-black, plumbago, and carbon in various forms, mixed with tar and kneaded so that the same may be rolled out into wires of various lengths and diameters. Each wire, however, is to be uniform in size throughout.

If the carbon thread is liable to be distorted during carbonization it is to be coiled between a helix of copper wire. The ends of the carbon or filament are secured to the platina leading-wires by plastic carbonizable material, and the whole placed in the carbonizing-chamber. The copper, which has served to prevent distortion of the carbon thread, is afterward eaten away by nitric acid, and the spiral soaked in water, and then dried and placed on the glass holder, and a glass bulb blown over the whole, with a leading-tube for exhaustion by a mercury-pump. This tube, when a high

vacuum has been reached, is hermetically sealed.

With substances which are not greatly distorted in carbonizing, they may be coated with a non-conducting non-carbonizable substance, which allows one coil or turn of the carbon to rest upon and be supported by the other.

In the drawings, Figure 1 shows the lamp sectionally. *a* is the carbon spiral or thread. *c c'* are the thickened ends of the spiral, formed of the plastic compound of lamp-black and tar. *d d'* are the platina wires. *h h* are the clamps, which serve to connect the platina wires, cemented in the carbon, with the leading-wires *x x*, sealed in the glass vacuum-bulb. *e e* are copper wires, connected just outside the bulb to the wires *x x*. *m* is the tube (shown by dotted lines) leading to the vacuum-pump, which, after exhaustion, is hermetically sealed and the surplus removed.

Fig. 2 represents the plastic material before being wound into a spiral.

Fig. 3 shows the spiral after carbonization, ready to have a bulb blown over it.

I claim as my invention—

1. An electric lamp for giving light by incandescence, consisting of a filament of carbon of high resistance, made as described, and secured to metallic wires, as set forth.

2. The combination of carbon filaments with a receiver made entirely of glass and conductors passing through the glass, and from which receiver the air is exhausted, for the purposes set forth.

3. A carbon filament or strip coiled and connected to electric conductors so that only a portion of the surface of such carbon conductors shall be exposed for radiating light, as set forth.

4. The method herein described of securing the platina contact-wires to the carbon filament and carbonizing of the whole in a closed chamber, substantially as set forth.

Signed by me this 1st day of November, A. D. 1879.

THOMAS A. EDISON.

Witnesses:
S. L. GRIFFIN,
JOHN F. RANDOLPH.

Advertisement by Edison Electric Light Company, Electrical Review, August 4, 1888, p. 17 (95: 233).

during which competing manufacturers might be driven out of business. Finally, there were many small manufacturers to seek injunctions against, and more entered the market as the clock was ticking on the patents.

The EELC did get effective injunctions against Westinghouse and some smaller manufacturers. But then Westinghouse and another smaller company developed non-infringing lamps that, although inferior, allowed competition to continue. Westinghouse then began manufacturing another infringing bulb. When Edison's Canadian patent expired, on November 14, 1894, the U.S. patent also expired. Shortly thereafter, General Electric entered into a market-stabilization and price-fixing agreement with other manufacturers, guaranteeing it the most but not all of the market.

Boundary disputes over ownership, such as those examined in this chapter, show how tenuously property arrangements are constructed and deconstructed. More cleanly stabilized arrangements are built from the same mechanisms, but their tenuousness is less visible. Speech acts, when they stand uncontested in one sphere, can carry powerful meanings into other spheres that are structurally related. Since property is a legal matter, legally defined and protected, a legal determination of property can have major impact on financial and corporate arrangements. Because Edison, despite contestation, was able to assert and maintain legal ownership of incandescent light technology during the period of development of the industry, his particular companies were able to maintain their dominance. The law provided a useful theater of determination that helped maintain some order on the many other stages on which the development of lighting was played out.

13

Charisma and Communication in Edison's Organizations

There is no organization; I am the organization.
—*Thomas Alva Edison to Maurice Holland, 1927*

Edison was the charismatic center of the organizations that formed around him. The first institutions to develop, manufacture, and disseminate his system of delivering light and power were built on the force of his authority and the trust he granted his close associates. This trust enabled his associates to reach beyond themselves while remaining loyal to their benefactor.

As the organizations moved beyond the reach of his personality and other organizational forces took over, however, Edison lost his role within the institutions he had founded. Only vestiges of his charismatic force remained in the successor institutional arrangements. The national Edison companies became consolidated into General Edison and then General Electric, and the local Edison-franchise power companies developed their separate and combined strengths. Nonetheless, to this day many of the regional power companies still draw on the symbolic force of Edison's name-for example, Commonwealth Edison, Consolidated Edison, and Southern California Edison (recently turned into Edison International).

Edison's personal power led to particular kinds of working and communicative relationships among the Edison pioneers. As his operations became larger, some of those relations became regularized in the forms of reporting within Edison organizations, and the curious proliferation of companies in the early 1880s was a further institutionalization of a personal style. These developments and other forces came to remold the original Edison relationships, establishing communicative patterns that left Edison institutionally irrelevant. Edison's charismatic relationship vanished as the communicative structure no longer accommodated it.

Charisma

Charisma is not an individual characteristic, but an effect of one person on another. The concept has its origins in the power of certain individuals, paradigmatically Jesus Christ, to inspire belief, commitment, devotion, and action in others. Contact with the charismatic draws others into an overwhelming vision of new orders of life, to which they then devote themselves. Revolutionary movements based on devotion have the potential to displace prior forms of order and authority.

Max Weber, for such reasons, considered charisma as one of the three pure types of legitimate authority, which lead people to obedience within a social order. People devote themselves to a way of life embodied in the charismatic figure "on the basis of a personal trust in the leader's revelation, his heroism or his exemplary character" (Weber 1994, p. 32).

Weber identified five typical characteristics of the relationship of the followers to the charismatic, which together constitute an ideal type of charismatic authority:

- The authority and power of charisma arises from the recognition of the ruled, although the charismatic appears to follow his or her own lights, not dependent on the approval of anyone else.

- Failure of the force of charisma is taken as a sign of divine abandonment.

- A group or community subject to charismatic authority is cohesive for reasons of emotional communal integration.

- Those under the authority of charisma are not subject to ordinary economic calculation and ordinary market strategies, although they may anticipate great gains. Charisma relies on patronage to a grand scale.

- Charisma is a revolutionary force that repudiates everyday routine.

The charismatic relationship gives special force to the word of the leader, so that the followers trust the leader's word even more than their own immediate experience of the everyday workings of the world and their ordinary calculation of self-interest. The charismatic's word is taken to lead them to another, more satisfactory experience or world. If, however, the word does not seem to lead to the anticipated benefits, the trust afforded the leader is withdrawn, and the charismatic power is lost.

Edison (bottom row, center) with Menlo Park workers.

Edison and His Disciples

We can see in the behavior, commitments, and statements of Edison's clos-est associates just such a relationship with him. Among the inner circle during the period 1878–1882 there was great belief in Edison's power and leadership, which evoked legendary effort on the part of the Menlo Park gang. These close associates worked for low pay but in anticipation of great windfall, the tokens of which were distributed through Edison's gifts of stock to those he deemed most valuable.[1] During the difficulties of 1879 there was wavering among the financial backers, and even some doubt among the Menlo Park team, but with the success that came in the autumn of that year they all became believers again—although the backers never again abandoned a rather cautious economic calculation. For the Menlo Park workers, however, the magic returned in full, everyday reality was sus-pended indefinitely, and those who managed to stay in the inner circle shared in enormous benefits.

A Magnetic Personality and Intensity of Work

It is no accident that the legend of Menlo Park lived long beyond the journalistic hyperbole that surrounded the first appearance of the phonograph and the electric light. This legend was rooted in the experience of those who lived and worked at Menlo Park and those who regularly visited. Consider the name of the organization of Edison alumni that met annually in the 1920s and the 1930s under the sponsorship of Henry Ford: The Edison Pioneers.

The experience of an inspired relationship with Edison is attested by Francis Jehl:

He was respected with a respect only great men can obtain, and he never showed by any word or act that he was their employer in a sense that would hurt the feelings. ... It was his winning ways and manners that attached us so loyally to his side, and made us ever ready with a boundless devotion to execute any request or desire.

He was the gentle master in everything—word, deed, and action; and we were but his disciples. (Jehl, quoted on p. 297 of Dyer and Martin 1910)

Later, Jehl was to say that what endeared all at Menlo Park to Edison was his "magnetic personality" (Jehl 1937, volume 2, p. 458).

That magnetic power is evident in the words of Edison's close friend George Barker, who even in the 1870s called Edison "a man of Herculean suggestiveness...he reaches out into the regions of the unknown, and brings back captive the requisites for his inventions" (McClure 1879).

Otto Moses, who only briefly entered into the inner circle of the Edison group, felt the attraction of its intense atmosphere even as he asked to join. In his letter responding to an Edison advertisement for an analytic chemist, he recalls his brief visit to Menlo Park in 1878, expressing "a longing to be surrounded by such an atmosphere of energy and genius as I had enjoyed in your laboratory."[2] Because he desires "contact with a fertile suggestion of work" and his "philosophy is to know and to conquer Nature rather than Fortune," he is willing to forgo a salary for the time being. Moses got the job over a number of other applicants who wrote more conventional letters and requested more conventional wages. Edison attracted and selected the one who was ready to suspend economic calculation, to escape everyday routine, and to participate in the revolutionary transformation promised by the charismatic leader. Later, after the Paris exhibition, when Moses was rejected by others in the group, he held to his personal affiliation with Edison with great tenacity, providing Edison with information and offering his service for private commissions.[3]

Samuel Insull, Edison's secretary, also felt the leader's force at their first meeting:

I finished my first night's business with Edison somewhere between four and five in the morning, feeling thoroughly imbued with the idea that I had met one of the master minds of the world. You must allow for my youthful enthusiasm, but you must also bear in mind Edison's peculiar gift of magnetism, which has enabled him during his career to attach so many men to him. I fell a victim to the spell at the first interview. (Insull, quoted in Dyer and Martin 1910, p. 332)

As secretary, Insull knew intimately Edison's suspension of everyday routine and business calculation:

I never attempted to systematize Edison's business life. Edison's whole method of work would upset the system of any office. . . . He cared not for the hours of the day or the days of the week. . . . Sometimes he would not go over his mail for days at a time. (Insull, quoted in Dyer and Martin 1910, p. 278)

Stories of all-night work sessions at Menlo Park, led and inspired by Edison's intensity, are legion in all the biographies and memoirs, as are tales of Edison's role in practical jokes and impromptu late-night entertainments, such as midnight sing-alongs around the organ he had installed in the laboratory. Jehl commented that Edison's personal suspension of everyday order put a great stress on others who followed his example: ". . . I have often felt that Mr. Edison could never comprehend the limitations of the strength of other men, as his own physical and mental strength have always seemed to be without limit." (Jehl, quoted in Dyer and Martin, p. 281)

In a 1931 series of newspaper reminiscences, William Hammer, a close associate of Edison from 1876 on, described the effect of Edison's unkempt intensity as he worked side by side with his associates in Menlo Park with no set lab hours:

On the second floor a small group of men laboring like Roman galley slaves. But they worked with a will, and they chuckled and played outrageous pranks.

Edison, like Napoleon and Grant, squandered men.

But like these he got results. . . . Edison himself was an inexorable pace setter. Francis Upton . . . once said that Edison used men like sponges and squeezed them until there was nothing left to come out of them.[4]

Although Hammer described Edison as driving his associates hard, he was not one to prescribe work in detail. Rather, he inspired them to draw on their own resources. "He expected those working for him to think ahead.

They were not to wait to be told." Hammer related that on the first morning he was working for Edison, when he asked the great man for an assignment, Edison growled back: "Find something to do."[5]

John Ott, a workman who first joined Edison at the Newark workshop in 1871 and who was to stay with Edison for the rest of his working life, told of the pleasure of working for Edison:

Edison made your work interesting. He made me feel I was making something with him. I wasn't just a workman. And then in those days, we all hoped to get rich with him. (Ott, in Nerney 1934, p. 64)

The Hope of Extraordinary Rewards

The most intimate and immediate account of Edison's charismatic relationship with his co-workers during the period of the development of incandescent lighting comes from the letters the young Francis Upton wrote to his father starting in November of 1878, when he joined Edison. Upton, having just completed years of schooling and earned a Ph.D., was at first delighted at having a real job and earning money. The following March, in place of a requested wage increase, he accepted Edison's promise of the copyrights to any publications on incandescent lighting that Upton might edit or ghostwrite, but Upton would have to pay for the drawings at his own expense. Anticipating the usual economic calculation that his father might make, Upton wrote:

You may smile at the prospect and think that promises will never buy bread and lodging, yet there is a certain degree of hope in such conditions.[6]

After writing of some recent advances, Upton hedged his bets:

I am fully satisfied with my prospects even if the light does not succeed for I shall have no trouble getting a job somewhere else on Edison's recommendation.

A month later, on April 27, 1879, he continued to hitch his wagon to Edison's somewhat sputtering star:

Mr. Edison will overcome them [difficulties] if any does. I have not in the least lost my faith in him for I see how wonderful the powers he has are for invention. He holds himself ready to make anything that he may be asked to make if it is not against any law of nature. He says he will either have what he wants or prove it impossible.[7]

After the success of Edison's light, Upton's letters were filled with talk of wealth. In a letter dated November 30, 1879 he wrote:

The Electric Light has taken a turn upward during the week and it looks as if I had drawn a big prize. I have 37 shares of the stock and during the last week I have heard of two shares selling for $650 each or $24,050 for the amount I hold. . . .

I think how timid you were at home about my giving up $600 a year and I have made at least $10,000. Mr. Edison is going to give me five per cent in England and the continent of what he receives. . . .

I am going to try for large pay, for I shall never have such a chance again in my life.[8]

A month later he wrote that shares were selling at $3500 and that he was selling options for sales of some of his shares at $5000 each.[9]

Edison relied on shares rather than high wages as a means of paying back his followers for their hard work and loyalty. Early in 1879, as was noted in chapter 8, he gave his close associates substantial gifts of stock in the newly formed EELC.

The files of Samuel Insull indicate how extensively Edison's Menlo Park associates invested in the Edison enterprises.[10] The list of stockholders in Edison Electric Light Company of Europe as of July 1881, for example, includes a large number of Edison's close associates and some of their relatives: James Banker, J. Bailey, Charles Batchelor Sigmund Bergmann, G. E. Carman and C. B. Carman, Charles Clarke, Charles Dean, Sherburne Eaton, Stockton Griffin, Samuel Insull, Edward Johnson, John Kruesi, Samuel Mott, Theodore Puskas, A. J. Saportas, Francis Upton. Many of the holdings range from 2 to 20 shares, roughly corresponding to Edison's degree of reliance on his associates at the time, so they were quite likely gifts.[11]

Charisma as Communication

The charismatic relation Edison had with his inner circle was enacted primarily in the laboratory and meetings, of which we now only have skeletal traces—most notably the lab notebooks, examined in chapter 4. The notebooks witness an intimate working communication between Edison and his colleagues that relied on several assumptions which affected what got communicated and how.

One assumption was that Edison's ideas were always worth being pursued with the full effort and imagination of all the people in the Menlo Park lab. This went far beyond the usual employer-employee relationship, in which employees carry out the boss's orders. In the notebooks we see each of Edison's suggestions and drawings setting in motion creative and energetic pursuit by collaborators like Batchelor, Kruesi, and Upton.

A second assumption was that the ideas ought to be understood in the way Edison understood them.

A third assumption, however, was of respect for the full expression and the candid perspective of each of the co-workers once they were aligned with the project. As individuals entered into closer relations with Edison, they were granted wider ranges of comments that served as records of the group judgment. From the outset, Batchelor and Kruesi were given wide judgment to make comments and to elaborate and modify plans in areas of their expertise, and Upton soon joined them.

These assumptions had further consequences for the nature of communications. There were few directive orders or elaborated arguments coming from Edison beyond requests to Kruesi to make up some parts according to sketches. He offered few general instructions or warnings, as though individuals were entrusted to know what they are doing beyond the specifics detailed in the sketch and annotations. Conversely, the judgments and additions of the key workers were not elaborated beyond necessary explanation, again relying on a trust and respect that needed no further persuasion.

Substantive information flowed rapidly among all participants. The fact that Edison was always at the center did not seem to dampen the information or to create a hierarchically cautious atmosphere. Rather, his oversight seemed to inspire the work of all. This oversight often involved participation in the smallest details, but few of the details reported by his subordinates seem designed to justify actions, create a self-protecting record, or meet arbitrary reporting requirements. Edison was micro-involved but not micro-managing.

This trusting but engaged relationship is evident in the letters that Moses, Puskas, Bailey, and Batchelor send back to Edison from Paris. In Paris they were on their own to make many decisions, including designing the exhibit, buying off reporters, and initiating negotiations for European operations, but they reported back in detail all they considered significant. They seemed to understand their mission quite well and to have little need to ask questions or respond to new instructions or reprimands. Yet they were detailed and candid in reporting to Edison, keeping him fully informed of all relevant occurrences.

In the aftermath of the Paris exhibition, Edison's conscious awareness of the need for trust and communicative ease among his team is evident. When friction arose from personality differences between Otto Moses and several others in the inner circle (especially Batchelor and Johnson),

Edison removed Moses from the group so as not to disrupt the atmosphere of ease and trust among his closest associates.[12]

An Enduring Habit of Relationships

When Maurice Holland, director of the Division of Engineering and Industrial research of the National Research Council, visited Edison's West Orange laboratory in 1927 to investigate the organization of Edison's labs, he noticed Edison's central role:

The entire organization revolves about Edison and progress recorded by personal contact with his "boys." He is constantly making the rounds of the laboratory and is entirely familiar with every detail of experimental projects. No important changes or development steps are undertaken without his personal approval. Mr. Edison has a large number of subordinates reporting to him personally. His remarkable memory serves him as a card index to apparatus and data used 20 years before in conducting experiments. No financial restraint hinders the development work since most of the funds come from personal royalties accruing from previous inventions. He is free to spend it as he likes—and he usually does.[13]

Edison expressed impatience with organization to Holland:

There is a tendency towards over organization in present day experimental work. The real measure of success is the number of experiments that can be crowded into 24 hours.

Despite the distaste for overregulation, Edison sought records and information, just as he did in 1878 through 1880. Holland reported:

Each member of his staff is required to keep detailed notes covering all experimental work in laboratory notebooks. Mr. Edison insists on an accurate and complete record being kept of each successive step in experimental development. A few years ago when fire threatened the valuable records in the laboratory Mr. Edison personally supervised the removal of the data books covering 7000 separate experiments on the storage battery.

Edison's personal participation as the source of planning and direction was accompanied by the same inspired relation with his workers that existed 50 years earlier, although he was then not old enough to be treated as a father figure. Holland reported:

An outstanding characteristic of this close knit, personally intimate organization, which might almost be called "a big family," is the evidence on every hand of an esprit de corps of the highest type. Loyalty and devotion to Mr. Edison is unswerving and borders on the intimate affection of children for their "Daddy." ... All his staff refer to him as "The Old Man."

This intimate, improvisatory, flexible social structure, with Edison at the directive center, would be put under strain as the enterprise of light and power required more extensive endeavors of more people at more locations.

The Attempt to Routinize Charisma in the Edison Companies

Among 20 people, all in direct contact with Edison, Edison's charisma could remain strong and at the center of all communications. However, once incandescent lighting was displayed, produced, and distributed from multiple distant sites, engaging larger organizations, Edison's personal force could not remain the only organizing principle.

At first, Edison attempted to spread out his personal relationships and maintain his centrality to all significant communications over greater distances and larger networks. Some of these attempts were consistent with developments in other organizations at the time; others were more idiosyncratically characteristic of Edison's operations. In both cases, however, the resulting organizations took on lives of their own in ways that got beyond the personal control of Edison, and Edison seemed to have neither the will nor the means to reassert authority at a higher level of managerial abstraction.

Forms and the Control of Communication

One major means of maintaining centrality in an extending informational web was to regularize the flow of information to the center, largely through the use of forms. The growth of forms in many businesses at this moment of history was part of the communication revolution that accompanied the rise of the modern corporation.[14] The required use of forms ensured that subordinates gathered just the information that management wanted, then collected and measured that information in ways consistent with information gathering at other times and places by different collectors. The standard form ensures that information is reported in a format that makes it easily comparable to other information filed on the same or a related form. As the forms organize and regularize the flow of information upward, they also regularize the work of subordinates. The forms themselves are implicitly a set of work orders sent downward to define what subordinates ought to be attending to. The work-order aspect of forms may be made explicit by accompanying instructions. The development of pre-printed forms for the reporting of information during this early period of expansion can be traced in the Edison papers. Most of the forms I have come across there deal particularly with the attempt to extend the production and marketing of Edison power to new domains.

The adoption of forms turns trust in personally transmitted information into narrowed regulation of subordinates. We can see this process at work by looking at the reports Alfred O. Tate sent to Charles Batchelor in 1884 while Tate was canvassing Michigan and Canada for sites for central power stations.

Batchelor, one of Edison's closest and most trusted partners, had just returned from two years in Paris overseeing the development of Edison Europe. During that time he had written regularly and extensively to Edison in personal letters. After his return, in the spring of 1884, he became manager of the Edison Machine Works and had many responsibilities throughout the Edison organizations, including overseeing the expansion of Edison power companies.

Alfred Tate too was moving into the realm of personal relations and trust in the Edison organization. Tate, a protégé of Samuel Insull, eventually replaced Insull as Edison's personal secretary when Insull moved up the organization.

During this period, as central plants were being established in small towns in Pennsylvania and Massachusetts, Tate was entrusted to check out the possibilities in Edison's home state, Michigan. Tate regularly reported back to Batchelor information about the agents contracted and the towns they were to develop. At first these reports were a mixture of business document and personal letter, but with the adoption of a printed form they became more like formal business records.

The reports began on September 23, 1884. They were on half-size stationery sheets, most with the letterheads of hotels Tate apparently stayed at. There were many formulaic elements to them, suggesting forms and contracts, but there were also elements typical of business correspondence. In about 10 days, Tate wrote 11 letters, totaling 18 sheets, reporting on 10 agents assigned to cover 13 towns. By October 4, however, Tate obtained printed forms, which he used thereafter. These forms, identified as "form 6," were imprinted "Edison Machine Works, Charles Batchelor gen'l manager" in large type. These half-size forms —with a horizontal format transforming them from a letter into something more like a note card—stabilized and fixed the formulaic elements and specific information to be provided. They also provided an undifferentiated space on the bottom half for "Remarks." At the bottom there was a dotted signature line. Tate filled out 49 of the forms between October 4 and November 6, each reporting on one agent.

Letter, Charles Tate to Charles Batchelor, September 23, 1884 (89: 591).

Flexibility of Letter Reports

Although much of the same information was reported in the standardized forms as in the letters, the introduction of the forms narrowed and routinized information and decreased the scope of the canvasser's judgment.

The earliest of Batchelor's letter reports, after the addresses, has the same key formula that would continue through all the documents:

D[ear] Sirs:
Subject to your approval I have appointed:

<u>John Hall</u>

<u>Agent</u>

–for–

<u>Woodstock</u>   <u>Ontario</u>
<u>Terms 5% on Gross Sales</u>[15]

As is shown here, the name of person and that of the town were in larger script, underscored, and centered. The terms were also underscored. These elements could be easily found when scanning a stack of such reports; the highlighting also gave the document the appearance of a legal contract.

Tate then continued the letters with whatever information appeared relevant to him. In the case of the first report on John Hall, Tate devoted several lines to the qualifications of the new agent:

Present occupation
Insurance Agent and Manager Great Northern Western Telegraph Office—was highly recommended by the member of Parliament for this County—Has much influence with Town Council and leading Spirits of the Town.[16]

Tate then detailed the town's prospects for electricity:

Woodstock: Pop 7500. Two or three mills—One of the Best towns in Western Ontario—streets lit with gas. No Electric light ever introduced but people anxious to get it. Agent will devote his energies to formation of stock company.

Tate exercised broad judgment as to what should be included in his reports. Some letters told little about the agent while elaborating on local business conditions, such as the recent actions of the town council. In his two-page letters, he sometimes turned to business strategy and instructions to the agents. In the case of Hamilton, Ontario, where a competing company had already been formed, Bachelor's narrative concluded as follows:

Their scheme is to run it for a while and then sell out to a company formed locally. Instructed our agent to watch them closely. To profit by any organization they might form, and when the proper time comes—before any sale is effected—to get in a bid. He understands thoroughly what he has to do.[17]

One letter was devoted to information Tate had neglected to report in a previous letter. In short, the letter reports offered flexibility despite their somewhat formulaic character and well-defined interests.

Standardization in Forms

The printed forms, while facilitating the reporting work, were to make the information more consistent and compulsory while reducing the extent of the information reported. The top of the form became a standard location line:

City of _____ County of _____ State of _____

The internal address then served as a prominent company identification. Instead of the letter of a traveling agent, this was a company document. The main text adopted the opening formula word for word, except with a major expansion to incorporate specific information about the agent and the contractual arrangement:

Dear Sir:

Subject to your approval, I have appointed Name Business, _____ , Address, _____ Terms, _____ Territory, _____ as your agent for _____ .

Although the text is similar, there are several interesting changes. The information required about the agent is particularized and constrained by the space available on the lines. The report on the agent is cast into the role of official contractual information rather than a more open-ended strategic business consideration of qualities, resources, and influence. Furthermore, the typography of the form, displaying the agent information at upper right and providing a large space for the region at lower left, made the official role of agent for a region more prominent than the name and particulars of the person fulfilling the role.

The lower half of the form was devoted to remarks. Though it followed the general format of the letters (which, after the formula, went off in directions restricted only by apparent relevancy), this simple placing of such a space on the form served to limit and change the content. Comments now tended to be only about the region and its economy; only

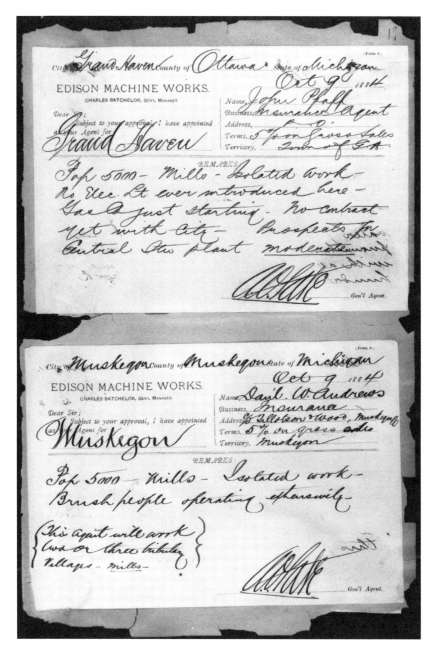

Letters, Charles Tate to Charles Batchelor, October 9, 1884 (89: 610).

in a few instances were particulars of the agent expanded upon, leaving unelaborated in most cases the restricted information in the preceding spaces. Comments now tended to be shorter and more telegraphic than they had been in the letters. Rarely did comments extend beyond six lines, and those lines contained less information. (The space seemed to invite larger handwriting.) No comments went over to a second page, as they were limited by the designated space. The comment on Grand Haven is typical:

Pop 5000- Mills- Isolated work No Elec. Lt ever introduced here—Gas Co just start-ing—No contract yet with City— Prospects for central lt plant moderate.[18]

Even the longest comment—that on the large, complicated market of Detroit—offers only cursory coverage, on a single sheet:

About every arc light in the market can be seen in Detroit. Tower lighting pro-nounced a failure. Cadman will devote all his time to pushing our business. He is an old bank manager and has good connections among the business men of the City, and is an excellent canvasser—Notwithstanding all the Elec Light Co's who are operating in the city, I anticipate a good deal of business for us.[19]

Despite the personal trust between Tate and Batchelor, the need for con-venience and comparability of information covering an extended region led to standardization; the standardized form, in turn, constrained and reg-ularized the gathering of information, thus decreasing the realm of trust.

Other Regularized Communication Flows

A similar process happened in regard to door-to-door canvassing of poten-tial customers within each projected lighting district. The canvassing of the original Pearl Street district developed ad hoc procedures, which crys-tallized into standard questions and handwritten tables for the entry of data. Further regularized formats were developed for the aggregation of the data and calculations back at the office. All these, however, were still handwritten.

As canvasses of new districts in new cities were planned, the ad hoc pro-cedural orderings became fixed in a printed canvassing book that opened with two pages of "General Instructions to Canvassers," which were fol-lowed by printed tables for the recording of information gained from each interview.[20] Less and less improvisatory leeway was given to each canvasser, who personally was more and more distant in the organization from Edison, Batchelor, or any of the inner circle.[21]

Other aspects of the operation underwent the same processes of formalization of procedures in reproduced documents. In the original installation and early operations of Pearl Street, Thomas Edison was regularly at the site, developing procedures of operation in conjunction with Charles Clarke (the chief engineer) and a few other employees. However, Edison or his closest associates could not always be present to supervise the construction and the operation of stations in Pennsylvania, Massachusetts, and overseas, so their expertise had to be transmitted textually. One early such document, titled Questions for Central Station Engineers, survives as a typescript dated November 21, 1883. The fact that it was produced only in typescript (and perhaps carbon) suggests that its circulation was not wide. However within that distribution it was still important that Edison was central to the document, even if he could not write it all himself. The title page explicitly establishes Edison's supervisory role in creating the document:

Instructions at the end of this Book were written by Mr. Edison.
All questions in this book were formulated by Mr. Edison.
Answers to meter questions were written by H. Ward Leonard.
Answers to questions on dynamos were given by W.S. Andrews
Questions on steam engines and boilers were answered by ?[22]

It is of particular relevance to Edison's personal role in ordering the social dynamics of building the business that Edison's instructions at the end pertained to the marketing of new stations and the handling of customers. For example, Edison passed on the following savvy advice:

In small towns, from 2,000 to 20,000 inhabitants, after the station has been started and three or four prominent places are lighted, stop soliciting orders for light, but run along quietly and wait. The customers will gradually come in and ask to be connected; their judgment and desires having been gained by seeing the light in the prominent places. Customers obtained in this manner give no trouble, and in time 75 per cent of the whole number counted on to take the light, will ask of their own accord to be connected.[23]

Technical matters of boiler operation could be entrusted to others, but Edison stayed close to the pulse of the human motivations that drew people into relation with Edison operations. This is something he clearly had thought about and worked on.

As operations became more extensive and distant, the increasingly impersonal corporate voice relied less on the personal charisma of Edison for guidance and less on the personal judgment of the employee. Even as early as 1881, the 76-page printed technical pamphlet titled Instructions for the Installation of Isolated Plants Published Exclusively for the Private Use of the Agents of the Edison Company for Isolated Lighting was presented entirely in an impersonal voice.[24] This pamphlet begins with a primer on electrical science. It then explains in great detail the features and reasoning of the Edison system so that any mechanically minded person can set up the machinery and repair it. The apparatus is presented as stable and intelligible and as not requiring personal improvisation or personal contact with the charismatic center of the organization.

In large modern organizations, such printed directives and work forms are associated with increasing distance, impersonality, and lack of improvisatory behaviors at lower levels. At upper levels, however, they are associated with the abstracted work of symbolic manipulation. Managers deal with information and paper as much as with people.

Managing People, Not Paper

It was not Edison's taste, habit, or practice to be a manager by paper. His lifelong habit was to work with trusted individuals and to spend time in laboratory and on the shop floor rather than in the managerial office. His antipathy toward scientific theory removed from practical mechanisms was paralleled by a dislike for managerial information removed from the persons in the organization. When Edison constructed the complex of Edison factories in West Orange, New Jersey, beginning in 1887, he located his library and offices in the same building as his lab and the machine shop. He never abstracted management from the development of products, inventions, and personal relations. Rather, he interposed individuals from his own personally trusted circle to manage the Edison enterprises, so he could maintain personal charismatic relations with them and let them deal with corporate abstractions of the more ordinary sort.

At first, Edison created a series of small manufacturing and regional power companies, each under the guidance of a personally trusted individual. With Edison remaining involved only in areas of major change (such as construction of the first group of central stations, in New York, Massachusetts, and Pennsylvania), the companies moved toward less charismatic forms of organization and toward orderly regimes of law and finance. As they did so, the separate manufacturing companies, within a few years, consolidated into Edison United Manufacturing and then Edison General. The regional franchises became stronger as independent corporations, and their bond to the central licenser, Edison Electric Light, became weaker. As Weber (1994, p. 43) points out, with respect to charisma in government operations, "a prerequisite for the routinization of charisma is the elimination of its noneconomic predisposition, its adaptation to financial forms of fulfilling needs, and therefore, to economic conditions necessary to raising taxes and duties."

The Formation of the Edison Companies

The second series of Menlo Park demonstrations, held in the winter of 1880–81, marked a major turn in the Edison operations. The system, now developed and workable, had to be brought into the marketplace and economically exploited. Fresh capital was needed to manufacture the machinery, bring power stations on line, and reach out to potential customers. The current backers of the Edison Electric Light Company, their patience having been stretched by Edison's promises, wanted to limit their risks. To protect the patents, which in a sense were the only substantial value created in the first few years, they made the EELC strictly a patent-holding operation. They were also willing to participate in two other parts of the business they viewed as immediately attractive and having long-run potential: the marketing of power to New York City and the marketing of small systems nationally. But these secondary investments involved no cash and were aggregated by the return of stocks to EELC in lieu of license fees (an arrangement that would continue for the development of other local central power companies). The current investors had to put up no further funds of their own, as new investors in the new companies would provide the needed capital.

The Edison Manufacturing Companies

The less immediately attractive investments in manufacturing had to be financed by Edison himself. Edison had a personal attachment to these workshop-like manufacturing businesses and was personally committed to the material development of the technology, whereas the financiers were moved more by the abstract value of patents and the commercial value of service being delivered to paying customers.

Yet, while Edison was committed to perfecting every mechanical and social detail necessary to make the system work, he was not much engaged in the daily repetitive operations of making a company work as a hierarchical financial and legal entity. Nor in running such an operation could he rely on the sense of the extraordinary that maintained his charismatic relations with the workers. Edison's solution was to hand over the day-today management of each manufacturing unit to a trusted colleague, with whom he could then coordinate informally. This cadre of Edison associates, in addition to being responsible for the ordinary operations of various Edison manufacturing companies, would be available for ad hoc collaboration, crossing the boundaries of the companies to maintain fluidity of the original small charismatic organization. In the meantime, each of the associates could establish mundane order and normal economic reasoning within his particular domain.

Edison had models for this practice in the ad hoc arrangements among inventors and machinists in the machine shop culture. A more specific model was provided by Sigmund Bergmann, who, after working in Edison's Newark Shops since 1870, had set out on his own in 1876 to manufacture a variety of devices, including a number of Edison inventions. Bergmann's letterhead of 1880 boasted, among other products, "Edison's Inventions—Phonographs, Telephones, Motographs, &c., &c."[25] Bergmann maintained a correspondence with Edison on both financial and technical matters. Early in 1881, Edison expanded their licensing arrangement to turn Sigmund Bergmann & Company into one of the core Edison manufacturing companies, forming a three-way partnership among Bergmann, himself, and another close associate, Edward Johnson. Bergmann & Company produced fixtures, sockets, connectors, and assorted other small devices for the Edison system.[26] Consistent with the personal trust at the center of the relationship, the partnership was not legally formalized for another year.[27]

Shortly before entering into the partnership with Bergmann, Edison had formed another manufacturing company to produce incandescent lamps. In November of 1880, the Edison Electric Lamp Company began manufacturing lamps at Menlo Park, but it was not yet officially incorporated.[28] In January, a contract was drafted formalizing the relation to the EELC, with Edison, Batchelor, Upton, and Johnson identified as principals. The Lamp Company was to have sole license to produce Edison lamps, which it would sell only to the EELC.[29] Thomas Edison owned the majority of this company and directly supervised early decisions about acquiring factory space and developing operations.[30] By February of 1881, however, Upton had taken over operations. He kept Edison posted in regular letters from 1882 through the middle of 1885.[31] Whereas the original stationery had identified all four of the principals, later stationery identified only Edison (as president) and Upton (as treasurer). The informality of the arrangements is suggested by a letter from a lawyer attempting to arrange the transfer of a minority share:

Upon making inquiry into the legal status of the so-called company, I was surprised to learn that its affairs were not clearly defined in writing, and that each of the partners depended on tradition, and the honor of his associates to establish his actual interest in the business; that the personal property was nominally owned by the company, while the real property is in your name, and in the case of death would descend to your heirs; that there is not even a declaration of trust in favor of your associates.[32]

Charismatic organization was beginning to come under the scrutiny of the orderly forces of economic rationality.[33] A Certificate of Organization drawn up shortly thereafter listed Edison as holding more than two-thirds of the ownership, Upton holding half of the remaining shares, Batchelor holding one-third of the remaining shares, and Johnson holding the rest.[34] By-laws for the company were finally drawn up in 1884.[35]

Early Tensions between Charisma and Routine

Even in the earliest, pre-formalization days of these two companies, one documented incident suggests, the new forms of organization created tensions with Edison's personal ad hoc style. A letter from Bergmann & Co. to Upton at the Lamp Company, dated October 12, 1881, contains the following request:

We want 100 lamps for our own use here, experimenting &c. Mr. Edison says we can get them from you for $35. If that is correct, please send them over immediately & oblige.[36]

Upton did not oblige. A week later, Sigmund Bergmann wrote to Thomas Edison:

Please send me by bearer 1 Dozen or 1/2 Dozen lamps. . . . I gave Mr Upton an order 3 or 4 days ago for 100 but have not received any yet.[37]

Edison did send a half-dozen lamps. Six days later, however, Bergmann & Co. received a formal typed letter stating:

Any lamps that you may require for experimental purposes must be ordered through the Edison Electric Light Company as the Lamp Company by their contract with the Light Company are only allowed to supply lamps on the order of the Light company.[38]

Earlier that year, in June, Edison himself had been formally instructed that even he would have to order lamps through appropriate channels.[39] Charisma seems to have been fading rapidly, being replaced by formal corporate entities with their own peculiar rationality. The files of the various

THE EDISON ELECTRIC LIGHT CO.,
65 FIFTH AVENUE.

New York, *June 3* 1881

T. A. Edison Esq
    Dear Sir
            The Lamp Company has called
upon us to approve numerous orders for lamps heretofore
drawn upon it by yourself. Mr Johnson and Dr Moses
and has requested that hereafter all orders shall be approved by this Company. under the terms of the contract between
the Lamp Co and this Company. this is quite right. and
I shall at once prepare the requisite blanks. and if you
will in future send your orders to me. I will see that
they are sent to the Lamp Co in proper form and will
thus be able to keep a record in this office. which is absol-
-utely necessary to the proper transaction of the business
            Be good Enough in each case to name
the purpose for which the lamps are required. that
we may know how to dispose of the bills
                Yours truly
                    C. Goddard
                            Secy

Letter, C. Goddard to TAE, June 3, 1881 (58: 34).

companies contain other similar correspondence concerning billing among the various companies and with Thomas Edison.

More Manufacturing Companies

The formation of the next company reveals the difficulties this mediated charismatic form of organization caused when Edison had no trusted personal associate in place in the company. The Edison Machine Works (largely financed by Thomas Edison, with some help from Charles Batchelor) manufactured dynamos and motors. Early in 1881, Edison himself managed the correspondence leading to the opening of the Goerck Street Works under the umbrella of the EELC.[40] Simultaneously, by-laws were prepared for the Edison Machine Works.[41] Though formal control was placed in the hands of Thomas Edison (as president), control of the day-to-day operations was to be placed in the hands of a manager. However, no manager was appointed for three years; rather, daily operations were left to the superintendent, Charles Dean, and a secretary, Charles Rocap. No letterhead bearing officers' names seems to have been prepared at this time, suggested that no permanent principals other than Edison were identified with the organization. Dean and Rocap, however, were allowed to enjoy a small percentage of the profits.[42]

With only loose supervision from Edison, Dean and Rocap apparently decided to help themselves to a bit more of the profits. In February of 1883, suspecting financial irregularities, Edison ordered all the financial matters to be channeled through his trusted personal secretary, Samuel Insull.[43] Rocap was ready to quit over this supervision,[44] but apparently stayed on for a while. Six months later, however, after an incriminating report,[45] both Dean and Rocap vanished, to be replaced by Gustav Soldan and E. J. Berggren.

At about this time, Batchelor, still in Europe, began to correspond with Insull about operations of the Machine Works.[46] The following February, at a meeting of the board of trustees (of which, unusually, detailed minutes were kept), the company was reorganized and new by-laws were adopted. Edison was again named president, and Insull was named secretary and treasurer; the office of manager remained vacant.[47] When the highly trusted Batchelor returned to New York in May of 1884, after a two-year stint in Paris, he immediately became general manager of the Machine Works.[48]

The Electric Tube Company (funded by Edison, with some distribution of shares to the other officers and board members) was formed in April of 1881 to become the last of this initial group of companies.[49] A couple of

years later, two more companies were added. The Thomas A. Edison
Central Station Construction Department was established in the spring of
1883 (out of the Engineering Department of the EELC) to construct central power stations for local power companies, which would then be
licensees of the Edison system. The Construction Department was established, owned, and managed by Thomas Edison, and his name alone
appeared on its stationery.[50]

The Edison Shafting Manufacturing Company (funded by Edison) was
established in July of 1884 to produce belts, pulleys, and shafts under a
contractual relationship with the Edison Machine Works.[51] Harry Livor,
who had proved his trustworthiness in the Machine Works, was named
general manager.[52] To avoid the kinds of problems that had arisen from
the informality of previous companies, by-laws were soon written[53] and
complex agreements were signed by the principals (Edison, Batchelor, and
Livor) to hold the stock unsold in trusteeship for three years.[54]

Consolidations and the End of Charisma

In the middle of 1884, by which time the development of new plants was
no longer an extraordinary matter, the responsibilities of the Construction
Division were incorporated into the Company for Isolated Lighting and
the Edison Machine Works.[55] In January of 1886, the Edison Shafting
Manufacturing Company was brought back into the Edison Machine
Works[56]; the Electric Tube Company followed suit around the same time.[57]
Kruesi became Batchelor's assistant general manager, and Livor continued to market the pulleys, shaftings, and related products now manufactured by the Machine Works. In December, with the opening of a new
factory complex in Schenectady,[58] the Machine Works was consolidated
with the Edison Lamp Company and Sigmund Bergmann & Co. under a
new umbrella corporation: the Edison United Manufacturing Company.[59]

Notes in Edison's hand from this period project Batchelor as the general manager and vice-president; however, he was only to visit occasionally,
as he was to return to the laboratory (by then in West Orange) to work
with Edison. The daily operations in Schenectady were to be run by Kruesi
(as superintendent) and Insull (as secretary-treasurer). Further, Insull was
to manage the Manufacturing Department, Kruesi was to manage the
Machine Works division, Upton was to manage the Lamp division, and
Insull was (temporarily) to manage Bergmann & Company. Others were
given responsibility for sales, engineering, construction, and management
of partially owned central stations.[60]

With this consolidation and reorganization, Edison's informal personal relationships finally turned into a hierarchic corporate structure. The economic rationality of a large organization became, in archetypal Weberian manner, a bureaucracy. It is worth observing, however, that Edison wrote up this organizational plan as personal notes, expanding on what individuals were to do, rather than as a formal organizational chart.

Others in the organization were more than ready to celebrate the emergence into rational economic order. A satiric document addressed to "To the unfortunate stockholders" makes fun of the earlier unbusinesslike practices and reflects a clearly defined new attitude. After some mock stock report boilerplate, it reads:

For those who are unacquainted with the aims of this company a brief resume of its object will be in order.

First. We wish to impress the fact indelibly on the minds of our stockholders that our object is not to make money, this fact once accepted will prevent much dissatisfaction and many unpleasant explanations.

The motive of the company is a pure and benevolent one; To disseminate light to suffering humanity at the least possible cost to them; To demonstrate to the business world and to competitors that a great business may be conducted with the best of feeling and brotherly love, and to provide fat situations for the numerous relatives of our genial and beloved Vice President.[61]

The confidence and humor of the early days were back, but this was company humor, reinforcing company values and rejecting the personal and the visionary. Edison was out of the picture, as he was pretty much out of the management and operations of the company of which he was nominally president. In 1889 this Edison United Manufacturing was to be further consolidated with the Edison Electric Light Company, the Canadian Edison Manufacturing Company, the Sprague Electric Railway Company, and Leonard and Izard (a small company formed earlier that year by Edison's long-time metering expert, H. Ward Leonard). This, in 1892, was to be merged with Thomson Houston (which in turn had consolidated many of Edison's competitors) to form General Electric.

The Local Power Companies as Extensions of Charisma

Establishing local central power stations presented a problem somewhat different from that of organizing the electric manufacturing companies, insofar as the power stations were consumer-oriented and were at many different locations around the country. Not only were capital and management needed; it was also necessary to have local familiarity and connections, so as to contend with city governments and state corporate laws. Edison's solution again started by distributing responsibility through personal charismatic relations, but the vehicle resembled modern franchises.

The already-established telegraph and telephone industries were moving to centrally organized, owned, and controlled management by the early 1880s. Edison had been deeply involved in these industries and knew them well. Telegraphy had developed in the 1840s through a large number of small regional companies; however, since the nature of the service was to connect distant points, regional and then national consolidation provided distinct advantages. By 1857 a cartel of six major companies controlled the national industry. By the end of the Civil War, Western Union dominated the national market, and within ten years it was a virtual monopoly run from the corporate and financial center in New York.[62] The telephone industry had a slightly different story but one with a similar ending. Three years of competition between Bell and Western Union had been resolved in 1880 by a financial agreement that had left Bell with essentially a monopoly on the new industry until patents expired in 1894. Since services were at that time delivered primarily within a local region, and since the company recognized that the value of the service increased with the number of customers connected to each local system, Bell wanted to expand service as rapidly as possible, but it was short of capital. To draw on local capital in each service area, Bell established license and lease arrangements with local companies, based on a fixed annual charge for each phone. The introduction of intercity toll service, however, became a strong force for regional consolidation. At that time, Bell renegotiated its initial arrangement with local companies to provide the central company with 35 to 50 percent ownership. By 1885, Bell was well down the path of centralization, a process that was to be completed in the first decade of the twentieth century (though with some authority remaining in the hands of local managers).[63]

Electric light and power, however, was and continues to be based on regional production of power, on the model of the local gas utility tied into local financial, business and political interests. Heavy capital investment was needed for local infrastructure, and local political clout was needed to obtain approvals, favorable regulation, municipal franchises, and contracts. Thomas Edison's interests in central control were limited to maintaining the Edison brand name, providing technical guidance and expertise to establish a standard of service and efficiency, and providing returns to himself, his colleagues, and his stockholders. Once the conversion to alternating current permitted the transmission of power over great distances, production-cost and load-sharing advantages provided an incentive to centralize, but even that impulse was carried out through a holding company rather than through centralized management.

In Edison's time there were only a few models of how to organize a national company that needed local service and delivery. As Chandler (1990) has noted, businesses were local until railroads and telegraphy made possible national marketing and shipping of goods produced in a single place. But these goods were largely sold as finished products, either directly to consumers or to local retail businesses. The Edison Company for Isolated Lighting followed this model, selling complete units for individual customers' use and maintenance. Edison's conception for the central production of power, on the other hand, bore more of a resemblance to the situation of two other companies that, early on, chose franchising arrangements over either full central control or direct sale: the McCormick Reaper Company and the Singer Sewing Machine Company (Dicke 1992, chapter 1).

Franchises

Cyrus McCormick first licensed the right to manufacture his mechanical reaper in the 1840s. Unsatisfied with the quality of production under this arrangement, at the end of the decade he became the primary manufacturer. Local agents and traveling salesman, contracted on an annual basis, sold and distributed the machines. However, this loosely organized and largely unregulated sales force did not generally have the technical competence to market the complex machinery or to provide the quality of service required by farmers who relied on these machines to get the harvest in. McCormick's need for tighter control over better-trained agents led to closer relationships between the local dealer agencies and the central office, achieved at first through a highly trained team of travelers who were close, trusted associates of Cyrus McCormick. Agencies, though independently owned, by the 1870s and the 1880s were brought under central control through company training, policies, and standards.

The Singer sewing machine carries a similar tale. During the early 1850s, Isaac Singer, in order to raise cash for his new company, sold exclusive regional contracts to use and sell his patented sewing machine. However, after he introduced a less expensive model in 1856, he began to rely more heavily on a mixture of company-owned branch houses and agencies. Again the need for competent service led to a closer relationship between agents and the central company. National advertising campaigns

established brand-name identification, uniform national marketing policies led to standards for the appearance of local agencies, and improved
accounting procedures made the agents less independent. By 1881, a system of traveling examiners enforced uniformity of policies in the agencies, which were still independently owned and operated.

When Edison was ready to create a national system of local power and
light companies, the agency franchise models of the McCormick Harvester
and Singer Sewing Machine Companies had recently been established.
Surely Edison, in view of his interest in developing a market for electric
power, had some familiarity with Singer (see chapter 7). Edison, in view of
the strategy he was developing for domestic lighting, would likely have
been particularly attentive to Singer's attempts to make his product attractive to middle-class housewives—indeed, it was through the sewing
machine that Edison hoped to bring the electric motor into the home.

In addition to the problem of inadequate capital, Edison shared with
Singer and McCormick the problems of maintaining technical competence and shared corporate vision across the entire network of local
providers. Whereas the incandescent lamp itself could be seen as a self-
contained product, central stations required craft skills and specialized
knowledge for their mechanical operations and maintenance, as well as
consumer service. Edison himself rolled up his shirtsleeves and got grease
on his hands to get the Pearl Street station up and running, and he kept a
close eye on the organization of the first regional power companies and
the construction of their power stations.

Networks of Charismatic Power

The solution to maintaining competence, standards, and vision while still
mobilizing local capital, leadership, and influence was to establish separate local power companies that would maintain close charismatic ties to
Edison. This tie was only minimally achieved by shareholding arrangements in which the regional companies typically returned a certain percentage of their ownership in lieu of a license fee—particularly since this
arrangement was mediated through the umbrella EELC patent-holding
company, the entity most controlled by the financial backers and the one
over which Edison had the least sway. Indeed, this financial arrangement
soon narrowed, with the identity of local companies becoming far more
significant both in operational choices and public perception than the tie
to a national corporation. The Edison Electric Light Company (by then a
part of General Electric) did not attempt to assert managerial control in

the regional companies until the 1905 formation of the holding company Electric Bond and Share Company (later to be known as Ebasco). As the transmission of power over great distances developed along with large generation projects and management of the national electric grid, Ebasco grew in managerial importance (Wallace and Christensen 1986).

A stronger but even more temporary tie to Edison was through individuals who had worked with Edison at Menlo Park. The personal communication, trust, and alignment that came out of Edison's charismatic relationships ensured that the Edison vision and competence would be communicated and maintained at each of the local sites during the early years.

Pearl Street, of course, was built and operated by men who had come directly from Menlo Park—most prominently the chief engineer, Charles Clarke. Clarke had come to Menlo Park in February of 1880, on the recommendation of his friend Francis Upton, and had carried out some of the earliest efficiency tests of the Edison system.

Other early plants followed that pattern of personal connection to Edison. After a small experimental station was open in Roselle (near Menlo Park) as a joint venture between EELC and the Edison Company for Isolated Lighting (Hellrigel 1989, pp. 87–89), the next central stations built were small plants developed for a series of small Pennsylvania and Ohio towns in 1883 and 1884. A regional agent, P. B. Shaw, helped organize a series of small power companies in Ashland, Bellefonte, Erie, Harrisburg, Hazleton, Mount Carmel, Shamokin, Sunbury, and Williamsport (in Pennsylvania) and in Circleville, General, Middleton, Piqua, and Tiffin (in Ohio).[64] The earliest of these were constructed by William Andrews and Franklin Sprague, both Menlo Park protégés, under the direct supervision of Edison, and Edison served on the board of the Sunbury station once it was in operation. Andrews served as the first manager of Sunbury and trained his local successor.[65] Moreover, Edison wrote a series of handwritten instructions for future station managers (Jehl 1937, volume 3: 1098, 1102–113).

During the same period, Edison constructed a series of plants in Massachusetts, the first at Brockton. The original contact for this plant was again personal. Henry Villard, the shipping and railroad magnate who had installed Edison's system on his ships and had become a financial principal in the Edison companies, had married the daughter of the abolitionist William Lloyd Garrison in 1883. His wife's brother, William Lloyd Garrison Jr., organized the Edison Electric Illuminating Company of Brockton. Edison sent Sprague to help install the plant and supervise its early operation. Edison came to supervise the opening, and he sent his own meter specialist from New York, H. Ward Leonard, to establish the metering system (Jehl 1937, volume 3: 1116–118). Several other central plants in the region—those at Lawrence and Fall River, Massachusetts, and Newburgh, New York—were developed with the support of trusted Edison associates.[66]

The first major city power plant after New York also was established through direct personal ties to Edison. The first attempt to establish a major urban power company in Boston was initiated by Henry Villard early in 1883, around the time he was encouraging his brother-in-law to establish the Brockton plant.[67] Villard immediately brought Edison into the negotiations, but they could not raise sufficient capital at that time. In 1885, Edward Johnson, then vice-president of the EELC and previously Edison's agent in London—an even closer associate of Edison—was still unable to raise Boston capital for the project, so he launched the Boston Edison company with New York capital. The company was at first run from New York by Edison's closest colleagues. Johnson was president, and Batchelor and Upton served on the board of directors. Frank Hastings, treasurer of the EELC, also served as treasurer of Boston Edison. From New York they managed by correspondence and occasional visit. The resident station manager, A.T. Moore Jr. had also been trained at Menlo Park. However, to establish local financial and political ties, several influential Bostonians were named to the board of directors.

After slightly more than a year of operation, Charles Edgar was named general superintendent of Boston Edison. Three years later he became its president, a post he was to hold until his death in 1932. Edgar had begun working for Edison in 1883 at the Machine Works. Having risen rapidly to serve as temporary head of that unit when his supervisor was out of town, he then was transferred to Bergmann & Co. After supervising the Edison display at the 1884 Philadelphia exhibit, he became one of the troubleshooters who traveled around the country installing power plants.

In Chicago the Edison involvement was a bit more attenuated but still substantial. Western Edison, opened in 1882, was an outgrowth of Edison's long-standing relationship with George Bliss (who also ran Western Electric, the major manufacturer for telephone and telegraph industries, and who also was the sole licensed manufacturer of Edison's Electric Pen). Bliss served as general superintendent, with the financier Ansons Stager as president, for the next five years. However, through 1887 Western Edison marketed only the isolated system. In 1887, the company reorganized and obtained a municipal franchise for a central station, which it opened the following year (Platt 1991, p. 54). The connection to the Edison tradition was reinvigorated when Samuel Insull took over the reigns of Western Edison in 1892. Insull, having served as Edison's personal secretary from 1881 to 1887, rose to a central position in the Edison companies. He came out to Chicago in 1892 (after the formation of General Electric) to run Chicago Edison; he then ran Commonwealth Edison until 1932.

The number of close Edison associates was small, however, and the growth of central stations was large. Further, once the manufacturing companies were united and Edison General was established, Edison had little role in the companies. Thus, when the first central stations were built in Detroit and Philadelphia in 1887 and 1888, the projects were backed almost entirely with local capital. As far as I can tell, no close Edison associates worked on the inside of those operations, although several came to inspect and consult (Marks 1895; Miller 1957).

In 1885, the regional power companies formed the Association of Edison Illuminating Companies. This organization, though it was undoubtedly encouraged and supported by the Edison Electric Light Company, provided the local companies with a channel of mutual communication and support that did not go through the corporate center and the remnants of Edison's personal charisma. The Association became a forum for the local companies to express their complaints against the Edison manufacturing companies (Passer 1953, pp. 122–123).

Although Edison's charisma lingered through his close associates who maintained substantial power in the Edison manufacturing and power companies, by the late 1880s it was replaced by typical organizational structures of large corporations. Edison had at first sought to create ways to mediate between his personal force in eliciting motivated, trusting work relationships and the kinds of organizations needed to run large and widely spread enterprises, but his presence quickly became stretched too thin, leaving little place for his own participation in the organizations he had created. As the standard story goes, by the second half of the 1880s the Edison electric enterprises were largely in the hands of the managers, and Edison went on to other endeavors where he could get back into the workshop, such as ore milling and cement manufacturing.

14

The Rhetoric of Capital Investment: Solvency, Profits, and Dividends

Other early entrants into the arena of electric lighting, including manufacturers of arc lighting, had modest ambitions compared to Edison. Elihu Thomson and Edwin Houston, working with arc lighting in Philadelphia, in New Britain, Connecticut, and in Lynn, Massachusetts, saw themselves for many years only as manufacturers of industrial products which they sold outright. Their ambition was to grow as a manufacturing firm tied to local economies (which explains their several moves as they sought conducive regional backing and development). Only in 1882 did Thomson and Houston enter the business of central generation and seek national market for their system, following the lead of the Edison companies. Local capitalization had been adequate for their earlier limited ambitions (Young 1941, pp. 400–401; Carlson 1991, pp. 176–182).

Charles Brush had envisaged central power stations to supply his arc lights, but the vision of central power he pursued was smaller than Edison's. By 1879, the California Electric Light Company of San Francisco powered 20 Brush arc lamps for private customers and the city (Passer 1953, pp. 19). However, because each arc lamp needed substantial power and because the dynamos were smaller and less efficient than Edison's, each of Brush's central stations could illuminate only a few lamps and serve only a few customers. Furthermore, Brush had no immediate plans to use electric power for anything other than lighting.[1]

The Edison companies, in order to raise capital for vertically integrated companies for equipment, production, delivery, consumer appliances, and service for widely distributed, centrally produced electricity, had to enlist many shareholders. During the early research stage, a small speculative market for Edison lighting formed for the few shares distributed to the original group of backers and rewarded associates; public financing of several ongoing corporations, however, meant that Edison enterprises had to support a more serious equities market. Initially three of the Edison companies were publicly held: the Edison Electric Light Company (with its initial capitalization of $300,000 increased in 1881 to $480,000, in 1882 to $720,000, and in 1883 to $1,080,000), the Edison Company for Isolated Lighting (first capitalized in 1881 at $500,000, which was doubled in 1882, with 51 percent of the stock in both instances returned to the EELC as a license fee for the Edison system), and the Edison Electric Illuminating Company of New York (capitalized in 1881 at $1 million, with 25 percent returned to the EELC as license fee and with first option on the remaining stock going to current EELC stockholders). All these companies remained under the direct control of financiers, unlike the manufacturing companies described in the previous chapter.

The standard arguments for investment concern security of the investment capital and potential returns through dividends and price increases. Arguments for company solvency and profitability are usually necessary corollaries, for they suggest that the investment will not vanish in bankruptcy and that profits exist to be returned to the investors. Thus, ultimately, to maintain capitalization, the Edison companies, like all publicly held investments, had to be able to represent themselves in terms of profits and dividends. But before then, as the Edison enterprises were getting off the ground, there were no sources of income, no earnings, and certainly no profits. There were only investment costs. Further, it was not clear that there was even a reliable, attractive product ready to be sold and delivered to consumers.

The first step in building investor interest was to introduce the nature of the business and the product, and then to establish the product's reliability, attractiveness, and current deliverability. The Edison companies had to convince investors that they were ready to do business. Later they could make the case for potential profitability. Only when the business had been built could the simple and direct proof of a solid, secure, profit-making investment be presented. Only then could the companies' value as investments be summed up in a few numbers in a stock market listing.

To make these successive cases to investors, the Edison companies distributed a series of explanatory publications. Although some of these pamphlets, stockholders reports, and bulletins had other audiences as well, they carried the information necessary for passive (offering financial backing only) and active (involved in the creation and management of regional power companies) investors to understand the attractiveness of the opportunity.

Exhibiting the Product: The Paris Exhibition Textualized

The first of the informational publications, a pamphlet consisting of reprints of four prominent descriptions of the Edison system's triumph at the Paris exhibition, appeared in 1882 under the imprint of the New York publisher van Nostrand, although it was clearly a piece of Edison publicity. The first chapter is a translation of the Count Theodore du Moncel's

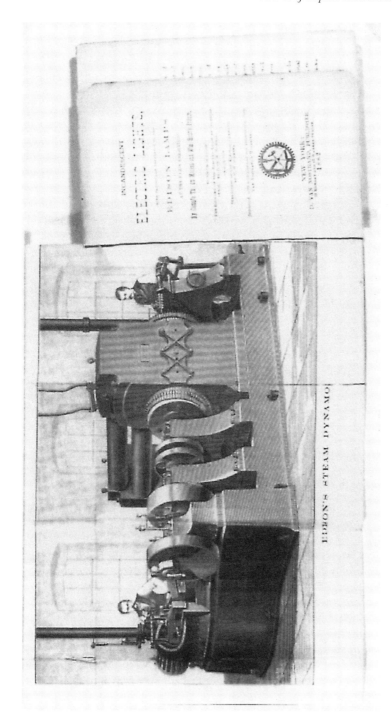

Theodore du Moncel and William Henry Preece, Incandescent Electric Lights (New York: D. Van Nostrand, 1882).

laudatory description from *La Lumière Electrique,* written while du Moncel was in the pay of Edison. Three articles from *The Engineering Journal* follow, all on the economics and technological accomplishment of the electric light; the authors are the outstanding electricians of Britain and Germany: William Preece, John Howell, and Charles Siemens.

As Henry Adams was to note shortly thereafter, the icon of the future was the dynamo. The pamphlet opens with a foldout frontispiece of Edison's steam dynamo, and much of the material within describes and evaluates it. After the title page, a photographic reproduction of the Diploma of Honor from the Paris exhibition appears, along with the texts of congratulatory letters from Swan and Barker.

Du Moncel's article begins with a brief history of the competition in electrical lighting, presenting a persistent and ingenious Edison making the crucial advance by developing the carbon filament. Descriptions of the illumination of the exhibition hall and of the electrical lamps again emphasize Edison's ingenuity. Edison lamps are presented as burning 1200 hours and being easy and inexpensive to replace. The lamps and the system are said to be simple, complete, and ready to go:

What constitutes Mr. Edison's system is not alone his lamps, it is the totality of the arrangements referring to them and which have attained such a degree of simplicity that henceforth nothing remains to be desired in practice. Generating machines, distribution of circuits, installation, indicating and regulating apparatus, meters, for measuring the amount of current employed are all combined for immediate application. As we have said, this application is about being made in a part of the City of New York, where a great number of houses are to be lighted by this system. (du Moncel and Preece 1882, p. 14)

Edison lighting is also described as aesthetically impressive:

The lighting of the two Salons of Mr. Edison at the Exposition is done by 16 small chandeliers like the above, two grand crystal chandeliers, and 80 brackets. The effect is very beautiful, the steadiness being complete as it could be desired. (du Moncel and Preece 1882, pp. 29–30)

A few pages later, an illustration shows a scene of middle-class domestic activity illuminated by three chandeliers and other incandescent fixtures. The room is appointed in an affluent Victorian clutter of columns, cloth, heavy furniture, moldings, mirrors, and high ceilings. Clean, pleasant, and attractive, electric light appears to be a part of the best of modern living that does not disturb traditional elegance or values. Du Moncel further comments on the attractiveness and versatility of incandescent lighting, emphasizing how it fits with many forms of activity without distortion.[2]

The other three papers are more technical, establishing that electric light technology is now viable and that Edison's system stands out from the others. Preece's opening comments on the brilliance of the light display at the Paris exposition establish the spectacular viability of electric lighting:

Those who saw it for the first time will never forget the vivid impression that the great blaze of splendor produced on their minds on entering the building. There never will be anything like it again, for as wisdom grows with experience, so no manager of a future Exhibition is likely to repeat that terrific *mélange* of lights that flooded the interior of the Palais de l'Industrie with great brilliancy. (du Moncel and Preece 1882, pp. 125–126)

The hall was flooded by such an immensity of light, it was hard to tell the benefits of one system over another. Yet, while giving due credit to the work of all inventors, Preece regularly singles Edison out for special praise as leading the field with his lamp, his dynamo, his meter, and his safety fuse. The overall system marks Edison's accomplishment. Preece begins a four-page discussion on that subject as follows:

The completeness of Mr. Edison's exhibit was certainly the most noteworthy object in the exhibition. Nothing seems to have been forgotten, no detail missed. There we saw not only the boilers, engine, and dynamo machine, but the pipes to contain the conductors; the conductors themselves, ... the insulation, the fixtures, the brackets, the safety catches, the lamps, devices to avoid the effects of expansion and contraction through changes of temperature, meters to measure the current used. regulators to control the consumption of fuel. (du Moncel and Preece 1882, p. 152)

Preece ends his discussion of Edison, culminating the essay, with these comments:

Mr. Edison's system has been worked out in detail with a thoroughness and a mastery of the subject that can extract nothing but eulogy from his bitterest opponents. Many unkind things have been said of Mr. Edison and his promises; perhaps no one has been severer in this direction than myself. It is some gratification for me to be able to announce my belief that he has at last solved the problem that he has set himself to solve. (du Moncel and Preece 1882, pp. 154–155)

The final two articles, one by Charles Siemens and one by John Howell, present technical tests establishing the efficiency and economy of Edison's generators and lights.

A Working System

After achieving presence and legitimacy through public demonstrations reported by public media, Edison's companies told their stories more directly through their own publications. The companies' early pamphlets were directed primarily toward potential customers, but for investors these pamphlets established the reality and marketability of the companies' products—an important step in establishing credibility as an investment. The earliest company pamphlet collected in the Edison papers, *The Edison Light,* dates from 1882. This 26-page pamphlet issued by the Isolated Lighting Company explains basic facts about the Edison system, presents the advantages of the system, describes applications, and offers testimonials. There are several large illustrations. The text consists of headlined sections, each only one or two short paragraphs in length. This pamphlet, or one similar to it, could have been distributed to potential customers of the Pearl Street station, which included many financial institutions and potential investors.

The straightforward text of *The Edison Light* first distinguishes Edison's incandescent system from arc lighting, which the readers are likely to have already experienced. The description of Edison lamps is similarly basic:

The lamp is screwed into a socket which is permanently attached to a gas or other chandelier or bracket, and contains a key whereby the light in the lamp may be turned on or off. The lamp, once screwed into the socket, needs no further attention or care until the carbon breaks, when the old lamp is unscrewed from the socket and a new one screwed in.[3]

This was indeed a totally new product and basic consumer education was needed. Advantages are enumerated in paragraphs with the headings "Absence of Heat," "Non-explosive," "Peculiar Adaptability," "Not Injurious to Eyesight, "Safety," and "Fires Impossible." Short descriptions of fixtures, the transmission of power, and the lighting of cities precede sections on economy and cost. The pamphlet ends with a mention of the Paris awards and ten pages of testimonials.

A number of pamphlets of this sort were produced over the next few years by the Edison companies based in New York and by various international Edison companies.[4] Over time, they become more sophisticated, elaborate, and better organized. Modified to meet local circumstances, they provided a textual mediation of the Edison product, so that people could understand what Edison had to sell.

An Extraordinary Investment Opportunity

With a product established, the Edison interests could turn more directly to establishing the company's value to stockholders. At first, extraordinary short-term gains were promised. On September 15, 1881, while work was proceeding on the Pearl Street station, Edward Johnson issued a remarkable typescript statement titled "Edison Electric Light Stock considered as a speculative holding for the ensuing quarter." At this time the capitalization of the Edison Electric Lighting Company was being increased and stock in the newly formed Edison Electric Illuminating Company of New

York was being sold; shortly thereafter, the Edison Company for Isolated Lighting was to be formed. This was the moment at which current investors had to be encouraged to increase holdings and new investors had to be brought on board. Accordingly, Johnson wrote this document for limited circulation, specifically "to enable the reader to form a judgment as to the result upon the public mind and consequently upon the stock of the Edison companies of the final opening up to the public gaze and use of this large district." He continued: "Will such a complete demonstration of Edison's success cause a material rise in the value of his shares—that is to say when Edison shall successfully supply some 20,000 lamps to the interiors from a single source on such terms as, while being below the price of gas, will still pay him handsome dividends will the public accept the problem as solved and seek investment in Edison stocks?"[5] Arguing that the enormous value of the investment was on the verge of being recognized and that anyone investing at that moment was likely to realize an enormous profit in the next several months, Johnson reviewed the history of the publicity and the public perception of the project, the current state of the technology, Edison's control of the market, and incandescent lighting's position relative to gas lighting. This seventeen-page stock promotional document offers a comprehensive view of how the multiple parts of the Edison endeavor contributed to establishing the value of the stock. Johnson recounted the extraordinary credibility and excitement that met Edison's announcement of having perfected incandescent lighting in September of 1878, attributing it to Edison's great celebrity for the invention of the phonograph. The escalating excitement fostered by the press and the skepticism of the technical experts, Johnson pointed out, had put Edison in an "embarrassing position." Johnson constructed Edison as a victim of an overexcited press and public, and as doing little to foster the overheated public excitement—contrary to the evidence I have presented of Edison's cultivation of publicity, his active deceptions, and his distribution of financial rewards. Johnson presented Edison, confident of his own abilities, as doing nothing worse than simply not deflating the public's beliefs. Johnson also inaccurately stated that Edison's confidence was based on his having had the basis of a solution from the beginning:

Well knowing however that he had found the true principle of accomplishing the work and thus ultimately justifying all that had been so prematurely claimed for him and confident of his ability to perform an immense amount of labor in an incredibly short time, he gainsayed nothing but went earnestly to work to evolve from out of the chaos of the un-known the order and proportion requisite to put his ideas into practical shape.[6]

Johnson recounted the effect of the Menlo Park demonstrations and anticipated the effect of the opening of the Pearl Street station. A description of the system and of the principles behind it, including the intentional modeling to be competitive to gas, was followed by an analysis of Edison's patent protections and trade secrets that would guarantee him market domination:

He has woven a web so compact that were it to be perforated in a hundred places it would still be an effective defence. The controlling features are already secured, so that Mr. Edison alone can operate a general system. Others even at the utmost can only do petty work in an isolated way.[7]

Assuming that Edison would gain all the business currently held by the gas business, Johnson valued shares of Edison (issued at $100 and then trading at $1000) to be worth between $6250 and $15,625. Johnson ended his optimistic hard sell with visions of imminent great wealth having its source in Edison's charismatic genius:

If it rose as it did to $3000 per share on his promise what will it go to on the fulfillment of the promise? This work will rank with the creation of the Telegraph, the Steam Engine, or the Printing Press, and would have taken the same number of years to perfect it but for the colossal brain, untiring energy and vast resources of the greatest inventor of this or any other age. Today the Edison Electric Light is popularly believed to be a failure; tomorrow its success will be recognized the World over.[8]

Although the technology had been demonstrated, it still seemed necessary to invoke Edison's charismatic force to support the projection of revolutionary transformation. Also worth noting is the importance attached to the word 'promise' throughout this document, which suggests that Edison's choice of words on the occasion of the opening of the Pearl Street station ("I have accomplished all that I promised") was no ad hoc invention.[9]

Routine Reports to Stockholders

Although a case can be made to selected early investors that they will be part of something extraordinary, over the long haul a company must establish itself in the more ordinary terms of solvency, profitability, and dividends. The primary communicative vehicle for establishing these standard investment values is the report to stockholders.

I have no information about the history of the stockholder's report as a form (and it may be a rather short history in the United States, in view of the limited history of securities markets in this country), but it is clear that by the 1880s annual reports were the primary method of communication with current and potential stockholders. Then, as now, reports certified the quality of the investment through two figures: profits and dividends. The best thing that can be reported is an increase in dividends; the second-best is an increase in profits. Failures to achieve profits (or profits as high as the stockholders would hope for) must be explained away, and threats to profitability must be dismissed as adequately defended against. A report typically closes with the treasurer's balance sheet. However, since the Edison Electric Light Company was a speculative venture that did not pay profits until eight years after its founding, its reports to stockholders had to rely on tropes of growth, promise, and investment for a number of years before they could rely on the normal rhetoric of successful investment.

I have found no original of the 1881 report of the Edison Electric Light Company; however, the *New York Times* presented its details.[10] The report to stockholders set out the increase in capitalization from $300,000 to $480,000 and the resulting issuance of new shares. The demonstration of the previous year was recounted, along with the opening of the New York office, the approvals for laying cable (in the name of the newly formed Edison Electric Illuminating Company of New York), the establishment of a separate company to develop railways, and the successes of the early isolated installations. The company was presented as a working entity, open for business, with a proven product and with plans that were being realized. Because the financial promise remained vague, no specific expectations were set.

Promises of Prudent Corporate Success

The report to the 1882 meeting of the Edison Electric Lighting Company continues the tale of a company expanding operations, organizing itself, and keeping its financial house in order, according to the excerpts from the report published in the *Bulletin of the EELC.*[11] Capitalization and new issues of stock were increased by an additional 50 percent. To maintain the character of the EELC as a patent-holding company, a new corporation was formed to develop and market isolating lighting, capitalized at $500,000, with 51 percent of the stock to be returned to the EELC for license and the rest to be sold for capital. One consolidation with a competitor had been completed; others were contemplated and rejected. The report projects cautious financial solidity, at the same time avoiding concrete promises of returns. No balance sheet appears in this or any of the other reports presented in the *Bulletin.*

Immediately following this report, in the same issue of the *Bulletin,* are excerpts from the reports of the two other publicly held companies spun off from the EELC. These are far more expansive, for there are substantial results to report. The directors of the Edison Electric Illuminating Company of New York tell of the opening of the Pearl Street station with great pride and promise:

This event has created wide-spread comment in both scientific and financial circles. In view of the fact that it constitutes the first attempt to distribute electrical current for lighting and other purposes, over a large area, from a central station, and in view of the further fact that it is but a short time since the highest scientific authorities in the world pronounced such an attempt not only impracticable but also impossible, the actual starting of the plant awakened the deepest interest. It is therefore with great satisfaction that your Board congratulate you upon the successful opening of the First District, and upon the bright prospects for the future of the Company's business of electric lighting in this city.[12]

The report continues with the particulars of the station, the wiring of the district, and the early customers. Though there is no income yet to report, it is in sight:

Your Directors were of the opinion that owing to the irregularity of the light, resulting from the imperfect governing of the engines, no charge should be made until the station was running light satisfactorily to our own engineers. For the last few weeks, however, this irregularity has been substantially removed, and we have accordingly now begun to charge.[13]

The consumers are reported to be very happy with the new service, and the directors report that their only mistake was in making the installation too small to serve all the potential customers. They are now ready to remedy that by installing more equipment in a second building. Additional isolated installations and plans for an uptown lighting district are reported. This report too closes with the great financial promise of the company:

Your Board are of the opinion that the present successful development of the Edison system of electric lighting . . . will open an attractive field of investment for the purpose of extending the Edison system of lighting, not only over the Twenty-Eighth District, but over the remaining districted sections of the city.[14]

The report of the Edison Company for Isolated Lighting is even more enthusiastic. In its first year of operation the company reports a profit and a substantial dividend. The report begins with a narrative of a company that started with little a year ago, has grown prudently but rapidly, and now needs more capital for expansion:

When this company was formed a year ago, the business of isolated lighting was entirely undeveloped, and no data existed whereby the future development of the business could be foretold. . . .

The policy adopted by our Company at its start was to call in only a small amount of money, and to develop the business, until it had passed its experimental state, only on a limited and economical basis. This policy was rigidly adhered to. Installments were called in only as fast as the necessities of the business, viewed from a conservative viewpoint, required. The growth of the business, however, has been so steady and reliable, that the entire capital has now been called in, and is being safely and profitably employed in the business, and additional capital is needed.[15]

Customers, customer satisfaction, pricing policy, the village lighting system, and agents throughout the country are then described, followed by a more detailed request for increased capitalization. The report ends with another round of congratulation:

We can truly say that we have actually been the pioneers in the practical introduction of incandescent electric lighting.... To accomplish this successfully on a large scale, within the short space of one year, and at the same time, to establish the business on a dividend earning basis during this, the first year of its existence, is, in our judgment, a matter for congratulation.[16]

These three reports, taken together, clearly indicate to stockholders and to potential stockholders that this is a prudent yet highly promising set of companies, with substantial successes already achieved.

The three companies' reports to stockholders for the next year, printed in full in the *Bulletin of the EELC*, continue in their optimism and promise but still provide no balance sheets. The Edison Company for Isolated Lighting emphasizes its rapid growth, the fact that it is now making a profit, the satisfaction of the customers, and its expanding markets. Its further issue of stock, to double the capitalization to $1 million, is warranted by current profitability and the need to fulfill orders already taken.[17] The New York City Illuminating Company recounts successful operations, growth of its customer base, customer satisfaction, expanding revenues from paying customers, plans for an uptown station, and the overall promise of central station lighting, on which it has a monopoly:

The problem of central station lighting on a large scale, in which we have no competitor, is solved. Our company controls the exclusive license for this valuable system of lighting for all New York City, the most profitable field for electric lighting in the country, and it now rests with your new board of directors to devise means for occupying this rich territory.[18]

Even the dour EELC takes a more expansive attitude of promise, retelling the successes of its subsidiaries and reporting the sale of additional stock to further increase the company's capitalization without the warrant of immediate profits. Reporting increases in its patent properties, it puts off for another year the issue of whether to consolidate manufacturing under its aegis. The report ends with some defensiveness about the amount the company has expended without yet producing profits, but asserts the success of the endeavor and makes promises for the future:

Large sums of money have been spent in experiments, new inventions and patents; also considerable amounts in developing the business in both Canada and South America. Your Board has considered it unwise to peremptorily limit the expenditures in new inventions, and to hamper the development of the business by unreasonable economy. On the whole, surveying the expenditures from the formation of the Company to the present time, the outlay has not been excessive, and your Board find adequate reason for congratulation in the fact that no expensive mistakes have been made, and that, generally speaking, full value has been received for all money spent.

In conclusion we will say that the ultimate success of our business is no longer an open question, and it is entirely within the bounds of safety to predict that before the close of another year the public at large will fully recognize the fact, now frankly admitted by all who have investigated it, that the Edison System of Incandescent Lighting is destined to supplant all other methods of illumination.[19]

Ordinary Profits and Dividends

In 1884, the second of the three companies, the Edison Electric Illuminating Company of New York, reached profitability, and its annual report, now a separately bound document including a treasurer's report and balance sheet, moved from promise to normal business considerations:

At the date of the last annual report, the First District had scarcely more than reached the point where its receipts were in excess of its expenses. During each and every month of the present year they have shown a handsome increase as compared with last year, and instead of a loss as in 1883, the operations of 1884 will leave a surplus of fully 3½ per cent. on the capital stock, after paying of expenses of every kind.[20]

Profit and sales table are incorporated in the report, along with further numerical data (such as lamp longevity and expansion costs). Other ordinary business information is presented, such as news that the introduction of motorized fans has counteracted the decreased need for electric light in the summer. The details of the uptown district are made more specific. Although there is some of the enthusiasm and self-congratulation of previous reports, as the operation becomes profitable, a more ordinary discourse of ordinary financial operations takes over from the anticipations of extraordinary futures.

Such a discourse of ordinary finances also pervades the 1884 report of the Edison Company for Isolated Lighting. Expenses, profits, detailed fiscal arrangements, and balance sheets dominate and justify various decisions made to increase profitability.[21]

The last of the three companies reached profitability in 1886, and the emergence of an ordinary discourse of solid investment is all the more prominent for its being delayed. The 1885 report to the stockholders of the Edison Electric Light Company presents a narrative of the advances made since the last annual report, reviewing the solid operations, improvements, new installations and increased capacity of the capital and capacity of the company. The New York Central Station is reported to be profitable and paying dividends, and thus to be "now on a firm financial footing." Reports of difficulties overcome and the amicable settlement of disputes indicate that this and other licensee units are now solid investments. This section of the report sums up as follows: "... your company has therefore striven earnestly to improve and develop the methods of central station installation on a *sound dividend-paying* basis."[22]

Reports of other regional licensees, demonstrated superiority over the competition, and expansion into street lighting and railway lead to the conclusion that the company now has a solid foundation for producing earnings and great potential for even larger earnings in the future. The investment has changed from a risk into a security. The report ends with the following summary:

The advance of our business in past years has been greatly retarded by our inability to refer to central stations that were earning money or paying fair dividends, for reasons given in reports of past years. Capitalists were unnaturally willing to invest money in any enterprise unless it was backed by good results, no matter how favorable the investment might appear prospectively. We can now point with pride to almost all of our central station companies where financial success is an accomplished fact, and a visit to them inspires such confidence that it is becoming much less difficult to enlist capital to organize similar companies in other cities and towns.

The 1886 report, similarly, announces that the prediction of a turning point in company history "has been fully verified" and that "all the expenses of the past twelve months, including heavy disbursements for patent litigation, have been provided for by the ordinary cash receipts of the company, and there have been accumulated in its treasury all the stocks received for license rights."[23]

Summaries of the reports of the ten best-established regional central station companies, including those in New York, Boston, and Harrisburg, provide further evidence that the company is now entering a phase of solid capital growth. Details of patent litigation indicate that the company will actively protect its now profitable and expanding territory.

The details of the profitable business promise a future of dividends by establishing the source of the profit, indicating future opportunities for expansion of profit, and recounting aggressive measures for protection of profits. The company is now headed toward the textual reduction of the daily stock pages, which characterize an investment by a few economic indicators: price, earnings, dividends, and the ratios among them.

A Franchise Opportunity

Edison's plans also called for another kind of investor: the franchise licensee. The investors in local power companies were a special group with special needs. In addition to considering the general promise of electricity, they had to evaluate a specific market and develop plans to satisfy local conditions. To encourage and support the efforts of these entrepreneurs, the Edison Electric Light Company circulated additional publications.

One such publication was the Edison Central Station Catalogue, the second edition of which is dated 1885.[24] The special nature of the circulation of such documents is made clear by the handwritten inscription of the copy that found its way back into the Edison Company files for inclusion in the filmed edition of the Edison papers:

L. Stieringer, Strictly personal and confidential
compliments of J. H. Vail, general agent.[25]

The introduction to this catalogue explains that it provides "such data as will enable [agents] at a distance ... to determine approximately the cost of Central Stations and the capitalization of Local companies of various sizes." Further, financial data "derived from *carefully compiled statistics...* afford *convincing proof* that *fair dividends* can be earned upon capital invested in a properly managed Edison company." Thus, the catalogue both sells potential investors on the profitability of the franchise and provides the means for evaluating local circumstances.

The body of the 59-page catalogue consists of a variety of materials. There is a kind of sales and public relations primer, consisting of arguments for an Edison plant, but it is not directed at the readers. The readers are treated as already committed to electrical illumination and to Edison's central station concept, but as needing arguments to present to civic authorities and consumers. To aid their rhetorical work as representatives of the Edison empire, the catalogue provides outlines of arguments and a list of basic facts to be used in arguing.

Planning procedures are provided for estimating local conditions influencing cost and profitability, for gathering data to send back to the central office, and for designing the electrical distribution system. Available real estate, water facilities, coal sources and transport, soil conditions, buildings, the advantages of underground versus pole wiring, and similar considerations are discussed from the point of view of evaluating choices in developing the installation. As an introduction to the technical side of the business, the catalogue provides illustrations of various devices, machines, and equipment for generation and distribution of power.

Stockholders needed to know little of the company beyond what produces or threatens profits, but agents establishing local companies needed a manual for the technical, financial, public relations, and consumer sales aspects of the business. The catalogue introduces the reader to a complex world which is knowable, orderly, and predictable—conducive to the planning and operating procedures provided in the franchisee manual. Running a company will produce profits. Estimating carefully will allow one to capitalize appropriately and set rates. Even the sections on machinery repairs, rather than detailing maintenance procedures, provide evidence of high reliability and reassure that breakdowns will not interfere with regular operations.

The text consistently aims at creating a sense of business as usual, guided by the strong hand of the national Edison companies. The catalogue creates the impression that, even though central station management is a new and profitable opportunity, the technology is proven and stable, and the business arrangements entail little uncertainty, risk, or innovation. The procedures for calculating potential profits and the reports of successful plants are recruitment devices, encouraging more capitalists to venture into licensing agreements.

In their early years, the Edison companies produced many similar documents directed toward passive and active investors. These publications typically assert the viability of the technology, the solvency of the company, the promise and solidity of the investments, and the advantage of creating a regional company or purchasing an isolated plant. Both national and local companies issued annual reports, catalogues, pamphlets, and bulletins.

A pamphlet produced by the Brockton Edison Company early in 1885 is of particular interest. It is a reprint of an article from *The Electrical Review* that seems directed toward investors interested in developing regional central stations. An introduction by the editor of the *Review* comments:

The interest manifested through the columns of journals which, like the *Review,* have for some years devoted their energies largely to popular education in electric lighting, has shown of late that the time is ripe for presenting the prospects for future investment in this direction, so far as they are indicated by the outcome in plants only recently established.[26]

This is followed by an account of the development of the Brockton plant, including details of numerous errors that were made in the design and early operations that serve as lessons for future plants. Despite these difficulties, the plant was self-supporting during the first year of operations and was projected to pay a dividend in the second. Statistics and accounts of the operations and equipment detail the success. The article ends with "hints to promoters" planning new central stations. By reprinting this article, Brockton was legitimizing its own viability as a business operation, encouraging Brockton-directed investment, and proselytizing new licensees for the Edison Lighting Company.[27]

Spreading Good Words and Maintaining the Faith

In the early years of the growth of central stations, continuing support for investors and licensees was also provided by the *Bulletin of the Edison Electric Light Company*. Published from January 1882 until April 1884,[28] the *Bulletin* was at first intended for a restricted audience. The first five issues are specified as "Confidential, and for the use of the Company's Agents only." The first four issues contained a wide variety of news useful for the operations and management of central stations—for example, anecdotes suggesting what kinds of lighting worked best in various circumstances, tips on installation, and warnings about keeping current regulated to minimize bulb breakage. This is the kind of practical information a new company just shaking down its product line needs to improve service. It also contained information useful to salesmen and managers looking to persuade customers and new investors: upbeat news about new installations around the world, testimonials, press accounts of successes, news of the international exhibits. Journalistic accounts were reprinted so that agents of local companies would have "objective" reports at hand, including comparative tests of various systems. Starting with the second issue, stories regularly appeared about the dangers of gas lighting, including fires, accidents, and gas-related deaths—and about the related concerns of insurance companies. Invidious comparisons with competing systems (gas, arc lighting, batteries, and the products of other incandescent manufacturers) appeared regularly. The first issue pointed out that the American Electric Light Company's patent was highly impractical, and the eighth issue (dated April 27, 1882) noted that company's demise.

Three technical items also had continuing presence, primarily for customer and investor reassurance. First, reports of continuing negotiations with the Fire Insurance Underwriters to develop standards indicated how safe the Edison system was in comparison with arc lighting, which used higher, lethal voltages, and with gas. Second, accounts of regional regulation of overhead wiring consistently pointed out that the Edison incandescent system had perfected underground wiring, whereas arc lighting systems required overhead because of the high voltages used. The third category of continuing items concerned the extent of Edison's patents and the weakness of his competitions' patents. The sixth issue (dated March 27, 1882) details all 81 of Edison's patents assigned to the company, 12 patents in the process of being assigned, and 101 pending applications. Shortly thereafter, the list was updated and further specified. The efforts to point out the weaknesses of the competitors' patents culminated in an eight-page analysis of Swan's patents that appeared in the penultimate issue (December 18, 1883). (Swan's patents were actually perceived as the gravest threat at that point, and Edison was soon to merge his British operations with Swan.)

All the good news about the Edison company and the bad news about its competitors made the *Bulletin* too useful to remain solely in the hands of company agents. By the fifth issue, the masthead was changed to indicate that the *Bulletin* was now to be distributed to all stockholders as well as agents, "to give them information of the progress of the company." The confidentiality warning was removed. The upbeat reports continued, but the more practical, troublingly realistic information immediately vanished. The masthead directed agents to communicate practical operation and installation suggestions to the vice-president. Operations were now made to appear seamlessly successful. Even stories about how to extend the life of lamps were transformed into upbeat pieces reporting the increasingly long life of Edison's lamps and presenting them as far better than those of the competitors. Every issue reported on new installations around the world. And there were accounts of the increasing anxiety of the gas interests and of fatal gas explosions and fires, attested to by fire insurance underwriters' statistics. Occasionally calumnies spread by the opposition were countered and corrected. A news report suggesting that "electric light cures short-sightedness" was reprinted. The *Bulletin* grew from a 16-page quick read to 50 and sometimes 60 pages of long technical reports. In the last half of its run, it also carried reports of stockholders' meetings.

After six months of biweekly publication, the *Bulletin* began to appear only once a month, then only once every several months. Reassurances of growing success were less necessary as the company took on solidity and faith in the investors' eyes. Every installation was no longer worthy of a report. Now only larger assurances about new markets, advantages over competitors, test results, and the like were important. However, detailed reports of gas deaths and accidents continued to appear. In the end, the *Bulletin* seems to have been replaced by annual reports, the first of which was independently bound and distributed within a year after the last issue of the *Bulletin* appeared.

The *Bulletin of the EELC* served as a model for bulletins issued by other Edison organizations, such as the Western Edison Light Company and the Compagnie Continentale Édison,[29] during their early days.

Interestingly, only one series of bulletins seems to have come from an already-established unit. The Isolated Lighting Company published these during its fourth and fifth years (1885 and 1886), when its solidity and the certainty of its prospects appeared questionable as a result of patent reverses in the courts. The *Bulletin of the Isolated Lighting Company* appeared for fourteen issues, addressed to "Agents and Friends" as a "weekly memorandum of information."[30] The first issue, however, was devoted solely to the patent litigation being mounted against Sawyer and Man and the last issue reported a court decision against Sawyer and Man and in favor of Edison. The last issue also recounted the nefarious history of the Sawyer-Man collaboration, in a story based on a highly interpreted account of the court testimony. The intermediate issues, in addition to reporting on the patent litigation, contained the same kind of investor-confidence-building material found in the earlier *Bulletin of the EELC.*

A Solid Corporation

The representation of the Edison System as a stable investment vehicle is fully realized in the 1888 pamphlet *The Edison System of Incandescent Light and Electromotive Power from Central Stations.*[31] This pamphlet offers stabilized representations and meanings of several sorts:

• the heroic accomplishment of the now-institutionalized genius Thomas Edison, presented as single-handedly responsible for the Edison System's technical success

• the public validation of that accomplishment through awards, demonstrations, and testimonials

• a perfected system

• a solid organization

• a widespread system and an established market

• a record of solvency, profitability, and dividend payment

• secure patents

• regularized procedures and clear roles for investors

• a company built on heroic foundations and incorporating the American myth of the rise of a man of humble origins through hard work, foresight, genius, and daring.

The unified accomplishment of all these things is symbolized by the frontispiece illustration of the large factory complex in Schenectady, its active smokestacks announcing its industrious activities. This accomplishment is contrasted with an illustration of humble origins in a farm-like Menlo Park with a few primitive electric poles and lights. The one-page introduction opens with the great accomplishment of "over 1,250,000 Edison lamps in the United States alone" but immediately translates this into an accomplishment of the imagination of the great man. The current system is presented as the successful solution to "the problem which presented itself to Mr. Edison's mind when he first announced the fact that a practical

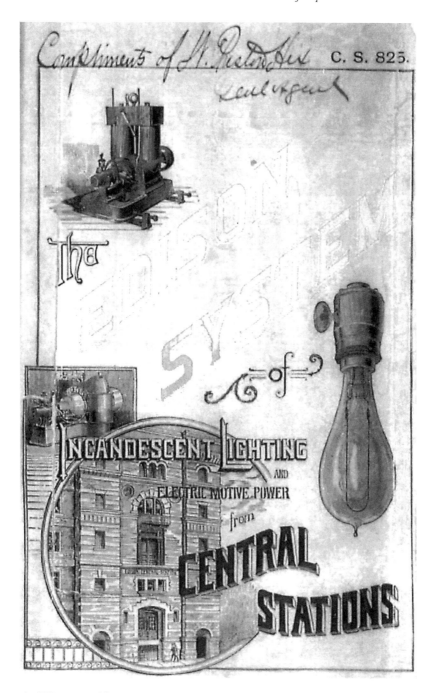

An Edison pamphlet.

incandescent lamp had been produced." Electrical lighting is presented as a heroic accomplishment of personal genius: the "system as originally laid down by Mr. Edison" is now "universally recognized as the only basis upon which the incandescent light can be successfully operated." Consequently, "as good profits can be made on the investments ... as with the best gas companies operating on a like scale." The public myth of the man Edison was part of what the Edison Companies offered. By investing in the Edison companies, one invested in Edison, much as Vanderbilt had a few years before; now, however, the faith in the man has been proved by actual success.

The next section turns from the imagination and achievement of Edison to the many prizes won by the Edison system. There is a foldout illustration of the award certificate from the Paris Exposition. The text quotes extensively from a Paris periodical attesting that Edison indeed had won over skeptical French scientists. Thus, the validation is scientific and international.

The ten-page description of the main components of the technical system includes a discussion of the criteria the system meets. The statement of criteria and the description of their fulfillment is a rhetorical tactic applied three times in this section, giving a sense of the perfected system. Criteria are offered for the general system, the electrical distribution system, and the characteristics of a good investment. The result of the perfection of this system is such that "it will prove more nearly universal and ultimately more profitable" than gas. The stability and profitability of utility systems are repeatedly emphasized, and the presentation of a perfected system qualifies the Edison system as a full-blown stable utility opportunity "which guarantees prosperity to the Vendor." Failure to meet these same criteria discounts other systems of electric lighting as valuable investments.

The description of the system's components echoes terms from the criteria, emphasizing that the components are reliable, productive, interchangeable, and adjustable. Illustrations of regulating devices, meters, and junction boxes give a concrete sense of the system's solidity, control, extensiveness, and delivery of a measurable product.

Five pages then detail the organizational structure of the company, various investment arrangements and procedures, and the advantages of these arrangements. Thus, electric light is presented as a well-thought-out, highly structured business opportunity controlled by well-considered licensing agreements.

Five of the major urban central stations are then described, primarily in terms of capitalization, profits, and dividends. Three of these (in New York, Detroit, and Boston are already running profitably); two (in Philadelphia and Chicago) are under construction. Facing the page devoted to each is an illustration of a solid-looking building. Together, the text and the illustrations form a portfolio of successes.

A summary of all operations follows, with additional substantial graphics of dynamo and boiler rooms and a list of other major capitalized franchises. A further description of central stations emphasizes how they have met criteria of reliability and economy, resulting in increasing public confidence and subscription growth:

> The general adoption of the incandescent electric light by the public is no longer a matter of uncertainty. Whenever an Edison Central Station has been established the light has become popular and has been generally adopted, no matter what other lights contested the field.

The books are closed and the ballots are in: Electric light is the technology, and Edison is the producer.

The case for investment having been made, the rest of the brochure is devoted to reassuring the investor that the patents are secure and that managing the companies is a well-structured task that—with much support from the central office—can be accomplished by men of planning, character, and energy. The tasks are elaborated, as is the promise of continuing support from the Edison companies and from the Association of Edison Illuminating Companies. The new opportunities in street lighting and electromotive power are described as further inducements. Billing procedures are spelled out in detail, reassuring the potential owner about the source of his income.

The value and security of the patents are detailed in a brief history of the lamp, Edison's advances, and the fundamental patent coverage. Two impressive illustrations of the lamp emphasize that this is what Edison owns. Quotations from the technical press emphasize the sweep and coverage of the Edison patents. Details of patent litigation indicate how vigorously the company is willing to protect the patent coverage and report the preliminary decisions in Edison's favor. Thus, it is well established both that the investor in an Edison company will have monopoly protection for the patents and that all attempts to exploit the electric light will have to go through the Edison system.

The pamphlet ends with a list of 169 Edison central station companies and an illustration of another large Edison factory (this one in Harrison, New Jersey).

Thus, the company represents itself as a solid industrial, corporate empire at work, making money at many brick-and-mortar locations throughout the country. This corporate empire rests upon many stabilized statuses, understandings, protections, and conditions that cannot be easily undone.

In ten years, the Edison companies had moved from an inventor with an idea who was able to evoke $50,000 worth of trust from a small group of financiers to a highly visible presence delivering light and power in major cities across the United States and around the world. The Edison companies could now represent themselves as secure long-term investments.

However, establishing a company as a publicly held investment frees it from its direct ties to individuals and to particular policies. Thomas Edison was already out of the management of the companies, and within a few years the national Edison companies were to be merged into General Electric and Edison's name was to be dropped from the corporate identity. The corporation, as an independent entity, representable in itself, on its own terms, could then go its own corporate way, free from the trust in any individual—a trust that, in the early days, had bought time for promises to be turned into deliverable light and for the commitment of a small group of close collaborators to be elicited.

In the course of these ten years, representations of the Edison companies came to rest on greater technological, legal, financial, productive, and marketing accomplishments, so that the claims of financial value could be increasingly warranted by reference to other textual and material realms (patents, exhibitions, awards, factories, central stations, balance sheets, and so on). The corporation's present and its future could now be represented by pointing to its multiple realities in the social, economic, legal and material worlds. This last brochure is a definitive collection and interpretation of the presences established by the Edison companies over ten years. It proclaims that the corporate empire cannot be undone without dismantling too much of the world. The Edison companies now can vanish only by being absorbed into something even larger.

The Language of Flowers: Domesticating Electric Light

Electric lighting arrived in the American home in a bouquet of flowers and ornaments. The fixtures were ornate, the illustrations of domestic scenes incorporating light were florid and elegant, and the descriptions of domestic lighting were full of the language of aesthetics. The domestic aesthetic established by these early representations of incandescent lighting was remarkably enduring, even today influencing the appearance of table lamps and ceiling fixtures. Until modern design invaded middle-class suburbs after World War II, lighting was almost always heavily adorned with ceramic and stained glass representations of flowers and classical art, creating an aesthetic of genteel affluence. Plain sockets and simple shades were well-known literary symbols of poverty, and one of the first acts of house-proud people trying to add decorative elegance was to purchase ornamented lamps and shades. One need only take a quick trip to a local discount store to be convinced that such still is the case.

The practical benefits of incandescent electric lighting for industry were quickly established. The earliest customers of incandescent lighting were industries that relied on flammable materials (such as newspapers and cotton and lumber mills) and public institutions that relied on surveillance and control (such as prisons, homes for the insane, and residential colleges). However incandescent lighting was not necessarily compatible with or necessary for domestic life, which had developed patterns of hearth and lamp that would be disrupted by this new brightness. Bathing the house in artificial light would require an aesthetic that conformed to acceptable values of family life.

An aesthetic of electric lighting was actively constructed by Edison companies and other members of the early incandescent industry in order to make electric light attractive to the domestic market. This aesthetic spoke to aspirations of the newly urbanized and increasingly prosperous American family. It spoke to a market highly inflected by gender and class.

Ornamental Floral Electrolier no. 35 in Catalogue and Price List of Edison Light Fixtures Manufactured by Messrs. Bergmann & Co. (New York: Bergmann & Co., 1883) (96: 227).

And it helped give cultural shape to a changing family life, formerly expressed in shared production in the day and gathering around the hearth at night but now dispersing in careers, shops, and leisure activities during day and night and, in the home, dispersing to separate corners and rooms.

Electric lighting was not only a tool of the expanded life of affluence; it also became a symbol of consumption, cultivation, and upward mobility. As such, it was intertwined with the growing aesthetics of a number of other aspects of late Victorian American life, such as advertising, lifestyle journalism, the department store, furnishing, and urban residential architecture.

This chapter will document the rise of this aesthetic, will identify the active participation of pioneers of the incandescent lighting industry in fostering it, will examine some of the social and cultural forces that underlay the appeal, and will tie it to other developments of the emerging culture of middle-class American life.

A Catalogue of Flowers

The 1883 Catalogue and Price List of Edison Light Fixtures Manufactured by Messrs. Bergmann & Co.[1] shows how thoroughly early incandescent lighting was wrapped in flowers. Even the plain electroliers and bracket fixtures presented on the first two pages configure the shades and lights as abstracted flowers, and in several cases the supporting tubing is given the sinuous bend of vine and branch. Starting with the third page of illustrations, the floral motif becomes more concrete; the shades become ribbed and fluted like morning glory petals. Two pages later, wrought iron and pressed metal vine and flower decorations begin to bedeck the fixtures. Over the next 20 pages, the designs become increasingly ornamental, with floral and plant motifs occasionally mixed with neo-classical vase and column figures. On page 24 a vase overflows with flowers; the stamens of the four largest are incandescent lamps. On the next several pages are majestically ornate floral chandeliers—some formal and geometrically regular, some with the random abundance of an English Garden. A few pages later, in the catalogue of individual parts, all the glass shades labeled "colored glass flowers" and "etched globes" have floral designs.

The Bergmann catalogue contains little print beyond names of items, catalogue numbers, and prices. However, the introductory page comments: "It will be observed that the use of the Edison Incandescent Light offers a wider field of ornamentation in Electroliers, Brackets, etc. than

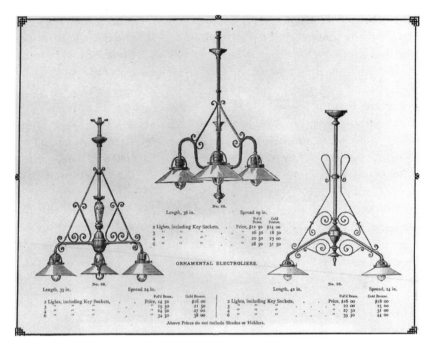

Ornamental Electroliers in 1883 catalogue of Bergmann & Co. (96: 190).

that of gas."[2] Contemporary catalogues of kerosene and oil lamps reveal a simple standard design based on a reservoir, a wick, and a chimney and ornamented only with patterns (sometimes floral) etched or painted on the glass or pressed into the metal.[3] Gas fixtures did venture into more ornamented floral designs, particularly on chandeliers, but nowhere to the extent that was pervasive with early domestic electric lighting. Electrical lighting proliferated non-functional, aesthetically inspired glass and metalwork. The incandescent lamp wasn't just decorated; it was embedded within a decorative ensemble.

Although many producers of lamps and fixtures were to cooperate in the development of lamp style, Bergmann & Company was one of the earliest and one of the most influential. Bergmann had been an employee of Edison in the Newark machine shop in 1871. As Edison's inventions grew in number, Bergmann went into business under his own name, but under well-publicized license to Edison, as a manufacturer of Edison's inventions in telegraphy, telephony, and sound reproduction. In 1881, as Edison set up the network of the Edison companies to market his now successful system, Bergmann & Company became the designated manufacturer of light-

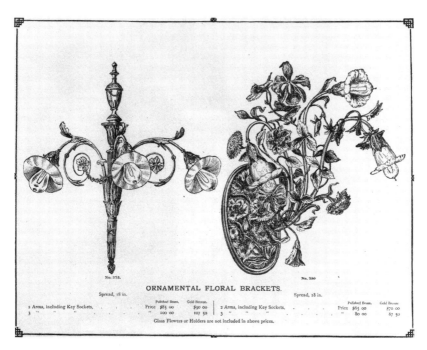

Ornamental Floral Brackets in 1883 catalogue of Bergmann & Co. (96: 223).

ing equipment, including fixtures, sockets, meters, and regulators. This was one of the few Edison companies that didn't have Edison's name on it, although Edison was a full partner. And when Edison set up his Fifth Avenue display room, he used Bergmann fixtures.

Woman's Place within the Trade Journal

The Electrical World of December 22, 1883 featured Bergmann & Company as "a representative american electrical manufactory."[4] Much of the text was devoted to a mechanical description of Bergmann's products, power system, and manufacturing processes, but nearly all of the ten illustrations were of floral design lighting fixtures, most of them (including an elaborate chandelier and a simulated bouquet in a vase) were reproduced directly from catalogue plates. The illustration of the display floor showed many of the designs of ceiling pendants and table fixtures from the catalogue being inspected by two couples. In the center, a man appears to be talking to his wife about the fixtures they are viewing, and another man is using a cane to point out a fixture to his wife. An unaccompanied man

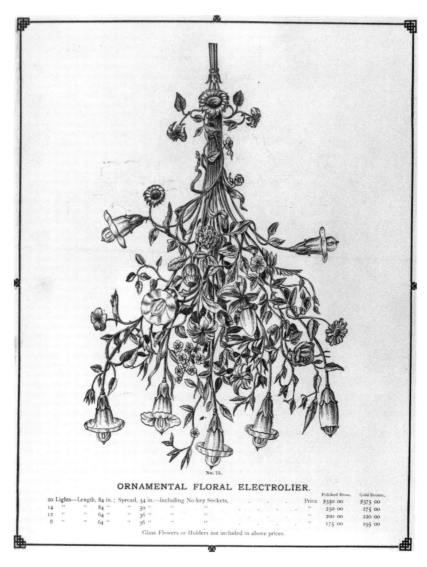

Ornamental Floral Electrolier no. 75 in 1883 catalogue of Bergmann & Co. (96: 226).

stares out the window, and another man is at the table, apparently looking over some equipment with a workman; no man in this illustration seems able to look at the fixtures without the aid of a woman.

Eventually, consumption decisions for lighting fixtures were to become predominantly the female homemaker's choice, being subsumed entirely into the role of middle-class women as ornaments of fashion and creators of domestic elegance. For example, in a 1925 company newsletter, the Beardslee Company condescendingly described the difficulty of working with the female consumer:

> When a woman buys a gown, or a hat, new furniture for the living room, or new lighting fixtures, she selects something unlike the wearing apparel, furniture and lighting equipment her neighbors have. Show her the most beautiful chandelier you have and she says: "Yes, that is a beautiful design, but one of my neighbors has a chandelier exactly like it in her dining room." That settles it! It's your move, and being wise in the ways of women you say:
>
> "Look at this piece. Here is something absolutely new—I just unpacked it this morning. You won't find another chandelier like this one anywhere in town."
>
> That's what she wants—something different. And so manufacturers must continue to produce new designs and dealers to stock them.[5]

This, however, was 40 years later, once houses and apartments were regularly wired.

During the early days of incandescent lighting, the consumption decision involved major utility choice and minor construction at the very least. Since the earliest central stations were in business and industrial districts, domestic electric lighting involved setting up a small isolated plant with a motor and a generator. Including wiring in new apartments was a major architectural construction decision, made with an eye toward marketing. All these decisions were, at the time, clearly men's decisions, made in a "man's world." Yet all these consumption decisions needed to be approved by women, who were in charge of the domestic space.

Handing women aesthetic control over the production of an elegant home was precisely suited to the ambitions of the male heads of middle-class and upper-class urban households. They wanted elegant and leisured wives presiding over elegant refuges from the world of commerce. This male construction of the female explains the curious allocation of content in the articles about the aesthetics of lighting that appeared in the male-dominated industrial journal *Electrical World*. For example, most of the aforementioned article on Bergmann & Company was technical, describing the construction of the product, the factory floor, and the assembly process. Though aesthetics were mentioned in passing as a business

VOL. II.—No. 17.   NEW YORK, DECEMBER 22, 1883. $3 PER ANNUM.
CENTS A COPY.

A Representative American Electrical Manufactory.

In our leading article of Sept. 8 ult., in which we studied the relations of "The Incandescent Light and Decorative Art," as exemplified at the Munich Electrical Exposition, we promised to show to our readers that American products in this line are quite equal to those of European art. To enable us to fulfill our promise, we paid a visit some time ago to the establishment of Messrs. Bergmann & Co., Avenue B and Seventeenth street, this city, who manufacture the fixtures and appliances used in installations of the Edison system of electric lighting. Our chief object, as just intimated, was to examine the elegant electrical chandeliers and brackets of artistic design and finish, for which the establishment is already well known, in order that we might point pen pictures of the beauties and merits of American decorative and ornamentative art for which the materials went through the various processes of transformation in their evolution toward a perfect finished article, were among the features which excited our interest. We were better prepared to appreciate the excellency of the productions turned out of this factory after we had had this opportunity to judge for ourselves of the facilities it affords for doing good work.

The factory of Messrs. Bergmann & Co. is a handsome brick building situated on the northwest corner of the junction between Seventeenth street and Avenue B. It occupies an area of 125 by 100 feet and is six stories high. The building is divided into two parts by a thick brick wall with open arched doorways connecting them. This mode of construction insures stable and solid floors, which is very essential where heavy machinery is to be used. The building is provided with elevators having self-opening and closing hatchways. On the roof is a tank which made in the factory, and it serves for experiments. Mr. Edison, whose laboratory now occupies the sixth floor of this building, not unfrequently avails himself of the use of this small testing-room. There is another small laboratory in the factory, with dark chamber photometer, etc. On the second floor we find the heavier machinery, such as engine lathes, machine lathes, die presses, drilling machines and screw machines. They form a diversified and highly interesting collection. We watched a workman making the neat little binding screws that add so much "electrical" grace to a piece of electrical apparatus and they were turned, threaded and finished almost as fast as we could count them. The workmen on this floor are among the most skilled of those employed in the whole factory, for it is here that the finest machine work is done. The lathes and heavy machinery are grouped on one side of the building as much as possible,

DISPLAY OF EDISON ELECTRIC LIGHT FIXTURES IN SHOW ROOM OF BERGMANN & CO.

our readers. But as it happened, our admiration was not evoked only by the beautiful chandeliers and fixtures which we saw; for as we passed from floor to floor, in our survey of this immense factory, Mr. Bergmann, who was kind enough to escort us, pointed out to us many interesting things, which have led us to enlarge the scope of our article. In addition to making the Edison fixtures, Messrs. Bergmann also make several other accessory appliances necessary with the Edison system, among which are the lamp sockets, current regulators, both hand and automatic, current meters and safety catches—in fact, we may say that everything but the Edison dynamos and lamps are made here. The manufacture of electrical instruments of precision, such as galvanometers, rheostats, etc., is also a specialty and besides this, work is done for various outside concerns on contract. While going on our round we saw work in progress for the Gold & Stock Telegraph Co., the Law Telegraph Co. and the Holmes Burglar Alarm Co.

The factory itself, with its improved machinery and large complement of help, the variety of work done in its different departments, the system and order with holds 3,000 gallons of water, to supply sprinkle pipes placed in the different parts of the building. The work to be done is mostly of a light character and does not require much power, probably not over 75 horse power for the whole establishment. This power is furnished by a 150 horse-power Corliss engine, located on the ground floor next to the offices. The boiler adjoins the engine room. On the ground floor is a brass foundry, where the brass and composition castings used for the chandeliers, switches, regulator frames, lamp sockets, etc., are made; and we must remark, en passant, that the casting for ornamental fixtures, with intaglio and relievo chased designs, were the smoothest and most perfect we have ever seen. There is also a cabinet shop adjoining the foundry, where the woodwork needed in the construction of the various appliances is made. The remaining space on the ground floor is given up to the shipping and storage departments, with the exception of a small testing-room. This room is provided with delicate measuring instruments, including a Thomson reflecting galvanometer. This apparatus is used to measure the constants of the wire purchased for use, and thus to determine its quality. It is also used to calibrate resistances and measuring apparatus and the other portion of this floor is devoted to the construction of fine apparatus. The Gold & Stock "tickers" are made in this portion, and the unequaled workmanship which we found bestowed on the wheels and pinions alone evinced, in our opinion, the high class of skill of the workmen who make them. The automatic regulators used with the Edison dynamo-electric machines described and illustrated in our issue of Sept. 1, are also made here. Besides these, galvanometers of all kinds, amperemeters, voltmeters are also made. Several galvanometers, both simple and astatic, and some reflecting galvanometers, of the Thomson single and double needle pattern and rheostats in the laboratory of Mr. Edison, are made by Messrs. Bergmann & Co., and they exhibited a workmanship quite equal to that of the best Elliott Brothers (London) make. On the third floor there is quite a distinction between the operations conducted on either side of the middle brick partition. One side is in a measure a continuation of that last described. Some Gold & Stock apparatus is made here, either wholly or partly, and in addition the work of this department includes burglar-alarm apparatus. Holmes' Burglar-Alarm Co.'s annunciators, signal-bells, telephone apparatus and similar contract work. On

The illustration from "A Representative American Electrical Manufactory." Source: Smithsonian Institution Libraries. Copyright: Smithsonian Institution.

necessity several times, only toward the end of the long article were the aesthetics of the fixtures discussed:

Polished metals, brilliant lacquers, outlining the framework of each fixture, glass globes, pendants, and ornaments of every conceivable shape and every possible color, and, withal brilliant flowers, all combined to surpass the rich munificence of the grandest pageantry and to remind us of a peep at fairy land.[6]

After this brief and grandiose discussion, the article returned to the technical. A few paragraphs later, another brief passage discussed the floral design:

One sees hanging vases with richest flowers, that seem to rival in freshness and gorgeousness those of the florist's conservatory, and amid the beautiful buds and blossoms one sees blooming tulips of glass all aglow with golden light, emitted within them by Edison lamps. Surely these are bouquets which are even more beautiful than those which Nature herself can supply.

Then the article again reverted to technical description. Yet this article was richly illustrated. Men may not have been interested in or able to talk about aesthetics; however, they could recognize pictures, and they could associate lighting with women. They might even have brought illustrations—such as those in the nearly wordless Bergmann catalogue—home to share with the family.

An Aesthetic Campaign

Just a month earlier, in the November 24 issue of *The Electrical World,* the connection between flowers, lighting and female aesthetics was made more explicitly in a brief note, unusually poetic in this technical, industrial journal, that was explicitly delivered through the voice of "a Lady":

POETRY OF THE ELECTRIC LIGHT. A Lady, writing to a Western journal of the recent exposition at Louisville, gives the following glowing description of the scene at night: "... All through the park the Brush lights glow like great animated lilies in the air. As the water-lily sways on the stream and is hidden sometimes by the swelling waves, so the current of air seems to bend and break over this great bed of electric flowers."[7]

And a month before that, the aesthetic reasoning behind the design of lighting fixtures was made even more explicit in an October 13, 1883 interview with Edward Johnson, vice-president of the Edison Electric Light Company:

At first the desire was to attract the popular fancy by means of elaborate and ornate designs. There, you see, is a specimen from our European exhibition, a flower-pot overgrown with a wilderness of foliage all done in polished brass. The lights spring from among leaves like flowers from their stems. While such forms were well enough for an occasional public exhibition, they were found to be unsuited to any other purpose. Men of wealth who wished to have the electric light in their houses wanted the supports to unite elegance and simplicity. The highly ornate plans of the English workers in brass had to be thrown aside. But the idea of treating the glowing bulb as the stamen, which was the first to occur to us, has, however, been retained through all the changes that were made to meet the requirements of popular taste. It was Mrs. Edison who suggested these floral forms as the best adapted to our purpose. We have not varied the discovery yet to the extent we shall have to do in the future, but have confined ourselves to flowers whose general shape is that of a cup, such as tulips and lilies. But we can as easily use the rose with its multitude of petals.... The handsomest device that has been thought of is a hanging framework of brass, to which the lamps are so placed that their stems form a basket that may be filled with artificial and plants in the natural colors.[8]

This statement reveals not only the conscious marketing thought that went into the choice of the floral motifs that turned out to be so pervasively successful but also several underlying themes to which the floral design appealed. Though presented simply as a matter of business choice, they are deeply revealing of the culture. In particular, domestic lighting was portrayed as appealing to men of wealth. It was being marketed as part of the affluent life. And at this point it was treated as a male decision. And the affluent male was portrayed as more than simply practical—as wanting elegance with his simplicity. Nonetheless, elegance was portrayed as provided by the feminine. This is the one time that Mrs. Edison (Mary, his first wife, who was to die of illness the next summer) is mentioned in any of the Edison business stories. The floral design is presented as the woman's touch. The sexuality of the flower imagery is further called to attention by the contemplation of the bulb as the stamen among the petals. Although one might not want to make too much of the psychic import of the imagery, it is certainly no stretch to see the female floral as softening the harshness of the technology when it is brought into the domestic nest, where the true beauty of the light may literally flower. After all, in the same interview Edward Johnson is quoted as follows:

Our aesthetic tendency did not receive much encouragement in the demand for electric lights in the factories. Utility and simplicity was the rule of those places.... But the beauties of the light itself are not quickly exhausted.

Another theme that underlay this interview was that of European taste, aesthetics, and craftsmanship. This was the time of the British Arts and Crafts movement, which attempted to wed aesthetics and industrial manufacture. It is not clear from the Johnson interview whether the British craftsmen or Mrs. Edison had came up with the floral idea, but clearly a connection is made between domestic elegance, the woman's touch, affluent appearances, and European taste, even if design needed to be made a bit more practical for the United States.

This theme of the United States rising to European standards of taste framed the aforementioned article about Bergmann and Company in the December 22, 1883 issue of *The Electrical World.* That article began:

In our leading article of September 8 ult., in which we studied the relations of "The Incandescent Light and Decorative Art," as exemplified at the Munich Electrical Exposition, we promised to show to our readers that American products in this line are quite the equal to those of European art. To enable us to fulfill our promise, we paid a visit some time ago to the establishment of Messrs. Bergmann & Co.[9]

The September 8 article had made an explicit link between aesthetics and the domestic market for lighting. It opened:

It was not enough for the incandescent light to have achieved the distinction of being practical. In order to supersede its rival and obtain a foothold in public buildings or private residences, it was obliged to accommodate itself to the requirements of decorative art.[10]

The Munich exhibition was described as moving beyond demonstrating the usefulness of incandescent lighting to displaying the light's aesthetic possibilities. Munich was portrayed as a center of art—"the Florence of Germany"—with museums that would inspire artisans. The article then described a series of six illustrations reproduced from the Paris journal *La Lumière Electrique,* one showing the illumination of an art room. The article ended with a nod toward Europe's artistic advantages over the United States, but with a promise to meet the challenge:

It must not be believed, however, that because we lack the advantages which Europe possesses in having able artists and skilled artisans in large numbers, we are unable to use the incandescent light to good advantage for producing tasty and brilliant effects.

In each of the four months that followed the appearance of this article, *The Electrical World* published an article that highlighted the aesthetics of incandescent lighting. Two of these four articles were illustrated features, and three of the four emphasized the floral as the center of the aesthetic.

Even earlier, Edison and *The Electrical World* had shown some concern for the aesthetics of lighting in their efforts to link European art, incandescent lighting, and an elegant affluent life.[11] At the 1881 Paris exhibition, Edison's representatives had gone to some expense to have fixtures produced by top craftsmen in order to produce an aesthetic effect, and had vied to illuminate art galleries and opera houses. Illustrations of these aesthetic placings of incandescent light within artistic settings (some of which originally had appeared in *La Lumière Electrique*) were a frequent feature of *The Electrical World.* These illustrations had something of the flavor of the latest Paris fashions for the home, carrying with them the model of civilized and affluent modern living.[12]

One illustration in particular, accompanying Theodore du Moncel's article in *La Lumière Electrique*, associated electrical lighting and modern active living.[13] The illustration shows a richly decorated late Victorian parlor, with decorated walls, thick curtains, and a large floral pineapple settee in the middle. Members of an elegantly dressed family are dispersed

Incandescent lighting portrayed as part of an affluent life: page 14 of du Moncel and Preece, **Incandescent Electric Lights** *(Van Nostrand, 1882).*

throughout the room, carrying on multiple activities with the aid of several chandeliers, wall fixtures, and lamps, all ornamented with floral designs. Aided by the incandescent light, a young woman plays the piano for an admiring male, a child plays with a cat, the mother looks after the child and holds a book, and the elder male looks on the entire scene with pride. This encapsulates the sense of European elegance tied to a new model of a family, where the father's reward as provider is to witness the elegance and prosperity of the life of his dependents, engaged in varied activities, aided by a wife who creates a beautiful domestic world by purchasing objects of taste and comfort that provide for an affluent life while simultaneously displaying to others the family's attainments. The decoration of the self and the home were core sites of consumption. This one picture shows how incandescent lighting was placed in the middle of these new cultural desires.

The Culture of Urban Domestic Life

Recent scholarship on the history of culture has been giving us a more precise look at the changing shape of domestic life in the emerging urban America of the late nineteenth century. Studies such as *Making America Corporate* (Zunz 1990), *The Light of the Home* (Green 1983), *At Home* (Garrett 1990), and *The Social Origins of the Private Life* (Coontz 1988) point toward the consequences for domestic life of the expansion of urban salaried middle-class and upper-middle-class labor, increasingly removed from the family and the home. That is, men went to work and made money. The value of this work lay less in the identity formed within the community or the enterprise one was part of than in that it provided the means to lead an affluent life and assured a class position. The home became the site for enjoyment of affluence and leisure, particularly through the consumption of the new products and aesthetic objects that were increasingly available. Moreover, the main way for the male to display class position, thereby reflexively engendering the trust that one indeed deserved a managerial position, was through the visible lifestyle of the family supported by his earnings.

It was hardly an accident that the Aesthetic Movement in the United States, with its particular emphasis on the decorative arts, arose at this cultural moment. This movement, centered in New York, helped distinguish the newly affluent urban elite from the urban poor. It also helped the new American elite to claim its place alongside the longer-standing European elites. Both phenomena are richly documented in the fiction and the social exposes of the late nineteenth century (Burke et al. 1986). Lighting fixtures were among the domestic furnishings that received aesthetic attention.

Among the most famous products of the Aesthetic Movement were the Tiffany lamps, first produced in 1899. Louis Tiffany had worked with Edison on electric lighting as early as 1885 (Koch 1971).

In 1888, Cincinnati—one of the secondary centers of the American Aesthetic Movement—was the site of an ornate display of floral lighting.[14]

As Trachtenberg (1982) and Barth (1980) have pointed out, the city became a place of differentiated locales as well as a place of differentiated cultures. It became important to mark out one's domestic space as a place of elegance and culture, lest it be assimilated into one of the city's less desirable locales that were proliferating, such as working-class tenement apartments.[15]

Another manifestation of the new elite urban feminized domestic consumerism to which electric lighting appealed was the department store, also a product of the 1870s and the 1880s. Department stores—centralized sites of consumerism—were designed to be meeting places for leisured urban women (Benson 1986; Hendrickson 1979; Loeb 1994; Norris 1990). Decorated in domestic elegance, they provided goods for the enhancement of the woman, the home, and the family. They regularly educated women on fashion and taste by means of merchandising events (such as fashion show) and trained sales personnel. European taste, portrayed in objects for sale, in illustrations, and in cultural symbols of European elegance, again set the standard. The same journals that set the tone for the Aesthetic Movement (*Scribner's Magazine,* for example) also carried news of the new department stores. These same department stores were among the leaders in interior electric lighting. John Wanamaker's store in Philadelphia installed arc lighting in 1878 and then switched to incandescent lighting as soon as that became available. Among the other department stores early illuminated by Edison light were Arnold Constable and Sterns (in New York), Marshall Field (in Chicago), and Jordan Marsh (in Boston).[16]

Thorstein Veblen examined this emergent urban consumer culture in his 1899 book *The Theory of the Leisure Class.* In particular, Veblen pointed that the home had become the site of conspicuous leisure, and that women had become the ornamental property assigned the task of displaying conspicuous leisure in themselves and in the domestic sphere. The men, of course, were too busy in the world of work to carry out the leisured occupation of consuming conspicuously. Thus, the elegant woman, as the ultimate property and extension of the prosperous male, became the vehicle of affluence, leisure, and distinction. Moreover, as Veblen pointed out, it was particularly important for the social leader to display that he engaged in leisured, non-industrial activities when he was out of public sight, and so the leisured, elegant home became a potent class symbol.

The incandescent light was also touted for its simplicity of care relative to oil lamps (which had to be trimmed and adjusted) or gas lamps. That a child could turn electric lights on and off safely with a switch was not only an appeal to safety but also an appeal to freedom from care and monitoring.

From Upper Class to Middle Class

It made marketing sense to attach incandescent lighting to the most desirable consuming lifestyle. Once an item of consumption became valued by a society's most prestigious members for its elegance and its lifestyle enhancement, it would also become valuable as a site for emulation for those aspiring to the status of the most eminent. J. Pierpont Morgan's installation of an isolated plant in his Madison Avenue home in 1882 was well publicized. Despite early problems (including a fire), Morgan's installation inspired his wealthy neighbors D. O. Mills and Whitelaw Reid to have their houses wired (Satterlee 1939, pp. 207–216; Hoyt 1966, pp. 164–165). Morgan then became instrumental in getting a central station built uptown to service his mansion and others. These were the very homes in which the Aesthetic Movement thrived.

In 1887 *The Electrical World* ran a feature story about the elegant light installation at the home of Edward Johnson, the Edison vice-president who had originally engineered the aesthetics of incandescence. Some of the fixtures in Johnson's home were, in fact, designed by Louis Tiffany:

> The main hallway is finished in quartered oak. The ceiling is composed of eight groined arches, each capped with a miniature sun made by Tiffany & Co. Behind these are Edison lamps, which, reflecting through the different thicknesses of glass, give the well-known outline of the man in the sun.[17]

Apartments and hotels were also important in bringing lighting into domestic life. They used the trappings of the affluent life to attract those who were transient in the city or those who could not afford the expense of their own mansions but who still aspired to the life of urban affluence. Lists of such hotels and apartment houses were prominent in the early publicity for the Edison Isolated System and for the Edison Central System. In New York the publicized installations as of 1885 included the most

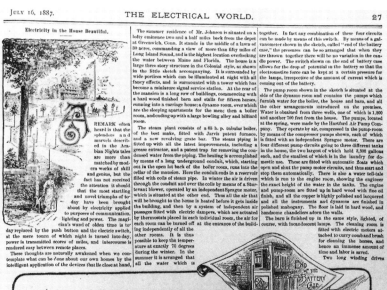

"Electricity in the House Beautiful," Electrical World, July 16, 1887: 27. Source: Smithsonian Institution Libraries. Copyright: Smithsonian Institution.

prominent apartments built in the period: the Dakota, the Osborne, the Barrington, the Franke, and the Steinhard apartments. Hotels that were early sites included the Buckingham, the Murray Hill, the two Everett Hotels, and the New York Athletic Club.

The 28-page Bulletin for Agents of the Isolated System dated July 25, 1885, was devoted entirely to hotel and apartment installations. It pointed out that "the popularity of a hotel is dependent, not only on the elegance of its appointments ... but also on the appearance of brilliancy it portrays."[18] Many of the installations were illustrated and described. There was an extensive description of the newest and most prestigious apartment house in New York, the Dakota, which was being wired for 10,000 lights—by far the largest residential installation of the time. According to the Bulletin: "Both the private houses and the Dakota rank among the very first buildings in the city for elegance of finish and substantial construction."[19]

Thus electric lighting early became one of the elegant appointments that made apartment living appear to be stylish and appropriate to the aspiring classes. Indeed, a 1913 commemorative volume produced by the New York Edison company, *Thirty Years of New York, 1882–1912*, devoted an entire chapter to the association between central light and power and the growth of apartment living in New York City.[20]

Although there had been a long history of flats in European cities, apartment life did not come to American cities until middle of the nineteenth century and did not expand rapidly until the closing decades of the century (Alpern 1975; Cromley 1990; Perks 1905), just when electric lighting became available. While incandescent lighting made more possible the kind of tenement living for which New York became notorious, the very presence of electrically illuminated tenements made it all the more important to distinguish the use of incandescent lighting in middle-class homes and apartments as a form of elegance through ornate fixtures—and therefore to place ornate fixtures among the amenities that would allow working-class families to express their middle-class ambitions. Lighting could then be perceived as one of the rewards of urban strife, producing upper-class elegance rather than the artificial machine conditions of the bare bulb.

The traditional associations of flowers with the leisured garden, feminine beauty, cultured elegance, and domestic charm; the pervasiveness of the floral in domestic decoration in Europe and Asia; and the Victorian enthusiasm for flowers as avocation, motif, and decoration all suggest that it was only natural that electric lighting would take on floral hues when entering the home. Yet, when we look into the cultural moment of urban

America in the 1880s, with its special drive to create the appearances of beauty, leisure, elegance, culture, and charm as marks of distinction within the new social and economic order and to create a place distinguished from the urban technological financial strife, we can understand why there was such a strong and immediate marriage between the technological marvel and the aesthetic of a cultivated nature. We can see why Johnson, Bergmann, and others in the Edison world were so attracted to flowers as the bouquet in which to bring lighting into the home.

Conclusion

Symbolic Invention: Making Meaning in Typified Activity Systems

During the development of incandescent lighting, Thomas Edison and his associates were busy people. They, of course, were making electricity and light; they too were making generators, meters, switches, and lamps; but they were also making meanings. They were creating value for their endeavor. They established relationships with people and influenced them to participate, collaborate, support, endorse, and purchase. Each meaning-making communication was within a complex communicative field that pre-existed Edison's entry, that was to some extent transformed by his participation, and that became an enduring site of meanings for incandescence. In each of these fields, Edison and his associates accomplished the magic of communication, saying the right things at the right time to keep the endeavor unfolding and creating the right representations to keep their project before the eyes and in the minds of the relevant parties. These representations were attractive enough to get the relevant audiences to grant more and more presence to Edison's project, to keep doing those things that would allow modern electric light and power to happen, and to make the decisions that would create an apparent inevitability and indisputable value to what was accomplished in those years.

The chapters of this book have told the stories of the symbolic presence of Edison's light in a variety of systems and of the active work that Edison and his collaborators carried out in each system to create those representations, presences, meanings, and values. These stories have been as various as the hints of backroom deals in politics and payoffs to journalists, the formal filing of patent applications, the cultural aesthetics of interior decoration, and the terse activities of the laboratory. Each chapter places Edison's communications within his immediate communicative needs and situations, framed within histories of Edison and of the larger systems Edison participated in. To understand these distinctive forms of meaning making, I found it useful to invoke everything from the legal concept of property to the sociological concept of charisma, everything from the activity theory of collaborative work to the impact of urbanization on journalism, and everything from the theory of speech acts to the theory of the leisure class. And I invoked phenomena as disparate as steam presses, world's fairs, department stores, the organization of the patent office, the stock markets, and the physical layout of Menlo Park.

Communication, as we have seen, is heterogeneous. The communication in each setting is responsive to and relies on myriad dynamic shaping forces. Edison Incorporated had to speak at each moment to complex dynamics and heterogeneous shaping in each unfolding setting.

Because each unique communicative configuration to which Edison spoke was located within an evolving communicative system, the incidents described in this book help us understand how larger social systems form the locale of immediate improvisatory action. Edison's newspaper interviews are remarkably novel and successful communications, but only because they took advantage of the possibilities offered by the press of the moment and Edison's developing public persona. Those interviews tell us something about the newspapers then and now, something about Edison as an emerging public figure, and something about Edison's acting at a specific time and place for immediate purposes. Edison's patents and patent litigation tell us about the patent system, about the emerging ownership of an emerging technology, and about Edison's pursuit of his interest through the system and his representations of the technology within the system.

In this way, the various chapters in this book display the evolution, structure, and current state circa 1880 of some of the most influential and powerful discursive systems in late-nineteenth-century America: the law and the courts, the newspapers, urban politics, class distinction, cultural events, consumption, finances, industry, corporations, and technology. We also develop a picture of a society as an aggregation of these systems in relation to one another, as news reports and financial dealings affect one another, as technological reports enter the press and boardrooms, as patent actions affect the stock markets, and as urban politics becomes embroiled with building technological infrastructure for commerce.

And the chapters also tell us something of Edison's history, illuminating dimensions only peripherally touched on in other kinds of accounts—his skills as a promoter, his knowledge of the press, his managerial talents and social organizational imagination, his political dealings and payoffs to journalists, his travails with backers, his spectacular displays. In the biographical literature on Edison, these are touched on as sidelights to his accomplishments rather than as major occupations that go hand in hand with the business of inventing. I hope this book makes it clear that these activities are part of what allowed Edison to create his accomplishments.

Beyond the particulars of Edison and the communicative systems he participated in, however, I hope the book raises more general issues about how communicative events unfold within communicative systems and how individuals participate in those events. I will now make those issues explicit, placing them within rhetorical theory, studies of science and technology, and social theory.

Heterogeneous Symbolic Engineering

For any technology to succeed (that is, to establish an enduring place within the world of human activities), it must not only succeed materially (that is, produce specified and reliably repeatable transformations of matter and energy); it must also succeed symbolically (that is, adopt significant and stable meanings within germane discourse systems in which the technology is identified, given value, and made the object of human attention and action).

New technologies are a matter of heterogeneous engineering—that is, the coordination and application of many kinds of knowledge and practice, all of which are united and instantiated in the final product (Bijker et al. 1987; see also Bijker 1995). Many of these knowledges and practices concern the social worlds that produce, support, and use the technologies as much as they concern the material world out of whose scraps and shards the technology is constructed. Those social worlds are worlds of transactions, most of which are symbolic, and even the material transactions—for example, cooking dinner for others and then sharing it with them, or firing up dynamos and delivering the consequent electricity into another's house—are surrounded by facilitating and meaning-attributing discourses.

Thus, it is useful to extend the notion of heterogeneous engineering to encompass symbolic engineering; that is, the development of symbols that will give presence, meaning, and value to a technological object or process within a discursive system. Moreover, as we have seen in the Edison story, symbols have to be created and managed in a number of different communicative systems, each with its own dynamics and considerations; thus, in building enduring meaning and values for the technology they wish to implant in our daily lives, the creators of that technology must engage in heterogeneous symbolic engineering.

Systems of Discourse Circulation

We can understand some of the principles of this heterogeneous symbolic engineering by considering how symbols are circulated. Discourses circulate within groups. Speech goes from one individual's mouth to another's ears. Texts go from hand to hand. The systems of discourse circulation define the realms in which symbolic forms exist and take on active life and meaning. They also form the realms within which representations surrounding a new technology, product, or concept must take on presence in order for the novelty to become part of some relevant human world—relevance being tautologically defined by the ability of the symbolic representation to be absorbed into the discourse stream. In a simple society, the social distribution of the language necessary to embed a technology in the human worlds may be relatively unarticulated, with the representation being uniformly accessible and acceptable to most individuals. Some simple technologies, such as using a rock to smash open clams, may make so much sense visually and kinetically for the common tasks of primates that they may be spread through the public observation of publicly performed tasks—that is, the technological process represented by itself without additional symbolic mediators. However, even with technologies that are so easy to understand and replicate it is not difficult to imagine that decreased grunting, increased sounds of pleasure, evaluative and ostensive comments ("Easy. Good. Do it this way") and requests ("Smash me a clam") will make the value of the technology evident and facilitate its spread. Moreover, improvement of stone tools will be facilitated by a language of tool evaluation, design, and production.

The movement from machine shops to Menlo Park and then to multiple central power stations is accompanied by increasingly complex involvement with extensive discursive networks necessary for the production, maintenance, and use of the technology: laboratory notebooks, installation and repair manuals, patents, monthly bills for service, publicity pamphlets, technical journals, city ordinances, and stock reports. The place of the technology must be established and maintained in many networks.

As technologies and societies become more complex, so do the symbolic accompaniments to the material technology. The technologies are tied into less obvious meaning systems for their development, appreciation, production, funding, operation, maintenance, social control, evaluation, and distribution. Moreover, these (and possibly other) functions are likely to be distributed among different groupings in the society, requiring differential distribution of representations to various social components that may or may not overlap. Papers must be filed with financial backers, government regulators, technical R&D departments, sales forces, materials suppliers, production machinery producers, and shop floor designers.

Furthermore, insofar as frequent experience with new technologies has led to highly regularized practices surrounding each of these functions, patterned, conventionalized, and generically recognizable discourses are likely to circulate in each area; the technology will have to establish adequate symbolic place within each relevant set of discourse conventions as those conventions are distributed among the relevant parties. It must achieve a patent. It must produce a corporate plan. It must generate an advertising strategy and appealing advertisements.

Differentiated Values

Not only do the kinds of documents and networks of discourse circulation become complex and differentiated; so do the values one needs to create in each arena. Easiest to perceive are the values of the stock market, as they are quantified and regularly traded as quantities; accordingly, we found the main tropes of value, as expressed in stock reports and related documents, to be barely removed from quantities: solvency, profit-making, dividend paying. All other representational and argumentative tropes were arrayed around these central values. The patent system, as a legal system creating property, finds its central expression of value as the patent grant, represented in a valid patent number. Within the patent system and the courts, all the representational argumentative work are legally framed discourses concerning the making, unmaking, or distribution of this valued property. This patent number, then, becomes valuable within the commercial and financial world as a sign of a uniquely valuable property realizable in a commodity or in the ability to profit from that commodity. In the financial and commercial markets, gas lighting was already established as a valuable and profitable commodity, so incandescent lighting and central power "piggybacked" on that established value.

Within the small group of elite financial backers that Edison drew his initial capital from, personal trust based on personal evaluation over years of contracted work established Edison's value as an investment gamble. In the Edison laboratories and companies, the charismatic faith that Edison elicited was the key value that elicited the work, the commitment, and the relationships that kept the project moving forward with energy and attention. This charisma, despite Edison's attempt to create organizations that would maintain its force, was to be replaced by ordinary institutional corporate values.

In the laboratory, the key consideration, the key value organizing all the work and representations, was a materially working technology; a secondary value was lowered cost. Both were necessary to fulfill the meaning and value positions in commercial and financial markets.

In public forums, the values of credibility are mobilized through the journalistic values of celebrity (with which Edison actively cooperated) and spectacle (which Edison and colleagues created). The ongoing tensions between what one could claim (and what resources one could draw on) as a celebrity and what one could produce as material spectacle (and then as a mundanely deliverable material technology) framed the entire period of development of the incandescent light, and they lay behind Edison's (and his colleagues') repeated manipulations of public representations through payoffs to journalists and deceptions. A similar but more complex tension existed within the expanding technical world, which, in its growing professionalism, found the discrediting of public celebrities and the establishment of professional credibility increasingly important. Edison's inability to establish solid credibility as a scientist and his suspect identities as a public celebrity, a mechanic, and an inventor led to professional resistance to accepting the value of his claims until the claims became backed up by material proofs carried out according to accepted professional standards by recognized credentialed professionals. Again we can see in Edison's sponsorship of certain scientists and their use as privileged testers, preferred judges, and professional advocates his awareness of the importance of establishing value in the spheres of official scientific knowledge.

In bringing lighting into the cities, Edison had to cooperate with the overt political values of economic and civic development and with the covert political values of patronage and individual advantage. In bringing light into the home, Edison had to align with and even advance the transforming values of family life—values driven by the new social positioning of upwardly mobile families within new urban economies.

The stunningly heterogeneous values that were constructed on this journey through value-constructing realms would have been hard to predict beforehand. Any enduring technology, or any other enduring aspect of human culture in the complex world of modern differentiated society, it seems, would have to establish itself within the values of a similar range of regimes. A movement for environmental consciousness and action, a pop band, a biotech therapy for Alzheimer's Disease, or a new-age religion must find terms on which to deal with the multiple discursive realms on which its survival depends.

Satisfactory Representations in Specific Discursive Systems

Because each communicative project intersects with different sets of communicative systems at different moments of the project's unfolding, each finding a different point and angle of intersection to pursue its own interests and trajectory, rhetoric cannot be a closed and fixed techne offering standard advice about a limited set of practices. Rather, the tools for analyzing and offering advice about communicative actions must be sensitive to the full range of forces, forms, and systems that influence each particular instance, and to the position and interests of each participant. If newspapers in the nineteenth century were responsive to economics, urbanization, printing technology, and news-gathering practices, a rhetoric for journalism must take such things into account. If the forms of patents are consequences of particular political, economic, regulatory, and legal arrangements, any rhetoric that would help guide one through the thickets of patent arguments must attend to those political, economic, regulatory, and legal arrangements. If charisma is a force within corporate life, rhetorical theory must be open enough to attend to charisma where it is relevant. This necessary openness and heterogeneity provides a challenge to rhetorical theory, which historically has tended toward unified, homogeneous accounts and advice based on particular social circumstances that have been taken to be more general than they are. (See Bazerman and Russell 1994.)

As a technology on its way to successful integration finds its satisfactory representation within each system, it takes on system-appropriate meanings and loses other meanings which are not germane to the discourse. In this way, the technology becomes something different in each representational realm: a legal entity, a financial entity, an end product of a production process, a series of costs, a social desirable, and so on. If it fails to take on any essential identity, its presence as a potentially successful technology is to that extent weakened, perhaps with negative consequences for the presence of the technology. Consider "orphan" drugs that have valuable meanings to government, chemists, production people, and consumers, whereas financial managers and accountants do not find the right numbers on the projected balance sheet. In this case, symbolic failure with one essential constituency has led to a proliferation of additional discourses drawing on new symbolic and rhetorical resources to make up for that failure—appeals to Congress for legislation based on social equity, appeals to private charities and the public based on sympathy, or appeals to pharmaceutical industry leaders under threat of regulation or negative publicity. In cases where just the one symbolic variable (in this case projected profitability) is different, large chunks of the symbolic profile surrounding the technology are changed.

Material Accountability

The success of representations occurs in particular circumstances and, often, in systems held directly accountable to manifestations of the material technology, as when consumers refuse to write checks to pay their bills when their lights do not turn on or when investors withdraw their funds upon perceiving that an inventor cannot produce a working prototype. Though some discourse-circulation systems can sustain a projected but unrealized technology for a period of time during development, even here material tokens are often needed as signs of faith—literally to maintain the faith in the ultimate production of a valuable material practice. Ultimately, however, the technology must be physically realized if it is to maintain discursive value, unless very strong factors hold together patient audiences waiting for an extraordinary appearance. Even cargo cults can fade when the cargo does not arrive.

A Rhetoric of Operations

Within action systems, the appropriate representations of technologies are active operants. People do things (including making their own representations) upon receiving the representations of others. That is, the formulation of a successful representation accepted as such by the relevant grouping within which the representation circulates becomes the basis for further symbolic and/or material action, within that discourse network or beyond. A patent application leads to a grant of a patent, which then sets the stage for further financial, corporate, and legal moves. A successful billing procedure results in cash receipts. An application to tear up the streets in order to install delivery systems for a new mode of power leads to the purchase of pickaxes and the employment of day laborers.

Each act of discourse is a move in a complex world of social and material activity. The texts and the spoken words are behaviors in unfolding dramas of moves, countermoves, and consequent moves. Many of these actions remain in the realm of symbols, but many have material consequences. Even imagined technologies that never make it to prototype still may cause trees to be cut down, inks to be trucked across country, and people to travel up and down elevators to attend lunch meetings. The aim, of course, is for the technology to become materially present, visible, and useful—to produce better and more enjoyable conditions of life for consumers, producers, and investors.

Persuasion as an Influence on Consequent Actions

Discourse, as conceived in this orientation of actions and networks, operationalizes the concept of persuasion. Persuasion, rather than being perceived only as a mental event happening invisibly within the recipient's consciousness, is also to be found in consequent actions to see whether and in what ways individuals have been influenced by receipt of the symbolic message. Persuasion then is a desired influence on the action of relevant parties within discourse networks. This definition extends persuasion beyond its usual associations of civic decision making to encompass the entire range of actions occurring across all discourse networks, while still recognizing the different kinds of influence and desired actions resulting from persuasion.

This definition also removes the confrontational and competitive aspects that seem deeply ingrained in most rhetorical concepts of persuasion. Although in some discourses influence may be exercised through confrontation, contention, and competition, in others cooperation, recognition, assent, accretion, or digression may be effective. With respect to the introduction of technologies, influence is needed to establish the place of the technology within the varied practices required for the acceptance, maintenance, and use of the technology within segments of the society. Financiers must be influenced to invest in stock offerings; patent examiners must be influenced to approve applications; reporters must be influenced to write positive stories; newspaper readers must be influenced to recognize Edison's accomplishment and authority; machinists must be influenced to produce a part according to new specifications; lab assistants must be influenced to coordinate their work with one another and with Edison. Though some of these may seem to offer instances of classic persuasive argumentation, others seem hardly contestatory. When examined in speech-act terms, even court and patent proceedings (which seem closest to canons of classical argumentation) reveal that their success depends more on the meeting of felicity conditions that perfect the speech acts of the patent process than on the mounting of powerful arguments.

Similarly, though in classical rhetorical terms one may say that the newspapers were a vehicle for establishing Edison's ethos, that hardly seems to capture the subtlety of the process by which credibility was created or the ways in which it came to bear on future discourses.

Extending the idea of discursive influence, we may view persuasion in light of the social and symbolic task to be carried out by the discourse. The other people involved in the discourse are appropriately influenced if they consequently act in a manner that respects the completion of the prior act. That is, competitors are influenced by a patent if their next design takes the patent as a constraint—as a social fact that they must respect. In some cases, this influence through the creation of social facts may be achieved through the affective or cognitive "heavy lifting" that is traditionally associated with argumentation. But sometimes this creation of social facts may simply be accomplished by filing the proper papers on time, or by transmitting a series of plans and instructions in a timely way. The criteria for successful utterance do not then come from general rhetorical rules for argumentation, but from analyses of the tasks to be accomplished by the utterance and the conditions that must be met if the actions are to be accomplished.

Conditions of Accomplishment

The above analysis is of the sort that Searle (1969) worked out for some of the more institutionalized, the more regularized, and the simpler linguistic activities of our society. However, the conditions that must be fulfilled in any case may be far more complex and far more local than the general guidelines that Searle established in abstracted philosophical terms for what he considered prototypical activities. For example, in the highly institutionalized setting of medical insurance it may be reasonably predictable what you must do and what conditions you must meet if your claim is to be filed appropriately, to be accepted, and to result in reimbursement. But these requirements are far more specific to the current insurance system and to the regulations and forms of your particular insurance company than would be suggested by Searle's general criteria for a successful request.

Furthermore, the environment for actions is not always easy to know and is not always institutionally regulated. We may not be aware of all the conditions that an utterance must meet. Perhaps you do not know all the arcana of your insurance company's procedures (which are in theory knowable). Perhaps its procedures are more haphazard than you suppose, and that this causes some variability in how claims are received and processed, thereby creating some randomness in regard to the chance that your claim will succeed. In such cases the mood, dispositions, or histories of the individual examiner may be relevant. If you knew your examiner and all the baggage she brought into the office that day, you might have a better probabilistic idea of what conditions your claim might have to meet to gain approval. Or perhaps a recent and unanticipated event, such as a scandal involving your doctor, has led to a change in the local procedures of examination.

If we step outside the most typified of discursive actions, the novelty or perceived novelty of our utterance acts as a catalyst for the formation of new reactions. Thus, there is no way we can know the conditions of success ahead of time, for they are revealed in the novel response to the utterance. Even in the regularized and institutionalized discourse of a court case, new arguments may be made on either side, new rulings may be made, and the judge may write an opinion establishing new legal precedents that reinterpret the meaning of all that has gone before and set new terms for all that will come after. This is the way in which rhetoric is constitutive and creative, as new actions evolve and gain social meaning in their consequences.

Symbolic Integration and the Process of Enlistment

The success of representations that are necessary for the social embedding of a new technology involves the process of enlistment. To Latour (1987, 1988), the enlistment of allies is a part of the process of making a finding or a technology strong by bringing more resources to its team. However, the specification suggested here goes beyond the general idea of an amassing of forces to consider the specific kinds of connections that must be established in each case, the particular symbolic means and discourses by which these enlistments or alliances can be made, the conventionalized social regularities to which each connection must speak, the operational force and operational mechanisms by which connections are made, and the ways in which the connections may be understood both conceptually and operationally by the participants. It also gives us means for considering how programs, concepts, or technologies become transformed as they move from forum to forum and for evaluating to what extent the various representations and the entire project remain coherent. The approach I present here specifies the discourse systems, the discourse practices, and the specific character of the representations that give some entity standing within each network. Further, this approach provides a view of all the networks within which an entity must gain standing if it is to attain social stability, and of how the various moves of influence hold together or become transformed in establishing the extensive complex of alliances. Thus, the approach I advocate provides a "structurationist" (Giddens 1984) account that allows us to go beyond the individualistic ad hoc networks suggested by Latour. Even as social arrangements evolve, actors operate within perceivable social orders to which they must attend.

Moreover, my approach draws attention to a wider array of players involved in the appearance, the maintenance, and the reproduction of technologies, not only to the overtly interested actors pursuing overt goals that the Latourian system highlights. Marginal participants (patent examiners and trained installers, for example) may have only a bureaucratic procedural stake, or no stake at all, until they are hired into a job. Influencing them into the appropriate action through proper representations, nonetheless, is as significant for the success of the technology as is influencing capitalists out to make a lot of money.

By focusing attention on situated, symbolic, discursive practices, the approach presented here moves us closer to the kind of micro-empirical study that is associated with ethnomethodology, conversational analysis, and sociolinguistics, allowing us to locate and examine the exact sites of social production and reproduction in particular discursive moments. It allows us to see in specific communicative moments the contextually constraining, scene-setting regularities of practice and expectations that produce and maintain social structure, as we perceive it from moment to moment. This approach also redirects discourse studies of technology out of the realm of disembodied persuasion and into the realm of sociomaterial practice, in line with the work of David Gooding (1990).

Representational Resting Points

At the micro-interactional level, where representations are accepted as adequate for the purpose at hand and speech acts are taken as accomplished, we see the formation of something like discursive black boxes as they empirically occur in moment-by-moment transactions. Originally from engineering, the term "black box," as it has been elaborated in science studies, refers to a technology or a concept that is taken as a stable and closed whole despite internal complexity or instability or a contended history. Once a concept or a technology is black-boxed, uncertainties surrounding it are put to rest and thereby forgotten, and the concept or technology is taken as a whole and unquestioned accomplishment.

In science and technology studies, the term "black box" has usually been applied only to accomplished scientific facts or accepted technologies. However, something similar to black-boxing occurs with representations and speech acts, which may at first appear unstable but then gain fixity and stability (say, in the form of a patent grant or a financing agreement for a now-solvent corporate entity), at least until some event (such as a patent suit or an extreme financial setback instigating bankruptcy) threatens the integrity of the patent or corporate entity. Until such a rupture occurs, the completed act and the various representations embodied in the act are taken as fixed. Thus, a stable patent fixes the property, but it also fixes a description of the innovation and the identity of the innovator. That is, a successful speech act, accepted as a social fact by others who respect the integrity of the speech act, fixes a social understanding of events and of the representations that are promulgated as part of the speech event.

These stable speech acts and the accompanying representations serve as resting points at which the interpretive action frameworks of various parties may meet. All parties, from their separate perspectives, can agree to a discursive resting point as a good-enough shared speech event that no longer is worth the trouble of contention. Thus, a patent case may end with no party fully satisfied but with each seeing in the decision as much as that party is likely to get, and therefore it is profitable to go forward on the basis of the agreed-upon representations. In the instance of the patent contention, the governmental authority of the patent office and the courts provide great forces for bringing closure and giving definitive power. (In the extreme case, if a dispute were to go all the way to the United States Supreme Court and one party were to remain unhappy with the result, gaining a constitutional amendment or creating a successful revolution would present a very high cost for destabilizing the discursive formation.) Even in less authoritative situations, such as an agreement over where a factory should be sited or a consumer's decision on where to spend his or her money, a stable representation then closes up former questions and becomes a ground for further action. Product loyalty, for example, has the consequence that one no longer worries about claims and counterclaims of the benefits of the contents of different boxes; one always reaches for (e.g.) the yellow box with the big red letters. If, however, the detergent in the yellow box with big red letters stops making the consumer's laundry clean, or the price doubles relative to that of a competitive detergent, he may want to go back to paying attention to advertisements, reading labels and price tags, and checking out consumer reports, thus reopening the consumer choice that had achieved discursive simplicity in a brand name.

It is, of course, in some people's interests to contest speech acts by undermining the completion or adequacy of any of the conditions and forms that go into their perfection, and to contest representations that accompany speech acts by attempting to hold the representations accountable to standards perceived as relevant by those whose judgment affects the social reality of the speech acts. That is, one can attempt to undo what has been done discursively. But if the undoing fails, or if its costs are too high, one has to recognize the social force of the speech acts and representations even if one personally refuses to accept their validity. For example, if I have failed to undermine the sales and the authority of a newspaper which I believe is illegitimate because it does not uphold journalistic standards and instead serves as the mouthpiece of malign interests, I nonetheless would be well advised to recognize that each day's issue of that paper will stand for most people as an adequate report of the news, and that the accounts of events represented in the paper will be accepted by most people as factual, even though I personally don't believe a word of it and continue to see most people as deluded.

A discursive resting point becomes a basis for common understandings. We agree to accept a contract or a patent. Most people are willing to accept as adequate for all practical purposes the claim that the United States landed people on the moon in 1969. We assent to these social facts and rely on most other people assenting to them.

Interpretive Variability

That one assents to a speech act or a representation as the resting point for social understanding does not mean that one understands, values, or uses the socially accepted representation in the same way that another individual does. To its holder, a patent grant is a potential source of profit. To a competing producer, it is a cost to be minimized or an obstacle to be gotten around by means of a substitute invention. To its holder, the patent still expands to the largest imaginable limits; to the competitor, it shrinks to the point where it is incontestable. To the consumer, it becomes a mark of uniqueness. The meaning of any representation is always open to personal and public interpretation.

Nonetheless, it is usually in most of our interests most of the time to keep most speech acts and representations fairly tightly constrained in their meanings, so that we can use them as the bases for mutual actions and agreements. Thus, mutually acceptable speech acts and representations have a conservative tendency—the more so the more widely they are shared. They make up the cognitive institutions of our society, and thus they are the grounds upon which we maintain the appearance of inter-subjectivity.

The stakes in these cognitive institutions are especially high for those favored by a representation, so they will struggle to maintain a favorable representation. A corporation, being both a legal and a fiscal entity, has among its primary goals the maintenance of its solvency and legal integrity. Without fiscal integrity, the corporation's legal status dissolves; and once the legal status dissolves, all the corporation's functions are dis-aggregated into separate actions, no longer held together and accom-plished under the corporate status. A product that has brand loyalty wants to maintain it, so that it will not have to freshly influence each purchaser at each point of sale. In this way, the brand-name product avoids constant detailed competition over price, quality, and convenience.

The discursive resting point, then, offers a starting point for new actions, because one can rely on common acceptance of social facts upon which to build new actions. One can seek new patents, negotiate new con-tracts, or attempt to increase sales, relying on a discursive world that can, in many respects, be taken for granted.

Structurationism and Social Learning

This structurationist account of individual discursive events respects enduring social understandings that provide a basis and a location for each communicative interaction and also respects our individual agencies and cognitive activities as we assert our selves, our desires, and our needs within the worlds that others structure around us by their actions. Each of our speech acts helps to establish and maintain structured discursive realms that others can orient toward in framing their own discourses as social actions.

In Vygotskian terms, in our interactions, as we orient toward one another's communications, we create discursive environments which serve as zones of communicative challenge and social learning for one another (Vygotsky 1978). That is, in desiring to communicate with one another, we orient toward one another's discursive universes and figure out what kind of utterance will be both intelligible and successful in accomplishing our ends—what kind of speech acts we can accomplish and how to go about accomplishing them. This means learning what others will find intelligible and then learning how we can frame our own inten-tions within what is assertable. In accomplishing speech acts and creating representations which others make sense of, we then fill out some of the social landscape which others will, in turn, speak to. In interacting with those who we can relate to and those who can relate to us, we keep entering into zones of mutual development—building both our own individual discursive skills and cognitive realms and our social worlds of intersubjective participation. We keep scaffolding the social, discursive world around one another.

The recognizable zones of interaction we keep creating in one another's presence form the social locales within which situated action takes place. A court or a seminar acts as a locale for interaction insofar as participants recognize and orient to the discursive space. More particularly, the issues at stake; the roles, attitudes, and resources of the various participants; the arguments on the table; and all the other elements that might be counted as the specifics of any rhetorical situation take shape insofar as they are asserted and recognized within the discursive forum. The situations in which situated action takes place are themselves social, discursive formations. Even the material components and constraints are made immediately relevant by our discursive recognition of or adjustment to their relevance.

Mutual scaffolding of the discursive landscape, then, provides a framework for conjoint activity. Edison, by enlisting others in his project in various ways through the successful completion of many symbolic and material actions, not only provided a focus for mutual attention; he also provided frameworks for mutual cooperation. His visions provided material for collaboration with news reporters looking for stories which people would be interested in; thus, Edison and the reporters together created interviews which generated news to fill newspapers. Similarly, Edison had to create domains of socially ordered discursive cooperation with others in order to create and maintain a laboratory, in order to gain and maintain financial backing, and in order to get other electricians to attend to and validate his inventive accomplishment.

Each of these complex, discourse-mediated social systems brought individuals from different positions into a shared project and distributed roles and possibilities for each, allowing them to bring their various resources to bear on the project at hand. For example, the unschooled Edison had the wherewithal and the savvy to hire men who held Ph.D. degrees in physics and chemistry to aid his work, just as he hired expert machinists, enlisted financiers with investment capital, and retained lawyers who had knowledge of the patent system and the courts. Common understandings, conjoint projects, social structures, accomplished social acts, and shared representations allowed Edison to turn himself into Edison Incorporated.

Ultimately, many less prominent individuals who were members of a corporate communicative network displaced the charismatic Edison from General Electric. A large social institution held together by situated, structured communication became more influential on discourse than the discourse of any individual. The corporation became the playing field upon which individuals constructed their lives and their careers.

The Individual and Society

Niklas Luhmann (1989) points out that society is not a group of individuals; rather, society is what passes between individuals. To engage in social action, each individual must take part in society through communication. The individual must project him- or herself, or his or her projects, onto a social stage—must enter his or her concerns into the discursively enacted concerns of society. Edison, by writing in his notebook, projected meanings into his laboratory and into the patent system. In carrying out demonstrations, either through chicanery or through honest accomplishments, Edison was projecting meanings out into the public, the financial markets, and the technological professions. Similarly, in talking to reporters, Edison projected himself into personal relationships with those reporters and into the virtual communicative space in each reader's newspaper.

Edison projected into various social spaces the possibility of a technology of power and incandescent light, making it intelligible in terms of prior communications concerning his own accomplishments and wizardry, prior gas lighting technologies, prior electrical accomplishments, and existing systems of property, patents, news, finances, politics, technological communication, and domestic life. He projected existing and possible meanings and values onto local and distant communicative forums. Perhaps most interesting is the way he projected a technology that had been developed in the male world of meanings and values into the domestic sphere, where values and aesthetics were constructed as female (though framed by male hierarchies of status and class).

Electrification as Revolutionary and Conservative

In retrospect, the new technology of electrical light (and electrification in general) brought about major ruptures in the way of life, creating new categories of thought and interaction, new organizational entities, new governmental procedures and bodies, new daily schedules, new products, and new dangers. Not only did it change meanings within existing discourses (see, e.g., Schivelbusch 1988); it also created new discourse networks with new sets of meanings. (For example, electrical engineering became a significant professional and educational network in the United States only with the generation of electricity, which brought with it all the other forms of applied electronics.)

Despite the great changes that came in the wake of the new technology, incandescent light and power had first to be built on historical continuities of meaning and value. It had to take a place within the discourse and the representational meaning systems of the time before it could transform them. Electrical light had to find representational terms that could be comprehended before it could create its own new world of experience and meaning.

Notes

Chapter 1

1. A patent and literature search for Edison, conducted by Francis Upton in November and December of 1878, turned up more than 200 patents (domestic and foreign) and articles relevant to Edison's project (Francis R. Upton, Literature Search Notebooks 1 and 2, Edison papers, microfilm edition, reel 95: frames 314–495). Edison seems to have been aware of the multiplicity of patents from the beginning, and to have soon informed himself in detail on the situation.

2. For example, on December 11, 1875, while Edison was away from his Newark workshop, his associate Charles Batchelor conducted some brief experiments with arc lighting in connection with the etheric force (PTAE 2: 688). Further experiments with sparks in January and February of 1877 are mentioned in Edison's notebooks (PTAE 3: 213–225, 245–247). In September of 1877 some of those experiments turned to preliminary explorations of incandescence (PTAE 3: 540–542, 546–548, 612–613, 651–652).

3. "Invention's Big Triumph," *Sun*, September 10, 1878; "A Great Triumph," *Mail*, September 10, 1878 (94: 349–351).

4. "Caveat for Electric Light Spirals," Menlo Park Notebook 16, beginning on p. 91, September 9, 1878 (4: 488). These designs were transcribed to Edison's notebook "Experimental Researches" beginning on September 27, 1878 (3: 376–380).

5. Telegram, TAE to William Wallace, September 13, 1878 (17: 925).

6. "Edison's Newest Marvel," *Sun*, September 16, 1878 (94: 354).

7. Private Communication, Keith Nier, September 29, 1997.

8. The correspondence examined in this chapter comes from the document file of the Edison papers, which consists largely of letters and cables written to Edison and only occasional indications of Edison's response. Almost all the documents in the file on the electrical light 1878 are handwritten.

9. Edison had had some preliminary correspondence on the incandescent light in the spring 1878, but the discussions had been general, and no experiments had

been done on what Edison recognized as "one or two obstacles" to the development of the light. See letter, E. Tillotson to TAE, April 30, 1978 (17: 916–917); letter, Charles Stovell to TAE, May 31, 1878 (17: 918–919); note of reply, TAE, June 30, 1878 (17: 916–917).

10. Letter, George Barker to Stockton Griffin, September 5, 1878 (17: 922–923); telegram, TAE to George Barker, September 6, 1878 (17: 924).

11. Letter, George Barker to TAE, September 16, 1878 (17: 927).

12. Letter, George Barker to TAE, October 23, 1878 (17: 979–981).

13. Letter, George Barker to TAE, November 10, 1878(17: 1031).

14. Letter, George Barker to TAE, November 22, 1878 (17: 1048).

15. On the British background of public demonstrations, see Simon Schaffer, "Natural History and Public Spectacle in the Eighteenth Century," *History of Science* 21 (1983): 1–43. Ebenezer Kinnersley (from Philadelphia, like Barker) was the most prominent public lecture demonstrator of electricity in the colonial period (Lemay 1964).

16. On the still-unsettled role of university professor of science at this time, see pp. 99–142 of Guralnick 1979.

17. Letter, George Barker to TAE, October 23, 1878 (17: 981).

18. Letter, Henry Morton to TAE, October 9, 1878 (17: 940); letter, John A. Blattau to TAE, October 22, 1878 (17: 976); letter, Professor Suley to TAE, November 21, 1878 (17: 1045–1047); letter, David Olegar to TAE, November 22, 1878 (17: 1051).

19. Letter, A. B. Williams to TAE, September 16, 1878 (17: 929–934).

20. Letter, Presbrey and Green to TAE, October 17, 1878 (17: 954); letter, A. H. Best to TAE, October 17 1878 (17: 955); letter, Charles S. Wells to TAE, October 26, 1878 (17: 1001–1002); letter, Willis Knickerbocker to TAE, October 28, 1878 (17: 1003–1004); letter, A. G. Nye to TAE, November 18, 1878 (17: 1040); telegram, TAE to Charles Siemens, December 3, 1878 (17: 1070); letter, Charles Siemens to TAE, December 4, 1878 (17: 1071).

21. Letter, Hugh Craig to TAE, September 23, 1878 (17: 935).

22. Letter, J. G. Kidder to TAE, October 15, 1878 (17: 949); letter, George Walker to TAE, October 16, 1878 (17: 951).

23. Letter, W. C. Miller to TAE, October 18, 1878 (17: 967–968).

24. Letter, Gerritt Smith to TAE, October 19, 1878 (17: 982).

25. Letter, Margaret Marsselis to TAE, November 5, 1878 (17: 1016–1017).

26. Letter, J. H. Pearce to TAE, November 9, 1878 (17: 1029–1030); letter, A. H. Best to TAE, October 17, 1878(17: 955); letter, W. R. Williams to TAE, December

22, 1878 (17: 1106), letter, William Morehouse to TAE, October 18, 1878(17: 959); letter, Frank W. Robinson to TAE, October 29, 1878 (17: 1008–1009).

27. Letter, Clinton Ball to TAE, October 23, 1878 (17: 986–991). A similar letter requesting employment at Menlo Park is H. S. Woodbury to TAE, November 15, 1878 (17: 1041–1043).

28. Letter, George W. Griffiths to TAE, October 24, 1878 (17: 992); follow-up letter, R. S. Scowder to TAE, November 9, 1878 (17: 1028).

29. Letter, A. J. Bryant to TAE, November 11, 1878 (17: 1034).

30. Letter, A. L. Fleury to TAE, December 23, 1878 (17: 1107–1109).

31. Letter, Henry Plimsoll to TAE, October 21, 1878 (17: 975).

32. See e.g. an adulatory letter from a British military widow offering Edison and his family a place to stay when they visit England: E. Shakespear to TAE, October 18, 1878 (17: 960–962).

33. On the rise of celebrity heroes in American mass culture, see Lowenthal 1984 and Greene 1970. On Edison's celebrity, see Wachhorst 1984.

34. Letter, George Barker to TAE, December 14, 1878 (17: 1081).

Chapter 2

1. The works of Michael Schudson, in particular *Discovering the News* (1978) and *The Power of the News* (1995), provide the most comprehensive and penetrating overview of the history of journalism in the United States. See also American Antiquarian Society 1991, Kobre 1969, and Mott 1950. On the history of the New York press, see Churchill 1958 and Wingate 1875. N.B.: Most of the articles considered in this chapter were collected in several scrapbooks in the Edison papers: the Batchelor Scrapbooks, the Menlo Park Scrapbooks, and the Miscellaneous Scrapbooks. When an article is available in the microfilm edition of the Edison papers, I will so indicate with a reel and frame number. I have, however, supplemented these with examinations of other newspaper archives.

2. The afternoon edition of the *Detroit Free Press*, which would have been the one Edison sold on his way back to Port Huron, ran this story under the headline "The Great Battle on the Tennessee." The early estimates of casualties in the story ranged from 53,000 to 60,000. The report was reprinted in the back of the next morning's edition.

3. See Edison's autobiographical notes, collected by Frank Lewis Dyer in 1908 and reprinted in PTAE 1: 627–646.

4. See e.g. "A Wonderful Invention: Thomas A. Edison's Startling Discovery," *Newark Morning Ledger*, November 29, 1875 (27: 282); "New Era in Science," *New York Daily Witness*, November 29, 1875 (27: 282); "Novel Electricity: The Remarkable Experiments of a Newark Inventor," *New York World*, November 30,

1875 (27: 282); "New Discovery in Electricity; Experiments of T. A. Edison," *New York Tribune,* November 30, 1875 (27: 282). See also "Etheric Force. Mr. Edison's discovery of a Supposed New Natural Force," *New York Herald,* December 2, 1875 (PTAE 2: 668).

5. Edison himself presented his claims. See TAE,"To the Editor of Scientific American," *Scientific American,* January 1, 1876 (letter responding to "The Discovery of Another Form of Electricity," *Scientific American,* December 25, 1875).

6. "Letter to the Editor," *Scientific American,* November 17, 1877 (94: 81).

7. "A Wonderful Invention," *Scientific American,* November 17, 1877 (94: 81).

8. "Is there a Law of Invention?" *Scientific American,* December 8, 1877 (94: 90).

9. "The Talking Phonograph," *Scientific American,* December 22, 1877 (94: 92).

10. See e.g. "The Telephone Outstripped," *North Wales Express,* November 31, 1877 (94: 93); "Phonograph," *The English Mechanic,* January 4, 1878 (94: 97); "The Phonograph," *Manchester Guardian,* January 18 (94: 102); "Speech Bottled Up," *New York Tribune,* December 26, 1877 (94: 93); "A Marvelous Invention," *Sun,* January 2, 1878 (94: 95).

11. "Transmitting Speech and Song," *New York Tribune,* January 18, 1878 (94: 100); "Preserving Sound," *New York Herald,* January 18, 1878 (94: 100); "Programme," *New York Herald,* January 18, 1878 (94: 114); "Phonograph and Telephone," *New York Herald,* February 22, 1878 (94: 114).

12. "Mr. Edison's Inventions. Hundreds of Curious Devices the Product of a Kaleidoscopic Brain," *New York World,* January 12, 1878 (27: 734; 94: 99).

13. "The Papa of the Phonograph: An Afternoon With Edison, The Inventor of the Talking Machine," *Graphic,* April 2, 1878 (94: 151); "Four Hours With Edison," *Sun,* August 29, 1878 (94: 339); "An Evening with Edison," *New York Times,* June 4, 1878 (94: 237); "Edison 'At Home.' A Ninety Minutes Interview by Telephone," *Philadelphia Record,* June 5, 1878 (94: 222). Another indicator of how vigorously Edison was sought by the press and how welcoming Edison was is the extensiveness of the articles folders of the Edison papers, carrying all the correspondence between journalists and Edison. In the microfilm edition, the first article folder in 1877 takes up fewer than 20 frames (14: 246–264); the 1878 folder is a dozen times that size (17: 1–274). The file was to be halved in 1879 (49: 651–759) and halved again in 1880 (53: 287–334). In 1881 the size held about the same (57: 395–452). It increased substantially in 1882 (60: 505–600); later, it gradually diminished. That is, correspondence with journalists hit its peak surrounding the phonograph and first announcement of the electric light, remained fairly strong throughout the period of development of light, then gradually tailed off.

14. "The Bores of Science," *Newark Sentinel,* May 21, 1878 (94: 200).

15. "Thomas Edison," *Phrenological Journal,* February 1878 (94: 105–108).

16. The first of these was McClure's *Edison's Life and Inventions* (1879).

17. "A Marvelous Discovery," *Sun*, February 22, 1878 (94: 115). This article was widely republished; see e.g. (27: 745–746, 749, 750–751, and 754–755).

18. "Edison the Magician," *Cincinnati Commercial*, April 1, 1878 (27: 790). This article may have been written by William A. Croffut, who wrote many favorable articles about Edison for the *New York Graphic*. See letter, W. A. Croffut to TAE, April 25, 1878 (17: 84–85).

19. "The Wizard of Menlo Park," *Graphic*, April 10, 1878 (27: 786–788; 94: 158–161).

20. "A Wonderful Genius," *Boston Herald*, April 14, 1878 (27: 799).

21. "The Inventor of the Age," *Sun*, April 29, 1878 (94: 186).

22. "The Man Who Invents," *Washington Post and Union*, April 19, 1878 (94: 170).

23. "Genius Before Science," *Washington Post and Union*, April 19, 1878 (94: 170).

24. "Astonished Congressmen," *Baltimore Gazette*, April 20, 1878 (94: 172). For more on the Edisonmania that seemed to overtake the press and the public, see letter, George H. Bliss to TAE, April 13, 1878 (17: 76).

25. "Awful Possibilities of the New Speaking Phonograph," *Graphic*, March 21, 1878 (27: 765). A month earlier, the *Sun* printed a poem about a stirring sermon delivered by a wax figure and a phonograph: Herwick Dodge, "The Speaking Phonograph," *Sun*, February 24, 1878 (94: 116).

26. "The Food Creator," *Graphic*, April 1, 1878 (94: 150). The aftermath of this hoax was gleefully reported in the *Graphic* of April 10 (94: 157).

27. "The Papa of the Phonograph," *Graphic*, April 2, 1878 (27: 776; 94: 151).

28. "Frank Leslie's Budget of Fun," *Frank Leslie's Illustrated Weekly*, May 1878 (27: 780).

29. For example, "How Edison Amuses Himself," *Graphic*, May 2, 1878 (republished from the Cincinnati Saturday Night) (94: 190); "Anecdotes About Edison," *Graphic*, May 3, 1878 (republished from the *Memphis Avalanche*) (94: 191); "Edisonia," *Graphic*, May 9, 1878 (94: 195).

30. "Edison's Frankenstein," *Graphic*, May 6, 1878 (94: 194–195).

31. "Mr. Edison's Reception," *New York Evening Post*, May 10, 1878. (94: 196).

32. See letters, W. Croffut to TAE, August 25, 1878 (17: 84–85), May 9, 1878 (17: 88), and May 10, 1878 (17: 91–93).

33. "To the Editor," *Graphic*, May 16, 1878 (94: 197).

34. On how telegraphy affected the gathering and dissemination of news, see Schwarzlose 1989 and Blondheim 1994.

35. See O'Brien 1968.

36. See Crouthamel 1989.

37. Edwin Marshall Fox characterized the editorial style of the *Herald* in a light-hearted letter to Edison (November 5, 1878 (17: 224)): "What the *Herald* wants is a vivid, succinct, fierce, graphic sketch not only about your past life but all the minor incidents of the present—how you act, talk, live, work, and look—the struggles and obstacles attending the completion of your chief inventions—your method of classification—your system of detail etc. etc.—the whole going to make up a lively and interesting pen picture."

38. Later the exoticism would provide a target for critical deflation by Edison's opponents, who, by making the place common, attempted to make Edison common, as in the *New York Times* later in 1878: "A man's first impression of Menlo Park is that of a misshapen little dwarf of a village trying to appear on terms of personal intimacy with a very large depot, and getting savagely snubbed whenever the depot happens to be in bad humor. One of the permanent fixtures on the platform is a young man with an overcoat four sizes too large for him, who has fits, and has to have plenty of room in his overcoat for their accommodation. As the representative of the *Times* alighted from the train, this young man executed a low whistle, and commenced to sing a song about a 'girl with a hole in the heel of her stocking'; then broke off abruptly, being seized with something resembling a convulsion, which he was pleased to call the 'silk and worsted double shuffle.' Considered as an ornament to the depot, the young man is not a desirable party, although success in life is open to him as a star on the variety stage." ("Edison's Baby," *New York Times*, October 27, 1878)

39. "Edison Whips Old Sol" (27: 932).

40. "The Gains of Genius," *Chicago Sunday Telegraph*, August 25, 1878 (27: 870).

41. "Tom Edison Back Again," *New York World*, August 27, 1878 (27: 861–862); "Edison's Trip and Inventions," *Graphic*, August 28, 1878 (94: 338); "The Elevated Roads," *New York Daily Tribune*, August 28, 1878 (27: 851); "Edison's Inventions," *New York Herald*, August 29, 1878 (27: 849; 94: 337); "Four Hours With Edison," *Sun*, August 29, 1878 (27: 850–851; 94: 339); "The Machine That Speaks," *Graphic*, August 30, 1878 (27: 862; 94: 340).

42. "Invention's Big Triumph," *Sun*, September 10, 1978. See also "A Great Triumph," *Mail*, September 10, 1878 (94: 349).

43. Edison apparently had intended to break the story to Edwin Marshall Fox of the *Herald*, but Fox was called away to another assignment. On September 11, only a day after the *Sun* story on the Wallace visit, Fox had sent a telegram asking if Edison had anything new (letter, Fox to TAE, September 11, 1878 (17: 168)). Edison wired back immediately: "All right. Something new in few days. " (17: 169).

On the September 14, Fox wired that he had been called to Connecticut (17: 173), and he introduced another reporter to cover the Edison story for the *Herald*. (17: 174). In the meantime, Ballard Smith, managing editor of the *Sun,* wired Edison to set up an interview for one of his reporters (Ballard Smith to TAE and TAE to Ballard Smith, September 13, 1878 (17: 170)). The *Sun* got the story. But the most surprising aspect of the incident is that Edison seemed to be ready to break the story within three days of his visit to Wallace.

Chapter 3

1. Here I follow the general accounts of the rise of American corporations given in Chandler 1990, Chandler et al. 1968, and Galambos and Pratt 1988. Histories of electrical industries include Passer 1953, Bright 1949, Howell and Schroeder 1927, Hammond 1941, Keating 1954, Nye 1990, Carlson 1991, and Millard 1990.

2. See letter, TAE to Grosvenor Lowrey, October 3, 1878 (28: 824): "Friend Lowrey, Go ahead. I shall agree to nothing, promise nothing, and say nothing to any person leaving the whole matter to you. All I want at present is to be provided with funds to push the light rapidly. Yours, Edison"

3. "Edison's Electric Light," *Sun,* October 20, 1878 (94: 382).

4. Letter, Stockton Griffin to Grosvenor Lowrey, November 1, 1878 (28: 893–896). Grosvenor Lowrey also used the argument that Edison need not pay the additional stockholder assessment because his talent was his contribution. See letters, Grosvenor Lowrey to TAE, October 24, 1879 (50: 280–285) and November 13, 1879 (50: 287–290).

5. Letter, Grosvenor Lowrey to TAE, November 2, 1878 (18: 59).

6. Letter, Grosvenor Lowrey to TAE, January 25, 1879 (50: 232–236). Tracy Edson, in another example, ended a letter to Edison (dated July 9, 1879) as follows: "I have every confidence in your ultimate success for I know your heart and your pride are in it, and I feel almost as much intent on it as you do, and I am proud of being able to assist you in accomplishing the desired result. " (50: 268–269)

Chapter 4

1. "Edison in His Workshop," *Harper's Weekly,* August 2, 1879: 607.

2. For a detailed study of employment at Menlo Park, see Finn 1989.

3. Pocket Notebook PN-69-08-08 (6: 757–793).

4. See e.g. (6: 767).

5. Pocket Notebook PN-70-10-03 (6: 794–819).

6. These drawing may be related to a revision of Patent 91,527 ("Improvement in Printing Telegraphs"), which was approved on October 25, 1870, as Reissue 4, 166, signed by Serrell as attorney (1: 43–47).

7. Pocket Notebook PN-70-10-03 (6: 819).

8. Newark Shop Notebook, Catalog 1174, p. 1 (3: 8).

9. Newark Shop Notebook, Catalog 1172, p. 1 (3: 78).

10. On Faraday's notebook-keeping practices, see Williams 1991; Tweney 1991a,b; Gooding 1990; Gooding and Tweney 1991.

11. Miscellaneous Notebooks, Catalog 1170 and 1171 (6: 216–298).

12. Pocket Notebook PN 73-00-00. 1 (6: 820–871) and PN 73-00-00. 2 (6: 872–906). Similar in character are Miscellaneous Notebooks 1176 (6: 1–65), 1175 (6: 144–215), and 299 (6: 66–143).

13. Newark Shop Notebooks, Catalog 1174, 1182, 1172, 1181 (3: 5–190).

14. Experimental Researches, volumes 1–6, Catalog 994–998, one volume missing (3: 192–427).

15. *A Guide to Thomas A. Edison Papers: A Selective Microfilm Edition*, parts I (1850–1878) and II (1879–1886) (University Publications of America, 1985): 7.

16. Charles Batchelor Notebook, Catalog 1308 (90: 695–701, 704).

17. In light of the narrative of Edison's involvement in the construction of his own fame described in the previous chapter, it is especially to be noted that these lists of publicity mailings date from April through June of 1878. April was when Edison's first fame for the phonograph was coalescing into a reputation as the Wizard of Menlo Park. June was when he was to set out in his expedition to observe the eclipse, using the tasimeter. Note, however, a telegram from TAE to Stockton Griffin, August 10, 1878 (97: 687), while Edison was in Virginia City, Nevada: "Better send Scientific Americans August tenth to regular list in Europe." That issue of *Scientific American* included an editorial favoring Edison's patent claims on the microphone. ("Letter from Professor Hughes," *Scientific American 39, August 10, 1878: 80)*

18. Literature Search Notebooks, volumes 1 and 2 (95: 314–495).

19. Experimental Researches, volume 5, Catalog 997: 49–56 (3: 376–383).

20. The Speaking Telephone Interferences, volume 1 (U.S. Patent Office, 1881): 11 (11: 26).

21. Ibid.: 60 (11: 51).

22. Menlo Park Notebook, volume 1 (29: 18–150).

23. Unbound Notebook, volume 16 (4: 478–873).

24. For example, Unbound Notebook, volume 16, pp. 24–25, September 17, 1878 (4: 0504) and pp. 107–108, October 5, 1878 (4: 568–569).

25. Unbound Notebook, volume 16, September 8, 1878, p. 8 (4: 487).

26. Ibid., September 9, p. 9 (4: 488).

27. See e.g. Menlo Park Notebook, volume 1, December 6, 1878 (29: 27).

28. Unbound Notebook, volume 16, p. 35, September 24, 1878 (4: 514).

29. Ibid., p. 36 (4: 515). For an even more complete statement of success, which seems to be there primarily for the historical legal record, see Unbound Notebook, volume 16, p. 40, September 25, 1878 (4: 518).

30. Unbound Notebook, volume 16, p. 17, September 15, 1878 (4: 499).

31. Ibid., p. 18, September 16 (4: 500).

32. Ibid., p. 12, September 11 (4: 492).

33. Menlo Park Notebook, volume 1, p. 32, December 15, 1878 (29: 35).

34. Unbound Notebook, volume 16, p. 97, October 3, 1878 (4: 561).

35. In 1947 the second Mrs. Edison (they had married in 1886, after the Menlo Park days) commented on the way TAE had used drawings to mediate work with objects that was largely carried out by others: "... He had marvelously delicate hands and was most skillful in sketching out his experiments. He drew very well. He kept notes with a lead pencil. Evenings, after dinner, he would come up to this room, sit down at his desk, and literally work out 50 or more things he was to do next day. In the morning, he would take his notes to the laboratory and apportion the work among his assistants. Evening after evening he would carry out this same procedure." ("Mrs. Edison Says Inventor Put Home First," Associated Press, *Sunday Star,* Washington, D.C., February 2, 1947: C.2)

36. Unbound Notebook, volume 16, p. 33, September 21, 1878 (4: 512).

37. Menlo Park Notebook, volume 1, p. 25, December 15, 1878 (29: 32).

38. Unbound Notebook, volume 16, pp. 47–48, September 27, 1878 (4: 525–526).

39. Ibid., p. 44, September 27 (4: 522).

40. Ibid., p. 60, September 30 (4: 539).

41. Ibid., p. 29, September 18 (4: 509).

42. Charles Batchelor Notebook, Catalog 1308 (90: 693–end).

43. Unbound Notebook, volume 16, p. 83, October 2, 1878 (4: 552).

44. Ibid., p. 40, September 25 (4: 518).

45. Ibid., p. 10, September 10 (4: 490).

46. Experimental Researches, volume 5, p. 54, September 27, 1878 (3: 379).

47. Unbound Notebook, volume 16, p. 26, September 17, 1878 (4: 505).

48. Ibid., p. 27, September 17 (4: 506).

49. "Improvement in Electric Lights," Patent 214,636, April 22, 1879.

Chapter 5

1. On whether this economic incentive has actually worked out well relative to other incentives, see Mokyr 1990, pp. 247–252; Weil and Snapper 1989; Gilfillan 1935; Horstman et al. 1985; Polanvyi 1944; Kitch 1977; Machlup and Penrose 1950; Wright 1983.

2. "Subdivision of current" is an obscure technical term of the time. It often suggests something like the operation of separate circuits using power from a single source. At other times, it suggests solving the problem of a single lamp breaking an entire series circuit. There are several other confusing usages too.

3. The complete file of Edison patents is available on reels 1 and 2 of the microfilm edition of the Edison papers. Edison's patents are also cataloged in Norwig 1954.

4. "Improvement in Electric Lights," patent 214,636, April 22, 1879.

5. "Safety Conductor for Electric-Lights," patent 227,226, May 4, 1880. This invention was, apparently an outgrowth of the work on regulation.

6. For outlines of the history and operations of the patent system in Britain and the United States, see Davenport 1979; Dutton 1984; Gomme 1946; Jones 1971); Klitzke 1959; Bugbee 1967; Fox 1947; MacLeod 1988; Sherwood 1983; U.S. Department of Commerce 1981; Vaughan 1956. P. J. Federico is the acknowledged dean of the history of American patent law and of the operations of the Patent Office; for a bibliography of his works on the subject, see Rich 1982. Also see the special issue of *Technology and Culture* devoted to patents (volume 32, no. 4, October 1991).

7. Copyright emerged out of comparable monopolies of printers and booksellers, in part to regulate and censor publications, and only later developed into a right granted to authors (Woodmansee and Jaszi 1994; Rose 1993).

8. The U.S. Department of Commerce (1981) puts the number at 55; Federico (1936) puts it at 57; Vaughan (1956) puts it at 67.

9. A further feature of the patent system at that time, the reissuance of patents to correct defects, created other opportunities to redefine the object being patented and the scope of the claim. Abuse arising from this opportunity to readjust patents on the basis of later knowledge about competition, workability of ideas, further developments of the product and marketplace considerations, led to curtailment of the reissue option in the late nineteenth century. See Dood 1991.

10. Lubar (1991) examines in detail the forces that led to changes of the law in 1836 and the consequences of the provisions. On the reorganization of the Patent Office in the wake of an 1836 fire, see Federico 1937 and Federico 1939. For more on the development and the current operations and structure of the Patent Office, see Jones 1971, Kursh 1959, and Mole 1978.

11. "Historical Patent Statistics," *Journal of the Patent Office Society* 46: 2 (1964): 112–116.

12. Edison was personally responsible for at least 1093 of those patents (Norwig 1954).

13. This form of address suggests a general public announcement, but the text is not made open to the general public until the patent has been approved and validated by the Patent Office. That is, the application, although submitted to the Patent Office, is not addressed to the Patent Office; rather it is as a public announcement awaiting official accreditation, after which it will carry the weight and legal standing of a government document.

14. Although instructions for replication imply that the invention is already successful in practice, such a determination of an operable invention could be made only by a court, litigation having been brought after the invention had proved to be of economic value.

15. For discussions of speech acts, see Austin 1962 and Searle 1969. On legal language as speech acts, see Bowers 1989, Kurzon 1986, and Kevelson 1990. It is not surprising that speech-act theory applies so well to the law, since Austin was influenced by the legal theorist H. L. A. Hart. (See Hart and Honore 1959.) For more on the role of language in patent formulation, see Stringham 1941.

16. These represent the five general kinds of speech acts Searle recognizes in chapter 1 of *Expression and Meaning* (1979).

17. E.g., perpetual-motion machines are not patentable.

18. On how heroic accounts of Edison were sustained after the fact, see Nye 1983 and Wachhorst 1984.

Chapter 6

1. Many of the early leaders of the electrical engineering profession were opponents of Edison (e.g., Henry Morton) or former co-workers who had fallen out with him (e.g., Franklin Pope, James Ashley, Otto Moses, Frank Sprague).

2. "An American Institute of Electrical Engineers," *Electrical World* 3 (April 5, 1884): 111.

3. "American Institute of Electrical Engineers," *Electrical World* 3 (May 17, 1884): 158.

4. Francis Hauksbee (c. 1666–1713) and Jean DeSaguliers (1683–1744), as official demonstrators for the Royal Society, displayed numerous electrical experiments before meetings of the society and developed several early devices for generating and measuring electricity.

5. The alliance with chemistry existed because batteries were the source of electrical energy for most applications at the time.

6. "Introductory," *The Electrician: A Weekly Journal of Telegraphy, Electricity and Applied Chemistry* 1: 1 (November 9, 1861): 1.

7. Another industry journal, *The Telegraphic Journal & Monthly Illustrated Review of Electrical Science,* was founded in 1872. As interest developed in a broader range of electrical technologies, it widened its scope and became weekly in 1881 under the new name *The Telegraphic Journal and Electrical Review.* In 1891 it became *Electrical Review.* See Strange 1985.

8. On the growth of the colonial telegraphic system, see Headrick 1988.

9. "Death of Count du Moncel," *Electrical World* 3 (February 23, 1884): 23.

10. The École supérieure de Télégraphie was founded in 1878. Because of the narrow civil service orientation of its program in relation to the state's telegraphic service, its graduates played little role in the development of incandescent light and central power. The Paris Municipal School of Industrial Physics and Chemistry (founded in 1883) and the Advanced School of Electricity (founded in 1893) were far more instrumental for power and light (Butrica 1987, p. 128).

11. In 1873 the more academic Physical Society was formed. See Gooday 1991 and Buchanan 1985.

12. Letter, Otto Moses to TAE, August 2, 1881 (58: 953–955)

13. Siemens served as president of the Society of Telegraphic Engineers in 1872 and again in 1878.

14. *The Electric Lighting Act, 1882 Minutes of Evidence given before the Select Committee of the House of Commons and Full Text of the Act* (London: The Scientific Publishing Company, Limited, 1882).

15. *The American Telegraphic Magazine,* 1.1 (October 15, 1852) through 1. 6 (July 15, 1853); *The National Telegraphic Review and Operator's Companion,* 1.1 (April 1853) through 1. 4 (January 1854). The only copies I was able to locate were in the Boston Public Library. See also Israel 1992, p. 63.

16. The piece in the May 9 issue described his "Combination Repeater"; that in the August 8 issue described a "Self-Adjusting Relay." In the issue of April 17, 1869, an anonymous paragraph—not by Edison—reports inaccurately on a trial of an improved version of Edison's Duplex telegraph. See PTAE 1: 63.

17. His advertisement for the double transmitter appeared from December 12, 1868 through January 9, 1869. The advertisement also appeared in *The Journal of the Telegraph* from December 15, 1868 through May 15, 1869.

18. *Telegrapher* 6, 1869: 401.

19. See PTAE 1: 98.

20. See PTAE 1: 139.

21. See PTAE 1: 141–142.

22. *The Journal of the Telegraph* mentioned the partnership with Pope in passing in a December 1, 1869. article on Leverett Bradley, a maker of electric clocks.

23. See PTAE II: 305.

24. First used on February 6, 1875. See PTAEII: 307.

25. "The 'Organ' Business," *Telegrapher,* September 26, 1874: 233.

26. "Edison About to Astonish the World Again. —Stand from Under!" *Telegrapher,* July 29, 1876: 184–185.

27. PTAE II, 287, n1. There is no support for the claim in any of the available documents.

28. *Electrical World* absorbed the *Electrical Railway Gazette* in 1896, the *Electrical Engineer* in 1899 (founded in 1882 as the *Electrician* and professionalizing itself as the *Electrician and Electrical Engineer* in 1884, then dropping the less professional *Electrician* from its title in 1887), and the *American Electrician* in 1906.

29. *Operator* 1: 1, March 1, 1874, p. 2.

30. *Operator* 1: 10, July 15, 1874, p. 5. In the same issue, immediately following, was this self-congratulatory editorial comment: "The telegraphic fraternity have an 'organ,' a little sheet rejoicing under the modest title of the OPERATOR. It seldom contains much of interest to an outside reader, so this can hardly be considered a 'puff' to the amateurs. Most of the matter is voluntary contributions from different parts of the country, as the boys like to see themselves in print. Among the correspondents is a young fellow in Portland Maine, who sports the nom de plume of Oney Gagin...."

31. PTAE II: 11.

32. Beach was a son of the editor and owner of the New York newspaper *The Sun.*

33. PTAE II: 330.

34. An exchange with the author of the paper Edison claims priority to, leads to a further published Edison letter on December 12.

35. PTAE II: 502.

36. "Supposed Discovery of a New Force," *Newark Daily Advertiser*, November 29, 1875 (27: 282); "A Wonderful Invention," *Newark Morning Register*, November 30, 1875 (27: 282); "New Discovery in Electricity," *New York Tribune*, November 30, 1875 (27: 282); "Novel Electricity," *New York World*, November 30, 1875 (27: 282); "New Era in Science," *New York Daily Witness*, November 29, 1875 (27: 282); several others are to be found on (27: 282) and on the following frames.

37. "Etheric Force," *New York Herald*, December 2, 1875 (27: 282); also PTAE II: 678.

38. "The Discovery of Another Form of Electricity," *Scientific American*, December 25, 1875: 400 (27: 289); "The New Phase of Electrical Force," *Scientific American*, December 25, 1875: 401 (27: 289).

39. "Etheric Force and Weak Electric Sparks," *Scientific American*, January 1, 1876: 2; W. E. Sawyer, "The Etheric Force," *Scientific American*, January 15, 1876: 36; George M. Beard, "The Nature of the Newly Discovered Force," *Scientific American*, January 22, 1876: 57; Electron (pen name), "Mr. Edison's New Force," *Scientific American*, January 29, 1876 (29: 333); see also PTAE II: 753

40. *Journal of the American Electrical Society*, volumes 1–5, 1875–1880.

41. The letter is quoted in chapter 2 of this volume.

42. "The Talking Phonograph," *Scientific American*, December 22, 1877: 384.

43. "Edison's Micro-Tasimeter," *Journal of the Franklin Institute* 106 (1878): 173–176.

44. "Edison's Electric Light," *Journal of the Franklin Institute* 107 (1879): 208.

45. This same paper was also published in the popular press: "Mr. Edison's Experiments," *Sun*, September 3, 1879 (94: 509); "Mr. Edison's Discoveries," *Graphic*, September 3, 1879 (94: 511). That this paper was understood by some as an attempt to maintain credibility for his struggling technical project of incandescent light is indicated in "The Brush Electric Light," *Engineering*, January 2, 1880: 13 (48: 605).

46. A letter from John Michels to Edison's secretary Samuel Insull, dated April 18, 1881, provides a subscription list with 175 names plus indicates an additional distribution of 100 copies to London and fifty through the American News Company (59: 502). Because sales were so few, copies of this journal are rare. I have located only two incomplete copies—one among the Edison papers (at the Edison National Historical Site, in West Orange, New Jersey) and one at the University of Pennsylvania. Consequently, there have been few studies of the material—as far as I have been able to tell, only one article appearing in the current journal *Science* in 1945 announcing the discovery of this material and a few pages in Baldwin 1995 (pp. 121–123). There are, however, substantial files of correspondence relating to the publication of the journal in the Edison papers (55: 360–492; 59: 451–608; 63: 661–671).

47. Based on Michels's letter to TAE of April 17, 1880, at a time when Edison was capitalizing on the New Year's demonstration to create the interlocking set of Edison companies: "I think you said that when you were further advanced you would cooperate with me. I notice you will shortly make another move forward...." (55: 361–363).

48. Prospectus, April 30, 1880 (55: 368–371).

49. Letter, John Michels to TAE, May 6, 1880 (55: 372–373).

50. Letter, John Michels to TAE, November 2, 1881 (59: 575–580).

51. Letter, John Michels to TAE, June 26, 1880 (55: 401–404). Although Edison kept his own role hidden, he kept Michels under strict financial and editorial control. According to the June 1880 "Memorandum of Understanding" (55: 412–414), Michels was to incur no debt, was to keep within a prescribed budget, and was to provide detailed weekly accounts. Michels was to work without capital or advances. The detailed records and financial correspondence in the Edison papers indicate that Michels was kept on a very short leash. Similarly, Edison kept Michel under strict editorial watch. Apparently his employee Otto Moses vetted the articles on physics and electricity at first, but once Moses left for Paris an article appeared that evoked Edison's ire. His response was quick and sharp: "In the last issue of Science (Oct 2d) I notice a great deal of space given to Weisendanger. Now I don't suppose you are aware of it but in Europe Weisendanger is looked upon as little less than an idiot and such stuff as the article referred to is nothing more or less than a disgrace to a scientific paper. If such a thing is possible I would like to see the proofs of what is to appear in Science before the paper is published and thereby avoid as much as possible that which would be detrimental to its interests. " (letter, TAE to John Michels, October 5, 1880 (80: 389))

52. "American Academy for the Advancement of Science," *Science* 1: 10 (September 4, 1880): 109–117. Barker was, of course, Edison's close friend. Although there is no indication in any of the correspondence that Barker was part of the planning or that there was any alliance of interests between Edison and the AAAS mediated by Barker, such cooperation is not unimaginable. Edison had attended the previous year's meeting of the AAAS, when Barker had been installed as president, in order to deliver a paper on "The Action of Heat in Vacuo on Metals."

53. "Editorial," *Science* 1: 27 (December 31, 1880): 321.

54. Similar editorials were to appear on September 25, 1880, and on April 30, August 20, and November 12, 1881. Half of the August 20 editorial on the Paris exhibition was devoted to Edison interests, including the seizure of the Maxim lamps (p. 389), and half of the November 12 editorial on the awards at the Paris exhibition was devoted to Edison's achievement (p. 533). Curiously, however, Michels was never successful in gaining information from the Edison team in Paris to be able to write the extensive report he desired of the exhibit. See letter, J. Michel to Samuel Insull, September 9, 1881 (59: 553–554).

55. "Edison's Method of Preserving Organic Substances," *Science* 2: 73 (November 19, 1881): 545.

56. See letters, John Michels to Stockton Griffin, December 1, 1880 (55: 479–480), to William Carman, February 24 (59: 475–476), and to Samuel Insull, November 18, 1881 (59: 581–582).

57. See letters, John Michels to TAE, February 10 (59: 461–464), February 14 (59: 469–470), March 18 (59: 481–482), and March 29 (59: 486–487), 1881.

58. Letter, TAE to John Michels, October 25, 1881 (81: 80).

59. "American Electrical Society. Seventh Annual Meeting," *Journal of the Telegraph* 14 (January 1, 1881): 2–3.

60. The final paragraph, nonetheless, mentions George Barker's comment from the floor that he was interested in Edison's work in this area.

61. "The Electrical Exhibition," *Journal of the Telegraph* 14 (August 16, 1881): 243.

62. "The Paris Exposition," *Journal of the Telegraph* 14 (October 16, 1881): 307.

63. "American Success at the Electrical Exhibition," *Journal of the Telegraph* 14 (December 16, 1881): 377.

64. "The Efficiency of Incandescent Electric Lamps," *Journal of the Telegraph* 15 (December 20, 1882): 224; see also the short notices, "Edison's Electric Meter" *Journal of the Telegraph* 15 (February 16, 1882): 55; "Consolidation of Electric Light Companies in the United States," *Journal of the Telegraph* 15 (November 20, 1882): 211.

65. "Mr. Edison and the Electric Light," *Operator* 9: 20 (October 15, 1878): 9.

66. "The Electric Light," *Operator* 9: 23 (December 1, 1878): 6–7.

67. "An English Opinion of Edison," *Operator* 10: 2 (January 15, 1879): 2.

68. "Thomas A. Edison," *Operator* 10: 3 (January 15, 1879): 1–2.

69. "Edison's Electric Light," *Operator* 10: 8 (April 15, 1879): 7.

70. "Mr. Edison's Friends," *Operator* 10: 9 (May 1, 1879): 7.

71. "How Edison Made a Bug," *Operator* 10: 12 (June 15, 1879): 5; "Progress of the Electric Light" and "Edison's Wall Speaker," *Operator* 10: 14 (July 15, 1879): 7; "Another Stride Onward," *Operator* 10: 17 (September 1, 1879): 8.

72. "The Electric Light," *Operator* 10: 19 (October 1, 1879): 8.

73. Untitled, *Operator* 10: 20 (October 15, 1879): 4–5.

74. "The Electric Light a Reality," *Operator* 11: 1 (January 1, 1880): 4–5.

75. "Public Exhibition of the Electric Light," *Operator* 11: 2 (January 15, 1880): 4–5.

76. "Edison's System of Electric Lighting," *Operator* 11: 4 (February 15, 1880): 4–5.

77. "Just What Mr. Edison Has Accomplished," *Operator* 11: 6 (March 15, 1880): 11.

78. "More Edisonian Reminiscences," *Operator* 11: 16 (August 15, 1880): 7; "Edison's Electric Railway," *Operator* 11: 14 (July 15, 1880): 4; "Edison's Electric Locomotive," *Operator* 11: 17 (September 1, 1880),"Mr. Edison on the Electric Light," *Operator* 11: 19 (October 1, 1880): 6.

79. "American Exhibit at the Paris Exhibition," *Operator* 12: 7 (April 1, 1881): 117; "The Paris Exhibition," *Operator* 12: 14 (July 15, 1881): 262; "The Electrical Exhibition," *Operator* 12: 16 (August 15, 1881): 297; "Edison's System of Electric Lighting," *Operator* 13: 2 (January 15, 1882): 25–28.

80. "Edison's First District Completed," *Operator* 13: 15 (August 1, 1882): 320; "The Edison System," *Operator* 13 (November 11, 1882): 528; "The Electric Light," *Operator* 13 (December 23, 1882): 690.

81. The boast first appeared in the issue of October 27, 1883, and continued in various forms throughout the decade.

82. See e.g. the stories about the Excelsior Electric Company (January 5, 1884), the Merchants' Electric Light and Power Company of Boston (March 29, 1884), the Sawyer Man Company (July 12, 1884), the Consolidated Electric Light Company (October 25, 1884), and the Weston Lamp Company (September 13, September 27, and October 11, 1884).

83. See e.g. "A Review of Electrical Events and Progress in 1884," *Electrical World* 5 (January 3, 10, and 24, 1885): 2, 12, 32; "Progress of Electric Light in New York City," *Electrical World* 5 (January 31, 1885): 42

84. "The Edison System on Board Steam Vessels," *Electrical World* 2 (September 1, 1883): 1–2; "The Edison Electrical Lighting Station, at Berlin, Germany," *Electrical World* 4 (December 27, 1884): 261. In addition, an illustration of Edison's plant in Milan accompanies "Electric Light in Theatres," *Electrical World* 3 (April 26, 1884): 135.

85. "A Representative American Manufacturer" *Electrical World* 2 (December 22, 1883): 275–278; "Decorative Fixtures For Incandescent Lighting," Electrical World 3 (March 15, 1884): 81–82. The latter praises the products of Mitchell, Vance & Company, which supplied the original fixtures for Menlo Park and several other early Edison installations.

86. "Aesthetics of Electric Lighting," *Electrical World* 2 (October 13, 1883): 100.

87. "The Electric Lighting of the Louisville Exposition," *Electrical World* 4 (September 13, 1884): 88; "Incandescent Lighting at the Exposition" *Electrical World* 5 (March 21, 1885): 115–116.

88. Of the 451 isolated plants lighting and 97,424 lamps in the United States, 368 of the plants and 78,432 of the lamps were Edison's. The article pointed out that since those statistics were gathered Edison had added more than 80 plants and

50,000 lamps. The Edison company also reported 88,000 lamps lighted by central stations, far in excess of any other incandescent supplier and showing greater growth than any of the arc light providers "Progress of Incandescent Electric Lighting in the United States," *Electrical World* 6 (November 7, 1885): 188.

89. "A Talk with Edison," *Electrical World* 1 (June 30, 1883): 404, 405; "Mr. Edison on the Situation," *Electrical World* 3 (April 19, 1884): 124: "Mr. Edison's Views and Work," *Electrical World* 4 (November 29, 1884): 220; "Mr. Edison as an Electrical Litterateur," *Electrical World* 5 (January 24, 1885): 38.

90. "Mr. Edison on Electric Lighting," *Electrical World* 1 (June 30, 1883): 407.

91. "The Edison Electric Light Company," *Electrical World* 10 (July 16, 1887): 11–12.

92. "A Day with Edison at Schenectady," *Supplement to the Electrical World* 12 (August 25, 1888): 1–12.

93. For an account of the forming of the AIEE that keeps Edison more on the inside, see McMahon 1976. For a detailed analysis of the development of the applied science/pure science distinction in the United States, see Kline 1995.

94. "National Electric Light Association," *Electrical World* 6 (August 22, 1885): 76–78; "First Annual Meeting of the National Electric Light Association," *Electrical World* 7 (February 20, 1886): 79–81.

95. See also Dyer and Martin 1910, p. 351.

Chapter 7

1. "Invention's Big Triumph: An Electric Machine that will Transmit Power by Wire," *Sun*, September 10, 1978, reprinted as "A Great Triumph. An Invention that will Revolutionize the Motive Power of the World," *Mail*, September 10, 1978 (27: 912–913; 94: 349); see also "Mr. Edison's Use of Electricity," *New York Tribune*, September 28, 1878 (27: 927).

2. "Edison's Newest Marvel," *Sun*, September 16, 1878 (27: 914; 94: 354).

3. "Power Flashed By Wire," *Sun*, September 17, 1878 (27: 915; 94: 354).

4. "Wonders Yet To Be," *New York Telegram*, September 18, 1878 (27: 915).

5. TAE, draft of letter to Presidents of Singer, Domestic, Howe, Wilson, Wheeler & Wilson, Remington, and Willcox & Gibbs sewing machine companies, September 8, 1879 (50: 115–116); letters of response to TAE from Willcox & Gibbs Sewing Machine Company, September 9, 1879 (50: 117–118); Wheeler & Wilson Manufacturing Company, September 10, 1879 (50: 119); E. Remington & Sons, September 12, 1879 (50: 122) and October 1, 1879 (50: 142); Wilson Sewing Machine Company, September 12, 1879 (50: 123), September 15, 1879 (50: 127), September 18, 1879 (50: 135), and October 21, 1879 (50: 148–149) [follow-ups: October 10, 1882 (60: 499) and October 18, 1882 (81: 948)]; Indiana Manufacturing Company, September 12, 1879 (50: 129); Singer Manufacturing

Company, September 15, 1879 (50: 127) [follow-ups: March 31, 1880 (53: 709), and March 18, 1881 (58: 244)].

6. Although he devoted substantial time and resources to this project, Edison never brought it to a marketable conclusion, leaving the field to the many competitors who crowded the field by 1883.

7. "Annual Meeting of Isolated Company," *Fifteenth Bulletin of the Edison Electric Light Company,* December 20, 1882, pp. 43–46 (96: 787–789).

8. Edison Electric Light Company, *Report of the Board of Trustees to the Stockholders at their Annual Meeting,* October 28, 1886 (96: 15–23).

9. "List of Edison Isolated Plants Installed Prior to October 1, 1885 in the United States," (96: 271–284).

10. "The Edison Light," *New York Times,* October 26, 1881.

11. "Annual Stockholder's Meeting, Edison Electric Light Company," *Twentieth Bulletin of the Edison Electric Light Company,* October 31, 1883: 47–48 (96: 890–891).

12. "The light" was the term used then, overwhelmingly. It implies the entire system, not just lamps.

13. The one notable exception is the article "Electricity as Power" written by Edison's employee Francis Upton, which appeared in the premiere issue of the Edison-funded journal *Science* (1: 1 (July 3, 1880: 5)). Here Edison was clearly promoting a different and grander vision of his project for an elite audience prepared, he might hope, to reach beyond the currently imaginable and the desires currently represented in the marketplace.

14. See also "Edison's Light," *New York Herald,* December 21, 1879 (94: 537–542); "Edison's Great Work," *New York Herald,* January 1, 1880 (94: 559).

15. "Plans for Using Electricity," *New York Tribune,* March 28, 1881. See also another statement by Eaton later the same year devoted primarily to the illumination of the Paris Opera and the completion of the Pearl Street station: "Electric Lights for Paris. The Edison Company to Light a Part of the Grand Opera House," *New York Tribune,* September 20, 1881 (94: 660). Buried near the end of this story is a comment about the lower Manhattan district: "But in this portion of the city the motor force is of more importance than the light. Machinery of all kinds, fly-wheels, printing presses, elevators, will be run by our electric engines."

16. "Edison's Newest Marvel," *Sun,* September 16, 1878 (94: 354).

17. Telegram, William Barrett to TAE, November 19, 1878 (17: 1044); TAE to William Barrett, November 22, 1878 (17: 1047).

18. See e.g. "The Darkness Dispelled. An Estimate of the Cheapness of Edison's Electric Light," *Sun,* December 14, 1881 (94: 0435); "The Light in Menlo Park," *Sun,* February 27, 1880 (94: 587).

19. The term "incandescent lamp" itself evokes the tradition of indoor light by oil and then gas lamps.

20. "Edison's Electric Light," *New York Herald*, October 12, 1878 (94: 375).

21. Two years later, when Edison was preparing to market his new utility to businesses of lower Manhattan, he made a detailed business by business survey of every gas lamp used, complaints against the utilities, and insurance rates, clearly gaining the information to sell his service in competition with gas. During the construction and early operation of the Pearl Street station, the Edison companies regularly issued a bulletin detailing the progress at this and other sites. The bulletin regularly reported, often with some glee but with little comment, every story of gas fire, explosion, or poisoning.

22. Henry Morton, "Lecture Upon the Electric Light," *American Gas Light Journal*, January 2, January 16, February 3, and February 17, 1879 (48: 534–551); "Edison's Electric Light," *New York Graphic*, October 14, 1878 (94: 377); "Electric and Gas Light," *Sun*, October 18, 1878 (94: 378); "Gas and Electricity," *New York Herald*, October 19, 1878 (94: 379); "Electric Light. The Invention Discussed by Members of the American Gaslight Association," *New York Herald*, October 17, 1878 (94: 379; 24: 81): "Illumination as a Science," *New York Times*, October 18, 1878.

23. See e.g. "Revenge is Sweet," *Brooklyn Daily Eagle*, December 1, 1878 (94: 438); "The New Light," *Brooklyn Daily Eagle*, December 18, 1878 (94: 436); "The Ninth Annual Meeting of the New England Association of Gas Engineers," *American Gas Light Journal*, March 3, 1879 (23: 518–519); "The Edison Light," *Journal of Gas Lighting, Water Supply & Sanitary Improvement*, February 18, 1879 (23: 521–522); "Edison's Electric Light," *Journal of Gas Lighting, Water Supply & Sanitary Improvement*, February 25, 1879 (23: 521); "Some Modern Lights," *Journal of Gas Lighting, Water Supply & Sanitary Improvement*, April 15, 1879 (23: 539); "The Electric Light," *Gas Light Journal*, June 1879; "The Great Edison Scare," *The Journal of Gas Lighting, Water Supply & Sanitary Improvement*, January 20, 1880 (94: 573–574); "Edison's Latest Device Reviewed by an Expert," *American Gas Light Journal*, February 2, 1880; "The Prospects of the Electrical Light," *American Gas Light Journal*, May 3, 1880 (23: 685); "Edison's 'Perfect Lamp' Perfected," *Gas Light Journal*, August 31, 1880 (24: 53); "The Edison Electric Light," *Gas Light Journal*, October 2, 1880 (89: 11); "Remarkable Report on the Cost of Incandescent Electric Lighting," *American Gas Light Journal*, February 16, 1882 (24: 115) "Gas versus Electric Lighting," *American Gas Light Journal*, August 16, 1882 (24: 112). "An Electric Leak," *American Gas Light Journal*, September 2, 1882 (24: 113); "The Edison System," *American Gas Light Journal*, October 16, 1882 (24: 110). "More About Mr. Edison and His System," *American Gas Light Journal*, November 16, 1882 (24: 105).

Chapter 8

1. "Edison's Light Turned On," *Sun*, September 5, 1882.

2. "Edison's Newest Marvel," *Sun*, September 16, 1878 (94: 354).

3. "Edison's Electric Light," *New York Herald*, October 12, 1878 (94: 375–376).

4. Ibid.

5. See letter, W. A. Croffut to TAE, October 28, 1878 (17: 212): "Don't let Cummings get ahead of me!" Fox also had competitive feelings about Cummings's reporting. See letter, E. M. Fox to TAE, October 20, 1878 (17: 206–207). See also letter, A. J. Cummings to TAE, December 29, 1882 (60: 834).

6. "Edison's Electric Light. The Inventor Declares it Perfect, and Makes an Exhibition," *Sun*, October 20, 1878 (94: 382).

7. "Edison's New Light," *Graphic*, October 21, 1878 (94: 380).

8. In November, Grosvenor Lowrey continued the deception with a letter to the Editor of the *Tribune* answering rumors that Edison had fallen seriously ill and had given up his work on light. After correctly pointing out that Edison's illness was minor and long past and that work on the light continued, Lowrey offered further excuses for the lack of public display: "In the meantime, the proper exhibition of what has already been invented, as well as the study of the economical questions involved, required the erection of large buildings, engines, etc. which is now going on with the utmost rapidity. Pending their completion, Mr. Edison, far from having given up his experiments, is pursuing a great variety of them with his customary energy, and even more than his customary good fortune." ("The Electric Light," *New York Daily Tribune*, November 15, 1878 (94: 0411))

9. Telegram, Grosvenor Lowrey to TAE, September 17, 1878 (18: 2).

10. Telegram, Tracy Edson to TAE, September 19, 1879. (18: 2). See also telegram, Tracy Edson to TAE, September 21, 1878 (18: 3) and reply telegram, TAE to Tracy Edson, September 21, 1878 (97: 695), canceling that day's meeting and suggesting one the following Monday or Tuesday, September 23 or 24. On September 24 Edson again canceled the meeting because he had been up all night working (telegram, TAE to Tracy Edson, September 24, 1878 (18: 7)). The rescheduled meeting was also to include Twombly and Lowrey (telegrams, Harrison Twombly to TAE, September 24, 1878 (18: 7) and Stockton Griffin to Grosvenor Lowrey, September 24, 1878 (18: 8)). That meeting also apparently never came off, and serious negotiations waited until the following week.

11. Letter, Hugh Craig to TAE, September 23, 1878 (17: 935); letter, J. G. Kidder to TAE, October 15, 1878 (17: 949); letter, George Walker to TAE, October 16, 1878 (17: 951); letter, W. C. Miller to TAE, October 18, 1878 (17: 967–968); letter, Gerritt Smith to TAE, October 19, 1878 (17: 982).

12. Letters, Grosvenor Lowrey to TAE, October 1, 1878 (18: 16–24); October 2, 1878 (18: 25–29); October 10, 1878 (18: 33–35); TAE to Grosvenor Lowrey, October 3, 1878 (28: 824).

13. "A Liberal Fund for Edison," *New York World*, October 18, 1878 (94: 0376). See also "Certificate of Incorporation of The Edison Electric Light Company," October 16, 1878 (18: 38–41).

14. "Fiat Lux!" *New York Herald*, October 14, 1878 (94: 0376); "Electric and Gas Light," *Sun*, October 18, 1878 (94: 0378); "Electric Light. The Invention Discussed by Members of the American Gaslight Association," *New York Herald*, October 17, 1878 (94: 0379); "Gas and Electricity," *New York Herald*, October 18, 1878 (94: 379).

15. "Edison's Electric Light," *Graphic*, October 14, 1878 (94: 377); "The New Inventions," *Graphic*, October 21, 1878 (94: 381).

16. Other optimistic stories out of Menlo Park included "Invention by Accident. Mr. Edison improves his electric burner by dropping a screwdriver on it" (*New York World*, November 17, 1878 (94: 411)) and "Edison Electric Light. The Latest Advances in Perfecting the Invention" (*New York Herald*, December 3, 1878 (94: 0431)).

17. In March 1879, the youth magazine *St. Nicholas*, in a story about a demonstration of an arc light ("A Wonderful Candle," *St. Nicholas* 6 (March 1879): 309–311), commented: "Let us all hope that Mr. Edison will succeed in making electricity the light that will, like the sun, 'shine for all.' "

18. Letter, Lowrey to TAE, October 31, 1878 (18: 56); see also letters, Lowrey to TAE, November 7 (18: 63) and November 25 (18: 83).

19. See letters, George Barker to TAE, September 16, 1878 (17: 927); October 23, 1878 (17: 979–981); November 10, 1878 (17: 1031); November 22, 1878 (17: 1048).

20. "Edison's Progress" *La Lumière Electrique* 1: 1 (May 15, 1879) (48: 564). See also Sylvanus P. Thompson, "Divisibility of Light from a Dynamical Point of View," *Engineering* 26 (October 25, 1878): 341 (48: 520–522); "The Electric Light," *Engineering* 26 (December 20, 1878): 499–500 (48: 522–533); William Preece, "The Criteria of the Electric Light," *Telegraphic Journal* 7 (February 15, 1879): 59–60 (48: 556–557).

21. "The Edison Light," *Journal of Gas Lighting, Water Supply & Sanitary Improvement*, February 18, 1879 (23: 521–522).

22. See chapter 5 above.

23. "Two Hours at Menlo Park," *Graphic*, December 28, 1878 (94: 444). The reporter (probably William Croffut) joins in the outrageous fun by ostentatiously refusing to name the competing reporters who fall into the sins of exaggeration—but clearly he implicates Edwin Fox of the *Herald*, his only peer in the Edison hype.

24. "Flight of the Airship 'Science' to the White Mountains," *New York Herald*, January 1, 1879 (94: 449).

25. "Punch's Almanack for 1879: Edison's Anti-Gravity Under-Clothing," *Punch*, December 9, 1878 (94: 445–447).

26. "Edison's Light. Already Perfected, but Requiring to be Cheapened," *New York Herald,* January 30, 1879 (94: 455).

27. "The Gas Stock Scare," *Graphic,* February 6, 1879. (94: 456).

28. Letter, Calvin Goddard to TAE, January 18, 1879 (50: 223).

29. Letter, Calvin Goddard to TAE, January 22, 1879 (50: 227)

30. "Edison Electric Light Stock," *Sun,* December 30, 1879 (94: 559).

31. Letters, Tracy Edson to TAE, January 24 (50: 229), Edgar Johnson to TAE, January 25 (50: 230–231), and James Banker to TAE, January 30 (50: 241), 1879.

32. Letter, Edwin Marshall Fox to TAE, January 26, 1879 (50: 237–238).

33. Letter, William Croffut to TAE, February 3, 1879 (50: 242–243).

34. See also letter, W. A. Croffut to Stockton Griffin, December 31, 1878 (17: 543).

35. Letter, William Croffut to TAE, December 8, 1879 (49: 748).

36. Edison's stock manipulations and financial relations with reporters did not go unnoticed at the time. See, among other passing comments on stock manipulations, the unidentified article from March or April 1880 "Detectives after Edison" (94: 596) as well as the unidentified cartoon from the same period, most likely from the *Graphic:* "The Wizard of Menlo Park—From the Phonograph to the Polyform" (94: 602).

37. Agreement, November 15, 1878 28: 1162–1170.

38. Ibid.

39. Letters, Grosvenor Lowrey to TAE, October 30, 1878 (18: 56); November 7, 1878 (18: 63); November 25, 1878 (18: 83); December 2, 1878 (18: 95–96). See also letter, Calvin Goddard to TAE, November 19, 1878 (18: 71).

40. See letters, Stockton Griffin to TAE, October 30, 1878 (18: 51–53); Grosvenor Lowrey to TAE, October 30, 1878 (18: 49–50); Grosvenor Lowrey to TAE, October 31, 1878 (18: 56–57); Edward Dickerson to Grosvenor Lowrey, December 11, 1878 (18: 116–118).

41. Letter, Francis Upton to Lucy Winchester Upton, November 7, 1878 (95: 497–499). In a telegram to H. R. Butler, November 12, 1878 (18: 67), TAE asks to see Upton. Upton's patent search notebook on the opening page is dated "Saturday Nov. 16. 1878" (95: 315). A letter from Francis Upton to Grosvenor Lowrey dated December 12, 1878 (18: 113) states that as of the next Monday, when Upton would relocate to Menlo Park, he would have been in employ of the Edison Electric Light Company for four weeks and two days.

42. Letter, Grosvenor Lowrey to Stockton Griffin, December 5, 1878 (18: 219).

43. Letter, Grosvenor Lowrey to TAE, December 10, 1878 (18: 226–235).

44. Letter, Grosvenor Lowrey to TAE, December 23, 1878 (18: 121–123).

45. Letter, Grosvenor Lowrey to TAE, January 18, 1879 (50: 224).

46. Letter, Grosvenor Lowrey to TAE, January 25, 1879 (50: 232).

47. Menlo Park Notebook, volume 2 (29: 190); volume 3 (286); volume 4 (29: 424); volume 7 (29: 778, 795–796); volume 22 (31: 12, 21); volume 23 (31: 182)

48. Menlo Park Notebook, volume 3 (286).

49. Earlier that month, Edgar Johnson gave an exhibition of Edison's other inventions in Boston, using the occasion to talk of the progress on the electric light. "Edison's Inventions," *Boston Daily Globe*, March 7, 1879 (94: 463); "The Edison Exhibition," *Boston Daily Advertiser*, March 8, 1879 (94: 462); "Edison's Inventions," *Boston Journal*, March 8, 1879 (94: 463).

50. More than 50 years later, at the time of Edison's death, the *New York Times* published a tale of disaster, with lights blowing out before reaching full incandescence and one of the directors, Robert F. Cutting, commenting that Edison ought to have learned the lesson from earlier failures of other inventors. "Light Bulb Balked Edison for Months," *New York Times*, October 19, 1931. See also Jehl 1937, pp. 246–247. I thank Keith Nier for suggesting in private correspondence that a more moderate interpretation of the events is called for.

51. "The Electric Light in the United States," *London Times*, March 22, 1879 (94: 464).

52. "Edison Altogether Serene," *New York World*, March 23, 1879 (94: 463).

53. "Edison's Electric Light. Several Recent Improvements in the System," *New York Herald*, March 27, 1879 (94: 464).

54. Letter, Francis Upton to Elijah Wood Upton, March 23, 1879 (95: 514–516).

55. TAE, annotation to letter from C. Goddard to TAE, March 29, 1879 (50: 251).

56. Letter, Francis Upton to Elijah Wood Upton, March 30, 1879 (95: 517–519).

57. Letter, Francis Upton to Elijah Wood Upton, April 4, 1879 (95: 520–521).

58. Letter, Francis Upton to Elijah Wood Upton, April 27, 1879 (95: 527–528). For further guarded optimism see letter, Francis Upton to Elijah Wood Upton, April 13, 1879 (95: 522–524).

59. "Edison's Electric Light. Several Recent Improvements in the System," *New York Herald*, March 27, 1879 (94: 464); "Edison Still Hard at Work," *Graphic*, April 11, 1879 (94: 474); "Edison's Electric Light. The Great Inventor Frankly Declares It Entirely Successful," *Sun*, April 12, 1879 (94: 475).

60. Letters, James O. Green to TAE, April 17, 1879 (50: 252); April 19, 1879 (50: 253).

61. "Subdivided Lightning," *New York Herald,* April 25, 1879 (94: 476–477).

62. "Edison's New Patents," *New York Herald,* April 25, 1879 (94: 477–478); "The Triumph of Electric Light," *New York Herald,* April 27, 1879 (94: 478); "Edison's Electric Light," *New York Herald,* April 27, 1879 (94: 479). The *World* also covered the story, although less extravagantly: "What Edison Has Done," *New York World,* April 30, 1879 (94: 480).

63. "Wanted: A Platinum Mine. Thomas Edison Willing to Spend $20,000 in the Search," *Graphic,* July 9, 1879 (94: 495–499).

64. See letters, Tracy Edson to TAE, June 30 (50: 263) and July 9, 1879 (50: 268–269).

65. Letter, TAE to Tracy Edson, July 7, 1879 (50: 264–266).

66. Letter, Tracy Edson to TAE, September 17, 1879 (50: 274–275).

67. "Mr. Edison's Experiments," *Sun,* September 3, 1879 (94: 509); "Mr. Edison's Discoveries," *Graphic,* September 3, 1879 (94: 511).

68. Letter, George W. De Long to TAE, April 21, 1879 (50: 44). See also letter, Jerome Collins to TAE, May 2, 1879 (50: 47); May 7, 1879 (50: 51); May 9, 1879 (50: 52); telegrams, Jerome Collins to TAE, May 7, 1879 (50: 50); May 16, 1879 (50: 58); May 22, 1879 (50: 69); May 22, 1879 (52: 57); TAE to Jerome Collins, May 15, 1879 (50: 58); May 16, 1879 (50: 58); May 20, 1879 (52: 54); TAE to George Walker, May 22, 1879 (50: 70).

Chapter 9

1. Letter, Francis Upton to Elijah Wood Upton, October 26, 1879 (95: 566–567).

2. Letter, Francis Upton to Elijah Wood Upton, November 2, 1879 (95: 568–569).

3. Letter, Francis Upton to Elijah Wood Upton, November 16, 1879 (95: 572–574).

4. Telegram, TAE to Norvin Green, November 4, 1879 (50: 158).

5. Exchange of telegrams, James Merrihew and TAE, November 7–8, 1879 (50: 159).

6. Letter, TAE to C. G. Wildreth, November 17, 1879, cited on p. 108 of Friedel and Israel 1986.

7. Letter, Grosvenor Lowrey to TAE, November 13, 1879 (50: 287–290).

8. Letter, Henry Hall to TAE, November 6, 1879 (49: 738–739).

9. Letter, Fred Webber to TAE, November 20, 1879 (49: 741–742).

10. Telegram, TAE to George Hopkins, November 7, 1879 (49: 740).

11. Exchange of telegrams, TAE and M. E. Stone, November 28, 1879 (49: 743–744).

12. Exchange of telegrams, TAE and George E. Gouraud, December 1, 1879 (49: 746).

13. Letters, John S. Barrow to TAE, annotated by TAE, December 4, 1879 (49: 747); December 15, 1879 (49: 748); December 20, 1879 (49: 750).

14. William Croffut, the other reporter who often benefited from Edison's largesse, sought the story: "Can't you let a fellow in, & give me a chance at the news?" (Letter, William Croffut to TAE, December 8, 1879 (49: 748)) See also letter, William Croffut to TAE, December 26, 1879 (49: 755).

15. "Edison's Light," *New York Herald,* December 21, 1879 (94: 537–542). See also the follow-up editorials and stories in the *Herald:* "A Lucky Horseshoe," December 22, 1879 (94: 542); "Proctor on Edison," December 22, 1879 (23: 538; 94: 542–543); "Edison's Triumph," December 23, 1879 (94: 543).

16. Letter, Egisto Fabbri to TAE, December 26, 1879 (50: 297–299).

17. The early public response to Edison's announcement of success is recounted in a letter from Peter A. Dowd to TAE dated December 27, 1879 (50: 192–195).

18. "Edison's Light," *New York Herald,* December 28, 1879 (94: 551).

19. Letter, Francis Upton to Elijah Wood Upton, December 28, 1879 (95: 586–588).

20. *New York Herald,* December 29, 1878, cited on pp. 411–412 of volume 1 of Jehl 1937.

21. "A Night With Edison," *New York Herald,* December 31, 1878 (94: 557–558). See also "Electricity and Gas. History Repeating Itself at Menlo Park," *New York Herald,* December 30, 1878 (94: 555–556).

22. "A New Light to the World," *Puck,* January 1, 1880 (94: 569).

23. "Edison's Great Work," *New York Herald,* January 1, 1880 (94: 559).

24. "Crowding Edison," *New York Herald,* January 2, 1880 (94: 560).

25. An interview published in the *Denver Tribune* on April 25, 1880 ("Edison at Home" (24: 11)) reported that fifty lamps "glowing here night and day for months [had] been put out." Eventually the laboratory was reconstructed as a public monument at Henry Ford's Greenfield Museum and Village in Dearborn, Michigan, for the purpose of presenting fiftieth-anniversary reenactments of the October 22 breakthrough and the New Year's Eve demonstration. On this, see Jehl 1929; see also Nye 1983 and *Ford News* 9 (October 15, 1929): 20.

26. "Edison and the Electric Light," *Nature,* February 12, 1880: 341–342 (48: 614–617). See also "The Great Edison Scare," *Saturday Review,* January 10, 1880 (23: 680–681).

27. See also "Edison's Critics," *New York Herald,* January 8, 1880 (94: 565); "Electricians on Edison," *New York Herald,* January 9, 1880 (94: 565).

28. "Mr. Edison's Opinion of a Superserviceable Professor" ("reprinted from the New York *Sun,* Dec. 27)," *New York Herald,* December 28, 1879 (94: 552).

29. "A Scientific View of It," *New York Times,* December 28, 1879 (58: 594–595). Other critical editorials, letters, and stories appeared in the *New York Times* on November 19, 1879 and on January 6, January 16, January 21, February 7, and February 16, 1880. The most positive coverage was the January 4 editorial "Edison and the Skeptics," which espoused a "wait and see" attitude.

30. "Electrician Sawyer's Challenge to Electrician Edison," *Sun,* December 22, 1879 (48: 596; 94: 549).

31. "Sawyer on Edison and the Herald," *New York Herald,* December 24, 1879 (48: 597).

32. "Edison's Electric Light," *New York World,* December 24, 1879 (48: 598–599); "Mr. Sawyer Skeptical," *New York Tribune,* January 2, 1880 (48: 599–601; 94: 564); "Mr. Edison Challenged by Mr. Sawyer," *Sun,* January 5, 1880 (48: 601); "To the Editor," *New York Herald,* January 5, 1880 (48: 602); "Mr. Sawyer's Exhibitorium," *New York Herald,* January 6, 1880 (48: 602); "W. Sawyer Declares He has Caused the Withdrawal of Edison's Patent for the Horseshoe Lamp and Why," *New York Herald,* January 22, 1880 (94: 578).

33. Letter, George Shaw to TAE, with annotations, February 3, 1880 (53: 350).

34. "Edison and the Electric Light," *Nature,* February 12, 1880: 341–342 (48: 614–617).

35. "Edison's Electrical Generator," *Scientific American,* October 18, 1879: 242.

36. "A Night with Edison," *New York Herald,* December 31, 1878 (94: 557–558).

37. "Edison's Latest Electric Light," *Scientific American,* January 10, 1880: 19.

38. "Edison's Great Work," *New York Herald,* January 1, 1880 (94: 559). Reportedly, a disappointed competitor from Baltimore was caught trying to create a short circuit (Association of Edison Illuminating Companies 1904: 86). See also "A Malicious Visitor," *New York Daily Tribune,* January 2, 1880 (94: 560).

39. "Edison's Secret," *New York Telegram,* January 23, 1880 (94: 577–578); "Edison and the Savants," *New York Telegram,* January 23, 1880 (94: 576–577).

40. "Edison's Latest Device Reviewed by an Expert," *American Gas Light Journal,* February 2, 1880.

41. "Lighting a Great City," *New York Times,* 7 February 1880. A similar technical critique appeared soon thereafter: F. G. Fairfield, "Menlo Park Laboratory. A Dispassionate Estimate of the Edison Light," *New York Times,* February 16, 1880.

42. "The Edison Electric Light," *Engineering*, January 9, 1880: 37–38.

43. "The Edison Electric Light," *Engineering*, February 6, 1880: 113–114 (23: 658).

44. See e.g. the interview "The Electric Light," *Washington Times and Expositor,* March 8, 1880 (23: 654).

45. Directive, TAE, February 19, 1880 (54: 368).

46. Telegram, C. F. Brackett to TAE, February 29, 1880 (53: 673).

47. The manuscript of the report is dated March 27, 1880 (53: 696–702).

48. Telegram, TAE to C. F. Brackett, March 30, 1880 (53: 703).

49. The results of the April 3 test were published in an Edison Company pamphlet (Brackett and Young 1883).

50. The report, without Upton's letter, was then published as C. F. Brackett and C. A. Young, "Notes of Experiments upon Mr. Edison's Dynamometer, Dynamo-machine, and Lamp," *American Journal of Science* 19 (June 1880) (48: 634–637).

51. See letter, H. A. Rowland to TAE, March 9, 1880 (95: 603); telegram, George Barker to TAE, March 20, 1880 (53: 688); see also "Testing Edison's Lights," *Sun,* March 21, 1880 (94: 589); "The Edison Electric Light," *New York Herald,* March 24, 1880 (94: 589).

52. Edison's Electric Light," *New York Herald,* March 25, 1880 (48: 625–626; 94: 590).

53. In its previous issue *The Electrician* had noted the success of both Rowland and Barker's and Brackett and Young's tests ("Electric Lighting in America," April 17, 1880 p. 261 (24: 16)). "Mr. Edison," the article observes, "is greatly vexed at the growing popular incredulity in regard to his reiterated assertion that a satisfactory system of illumination has been perfected."

54. A critique of this article by one of Barker's colleagues at the University of Pennsylvania, William D. Marks, appears as a letter to the editor titled "The Edison Electric Light" (*Engineering and Mining Journal*, April 24, 1880 (24: 26)).

55. Edwin Marshall Fox wrote Edison earlier inquiring about the report, saying he wanted to publish it "if it's favorable " (letter, Fox to TAE, March 17, 1880 (53: 686)).

56. The Electrician republished the results as "Edisonia" (May 8, 1880: 292 (24: 34)).

57. "Edison's Horseshoe Lamp," *Engineering*, May 14, 1880.

58. Contracts were signed in January, and J. C. Henderson of the navigation company went to Menlo Park to consult as work proceeded in the early months of the year. The ship, after its launch, was brought to New York Harbor to be outfitted

with lights during the months of March and April. See 53: 639, 53: 675, 53: 689, 53: 712, 53: 715, 53: 724, 53: 725, 53: 731, 53: 732, 53: 734, and 53: 735.

59. See letter, Francis Upton to Elijah Wood Upton, May 9, 1880 (95: 605–607); "The Columbia," *Scientific American,* May 22, 1880: 326; Jehl 1937, volume 2: 557–564.

60. Letters, J. C. Henderson to TAE, January 5 (60: 754) and February 24, 1882 (96: 676)

61. See e.g. "Into Brazil for Edison," *New York Times,* November 27, 1880.

62. See e.g. "Edison's Work," *New York Daily Tribune,* November 26, 1880 (94: 618; 89: 18); "How Far Edison Has Got," *New York World,* November 29, 1880 (89: 18).

63. "Bernhardt and Edison," *New York Herald,* November 29, 1880; "Bernhardt at Menlo Park," *Sun,* November 29, 1880.

64. "Aldermen at Menlo Park," *New York Truth,* December 21, 1880 (94: 623); "The Wizard of Menlo Park," *New York Herald,* December 21, 1880 (94: 623).

65. Association of Edison Illuminating Companies 1904, p. 166.

66. "Economy Test of the Edison Light at Menlo Park, 1881," in Association of Edison Illuminating Companies 1904.

67. Letter, Grosvenor Lowrey to TAE, December 17, 1880 (54: 12–122).

Chapter 10

1. For the history of fairs and exhibitions, see Altick 1978; Benedict 1983; Greenhaigh 1988; Luckhurst 1950; Rydell 1984; Smithsonian Institution Libraries 1992.

2. Letter, George E. Gouraud to TAE, December 21, 1880 (54: 359–362).

3. Letter, George E. Gouraud to TAE, April 23, 1881 (58: 871).

4. "Estimated Cost of Exhibition of Electric Light at Paris," June 24, 1881 (58: 901).

5. Letter, Charles Batchelor to Sherburne Eaton, October 27, 1881 (58: 1109–1111).

6. Letter, Charles Batchelor to Theodore Puskas, May 31, 1881 (58: 882).

7. Letter, Theodore Puskas and Joshua Bailey to TAE, June 14, 1881 (58: 892–896).

8. "Electric Lights for Paris. The Edison Company to Light a Part of the Grand Opera House," *New York Tribune,* September 20, 1881 (94: 660). "L'Éclairage Électrique du Foyer de l'Opéra," *L'Illustration,* November 19, 1881 (94: 678; 89: 136–137); "The Electric Lighting of the Grand Opera at Paris," *Engineering,* October 21, 1881: 417–419 (89: 161–162). See also letter, Theodore Puskas and

Joshua Bailey to TAE, September 16, 1881 (58: 1067–1068); letter, Otto Moses to TAE, September 22, 1881 (58: 1075–1077).

9. Letter, Otto Moses to TAE, July 10, 1881 (58: 908–916).

10. Letter, Otto Moses to TAE, July 25, 1881 (58: 947–952).

11. The offices of the USELC were at 25 Avenue de l'Opera.

12. Letter, Charles Batchelor to TAE, August 30, 1881 (58: 1020–1024). See also letter, Theodore Puskas and Joshua Bailey to TAE, August 17, 1881 (58: 985–988).

13. Letter, Otto Moses to TAE, September 7, 1881 (58: 1038–1040).

14. Letters, Otto Moses to TAE, July 10, 1881 (58: 908–916) and July 12, 1881 (58: 920–923).

15. See enclosure of letter from Puskas to TAE, July 26, 1881 (58: 938–941).

16. See e.g. letters, Theodore Puskas and Joshua Bailey to TAE, July 22, 1881 (58: 930–934) and September 9, 1881 (58: 1056–1058).

17. Letter, Otto Moses to TAE, July 12, 1881 (58: 920–923).

18. Letter, Theodore Puskas and Joshua Bailey to TAE, July 29, 1881 (58: 939–941).

19. Letter, Otto Moses to TAE, August 11, 1881 (58: 978–981).

20. Letters, Theodore Puskas and Joshua Bailey to TAE, August 17, 1881 (58: 985–988); August 19, 1881 (58: 992–995). See also letter of agreement with du Moncel from Joshua Bailey, August 22, 1881 (58: 1015–1018).

21. Letter, Otto Moses to TAE, August 2, 1881 (58: 947–951).

22. Letter, Theodore Puskas and Joshua Bailey to TAE, August 8, 1881 (58: 956).

23. Letter, Theodore Puskas and Joshua Bailey to TAE, July 11, 1881 (58: 917–919).

24. Letter, Theodore Puskas and Joshua Bailey to TAE, July 22, 1881 (58: 930–934).

25. Letter, Theodore Puskas and Joshua Bailey to TAE, July 22, 1881 (58: 931).

26. Letter, Theodore Puskas and Joshua Bailey to TAE, July 14, 1881 (58: 924–926).

27. Letters, Theodore Puskas and Joshua Bailey to TAE, August 19, 1881 (58: 992–995) and August 26, 1881 (58: 1008–1011). *Moniteur Officiel de l'Électricité* 1: 1 indeed features a three-page story plus a cover portrait of Edison ("Thomas-Alva Edison," *Moniteur Officiel de l'Électricité* 1: 1 (August 25, 1881) (89: 86)). The second issue follows, after the general stories of the exposition and organizers, with two more pages of continuation, plus an account of the Edison Light ("La Lumière Edison," ibid. 1: 2 (September 3, 1881) (89: 79–82)).

28. Letter, Theodore Puskas and Joshua Bailey to TAE, August 19, 1881 (58: 996).

29. Letter, Theodore Puskas and Joshua Bailey to TAE, August 26, 1881 (58: 1008–1011).

30. Ibid. An article by du Moncel already had been planted in *Journal des Debats.*

31. Ibid.

32. Letter, Otto Moses to TAE, August 11, 1881 (58: 979).

33. Letter, Charles Batchelor to TAE, August 9, 1881 (58: 961–962).

34. Letter, Charles Batchelor to TAE, August 12, 1881 (58: 981–982).

35. Letter, Theodore Puskas and Joshua Bailey to TAE, July 29, 1881 (58: 939–941); see also August 17 (58: 985–988).

36. Letter, Theodore Puskas and Joshua Bailey to TAE, August 8, 1881 (58: 956–958).

37. Letter, Otto Moses to TAE, August 11, 1881 (58: 978–980).

38. Telegram, Heraclite to TAE, August 21, 1881 (83: 885).

39. Letter, Otto Moses to TAE, August 31, 1881 (58: 1025).

40. Letter, Otto Moses to TAE, August 26, 1881 (58: 1005).

41. Letter, Otto Moses to TAE, August 19, 1881 (58: 990).

42. Letter, Theodore Puskas and Joshua Bailey to TAE, July 22, 1881 (58: 930–934).

43. Letter, Theodore Puskas and Joshua Bailey to TAE, August 17, 1881 (58: 985–988).

44. Letter, Otto Moses to TAE, August 26, 1881 (58: 1005–1007).

45. On the close relation between the AAAS and the Edison-sponsored journal *Science* during this period, see chapter 6 above.

46. Letter, Otto Moses to TAE, September 7, 1881 (58: 1038–1040). Moses also mentioned a European ally who had been approached to be on the jury, who claimed to have an agreement with Edison for Swiss rights.

47. Letter, George Barker to TAE, June 10, 1881 (58: 888–890).

48. This was Sir William Crookes, a fellow of the Royal Society.

49. Letter, Charles Batchelor to TAE, October 1, 1881 (58: 1082).

50. Letter, Charles Batchelor to TAE, October 4, 1881 (58: 1085).

51. Letter, Grosvenor Lowrey to TAE, October 23, 1881 (58: 1098–1104). According to Hounshell (1980, p. 615), who cites the minutes of a special meeting of the Executive Committee of the Board of Directors of the Edison Electric Light Company, January 3, 1889, Barker was kept on an annual retainer of $500

plus guaranteed a hundred days per year consultation, at a rate of $50 a day, for a minimum annual recompense of $5500.

52. "The Electrical Exhibition," *London Times,* October 22, 1881 (89: 0139) provides the most detailed account of the awards.

53. In the 1882 republication of Theodore du Moncel's 1880 article in *La Lumière Electrique,* for example, the diploma was reproduced in a fold-out illustration with an accompanying text attributing all five medals and the diploma specifically to the "Complete System of Electric Lighting."

54. All three cables are reported in a letter from Grosvenor Lowrey to TAE dated October 23 1881 (58: 1098).

55. "Report on the Incandescent Lamps Exhibited at the International Exposition of Electricity, Paris, 1881," *Electrician,* June 17, 1882: 104–107.

56. "Les Lampes Électriques," *La Lumière Électrique,* October 1, 1881 (94: 662–666).

57. *La Lumière Edison* (Paris: A. Lahure, 1882); *Das Licht Edison* (Berlin: W. Buxenstein, 1882); Theodore du Moncel and William Henry Preece, *Incandescent Electric Lights* (Van Nostrand, 1882).

58. See e.g. "The International Exhibition of Electricity," *Scientific American Supplement* 309, December 3, 1881; "Edison's System of Electric Lighting," *Operator* 13: 2, January 15, 1882: 25–29; "Exposicion Internacional De Electricidad," *La Illustracion Española y Americana,* October 22, 1881 (94: 675).

59. "Exposition Internationale de l'Électricité," *Journal des Debats,* September 8, 1881 (94: 659); Henri de Parville, "Exposition d'électricité," *Journal des Debats,* October 22, 1881 (94: 667–670; 89: 68); October 27, 1881 (94: 680–684 89: 69–71); Olympe Audouard, "Edison," *Le Papillon,* September 4, 1881 (94: 652–653; 89: 108–109); A. Vernier, "L'exposition d'électricité" *Du Temps,* September 6, 1881 (94: 656–658); "Edison ou la Lampe Merveilleuse," *Le Figaro,* October 29, 1881 (94: 686–687; 89: 102–103); "Exposition Internationale d'électricité," *L'Illustration,* October 1, 1881 (94: 676; 89: 115); "M. Edison," *L'Illustration,* October 15, 1881 (94: 666–667); "Le Système d'Éclairage Edison," *L'Illustration,* October 22, 1881 (94: 672–674); "L'Éclairage Électrique du Foyer de l'Opéra," *L'Illustration,* November 19, 1881 (94: 678); F. A., "Exposition d'Électricité," *Republique Français,* September 4, 1881 (94: 647–649); "L'Éclairage Électrique," *Le Voltaire,* October 13, 1881 (94: 671; 89: 134); E. Vignes, "Un systeme pratique complet d'Éclairage Électrique," *La Science,* no date (89: 112–113); "Exposition d'électricité," *La Correspondence de Paris,* September 26 1881 (89: 91–92); "L'Éclairage Électrique Edison," *L'Écho Industriel,* no date: 434–437 (89: 72–73).

60. "The Exhibition of Electricity at Paris—No. VII," *Engineering,* September 2, 1881, p. 242.

61. "The Electric Lighting of the Grand Opera at Paris," *Engineering,* October 21, 1881: 417–419.

62. "The Large Edison Dynamo-Electrical Machine at the Paris Exhibition of Electricity," *Engineering*, October 21, 1881: 419.

63. "The Edison System of Electrical Illumination," *Engineering*, March 10, 1882: 226–228 and March 31, 1882: 305–307.

64. "The Electrical Exposition at Paris," *Electrician*, October 22, 1881: 361.

65. See *Scientific American*, August 13, September 10, September 17, September 24, November 5, November 12, November 19, December 3, December 10, and December 17, 1881.

66. See *The Operator*, July 15 and August 15, 1881, and January 15, 1882.

67. See letter, Grosvenor Lowrey to TAE, October 23, 1881 (58: 1098–1104).

68. "The Crystal Palace Exhibition," *Electrician*, March 4, 1882: 245.

69. "The Crystal Palace Exhibition," *Electrician*, June 10, 1882: 89–91.

70. "The Munich Exhibition," *Electrician*, October 7, 1882: 498.

71. "The Munich Electric Exhibition," *Engineering*, November 3, 1883: 431–432 (89: 549–550).

72. "Munich Exhibition," *Electrical Review*, October 25, 1883; "The Incandescent Light and Decorative Art," *Electrical World* II: 2 (September 8, 1883): 17.

73. See letter, Francis Jehl to TAE, August 4, 1883 (70: 761).

Chapter 11

1. See chapter 1 above.

2. Hellrigel, *The Quest to Be Modern* (dissertation in progress, Case Western Reserve University).

3. Tammany Hall was a political club, often associated with corruption, that dominated New York City's Democratic Party and its city government from the Civil War through the 1920s.

4. Kelly was never caught in a scandal.

5. This can be seen concretely in the Edison papers: typed documents displace handwritten ones as the 1870s progress.

6. In another incident, in October 1881, Edison had to have the streets cleared to transport the Jumbo dynamo to the docks for it to be shipped to the Paris exhibition. Speaking to the Tammany leader ensured that the route was lined with police. (Dyer and Martin 1910, pp. 326–327)

7. Two years later, in 1882, Campbell was to make an unsuccessful run for mayor on the Republican line against a Tammany-Swallowtail coalition candidate.

8. Letter, Tracy Edson to TAE, November 27, 1880 (54: 107–108).

9. Letter, Tracy Edson to TAE, November 27, 1880 (54: 108).

10. Letter, Calvin Goddard to TAE, December 15 (54: 118). On December 15, Calvin Goddard, secretary of the Edison Electric Light Company, sent Edison this draft for Edison to copy and sign in his own hand so it could be delivered the next day at an apparently crucial meeting. Just a few days before, Edison had received an application, dated December 6, from a G. Morris for an engineer's position in London. I cannot determine whether G. Morris, then an Inspector for Metropolitan Telephone & Telegraph Company in New York, was related to Joseph Morris, the president of the Board of Aldermen; however, the letter was annotated by Edison to be sent to Edward Johnson, who was in charge of Edison's London Electric Light and Power operations. Letter, G. Morris to TAE, December 6, 1880 (56: 814).

11. TAE, draft of letter to members of New York City government, December 18, 1880 (54: 123).

12. Telegram, Calvin Goddard to TAE, December 20, 1880 (54: 125).

13. "Aldermen at Menlo Park," *New York Truth,* December 21, 1880 (94: 623); The Wizard of Menlo Park, *New York Herald,* December 21, 1880 (94: 623); "The Aldermen Visit Edison," *New York Times,* December 21, 1880.

14. "Mr. Edison's Melancholy," *New York World,* January 1, 1881; Jehl 1937, p. 785.

15. "Lightning over Snow" *New York Herald,* January 20, 1881 (94: 627–628).

16. "Mr. Edison's Melancholy," *New York World,* January 1, 1881; "The Edison Light," *Newark Daily Advertiser,* January 7, 1881; "Edison's Electric Light," *Sun,* January 7, 1881.

17. "Visiting Edison's Home," *Newark Morning Register,* February 9, 1881 (94: 631); see also "The Edison Electric Light," *Newark Daily Journal,* February 9, 1881; "Inspecting Edison's Invention," *Newark Daily Advertiser,* February 9, 1881.

18. "The Edison Light," *Newark Daily Advertiser,* January 7, 1881; "Edison's Electric Light," *Sun,* January 7, 1881.

19. *Proceedings of the Board of Alderman of the City of New York,* volume 151 (New York: Martin B. Brown, 1881): 403–404.

20. *Proceedings of the Board of Alderman of the City of New York,* volume 152 (New York: Martin B. Brown, 1881): 40–43; "Mayor Grace Vetoes," *New York Herald,* April 6, 1881; "Vetoed by Mayor," *Sun,* April 6, 1881.

21. There is no record in the municipal archives of his having received any recommendation on this matter from his new Commissioner of Public Works, Robert Thompson, despite the many detailed recommendations on the most minor of aldermanic resolutions, including ones for the construction of single awnings in front of houses.

22. *Proceedings of the Board of Alderman of the City of New York,* volume 152: 95–97, 101–103.

23. Ibid.: 202–207.

24. Franklin Edson is not known to have been a relative of the by-then-deceased Tracy Edson.

25. See also Maltbie 1911 and Arent 1919.

26. *Iconography of Manhattan Island,* volume 5 (New York: Robert H. Dodd, 1928): 1986. The most complete account is to be found on pp. 139–144 of Myers 1974.

27. Similar stories concerning the laying of conductors under the streets of Philadelphia are recounted on p. 124 of *A Course of Thirteen Practical Talks to the Working Men of the Edison Electric Light Company of Philadelphia* (Philadelphia: Edison Co., 1895).

28. Among the steamers were the *Pilgrim,* the *Albatross,* the *Carolina,* and the *Virginia.* Sources: "List of Edison Plants Installed in the United States," *Fourteenth Bulletin of the Edison Electric Light Company,* October 14, 1882: 19–21 (96: 763–764); "Edison Plants in Europe," *Sixteenth Bulletin of the Edison Electric Light Company,* February 2, 1883: 27–30 (96: 802–804); "Edison Isolated Plants. Full List," *Eighteenth Bulletin of the Edison Electric Light Company,* May 31, 1883: 30–38 (96: 842–846); "Plants Sold Since May 31st, 1883," *Twenty-Second Bulletin of the Edison Electric Light Company,* April 9, 1884: 5–38 (96: 926–928).

29. *Bulletin of the Edison Electric Light Company,* nos. 1–22 (January 26, 1882–April 9, 1884) (96: 667–840).

30. Menlo Park Notebook, volume 120 (37: 586–632). See also "The Light of the Future," *New York Truth,* March 6, 1881 (94: 629).

31. Menlo Park Notebook, volume 132 (38: 315–413). Claudius, a former officer in the Austrian Telegraph Corps, began working for Edison in August of 1880.

32. "First District, New York City," *Seventeenth Bulletin of the Edison Electric Light Company,* April 6, 1883: 3–4 (96: 810–811); "First District, New York City," *Twentieth Bulletin of the Edison Electric Light Company,* October 31, 1883: 3–6 (96: 868–870).

33. "The Distribution of Light and Heat in New York City," *Scientific American* 45, November 19, 1881: 319–320. See also "Edison's New Steam Dynamo," *Scientific American* 45, December 10, 1881: 367.

34. "A Glance Backward," *Scientific American* 45, December 24, 1881: 400.

35. "First District, New York City," *Fourteenth Bulletin of the Edison Electric Light Company,* October 14, 1882: 1–3 (96: 754–755).

36. "Annual Stockholder's Meeting, Edison Electric Light Company," *Twentieth Bulletin of the Edison Electric Light Company,* October 31, 1883: 41–50 (96: 887–892).

37. "Edison's Incandescent Light," *New York World,* September 5, 1882 (24: 248).

38. "Edison's Illuminators," *New York Herald,* September 5, 1882 (24: 82).

39. "Edison's Light Turned on," *Sun,* September 5, 1882.

40. "Electricity Instead of Gas," *New York Tribune,* September 5, 1882.

41. "Edison's Electric Light," *New York Times,* September 5, 1882.

Chapter 12

1. "Index, Abandoned Patent Applications, Case 202," (45: 434).

2. Figure, (45: 435–436).

3. Charles Batchelor Notebook, Catalogue 1304, entry of November 28, 1878 (91: 4).

4. *Edison Electric Light Company v. United States Electric Light Company,* U.S. Circuit Court for the Second District: 2 (46: 257).

5. Letter, H. E. Paine to TAE, March 30, 1880 (45: 437–438).

6. Letter, TAE to Commissioner of Patents, April 20, 1880 (45: 439–442).

7. Letter, E. M. Marble to TAE, September 20, 1880 (45: 443–445).

8. Letter, TAE to Commissioner of Patents, December 9, 1880 (45: 446–447).

9. Letter, E. M. Marble to TAE, December 15, 1880 (45: 448).

10. In recent law, because of the difficulties of the concept of novelty, the criterion has become defined as non-obviousness to one versed in the art. See Witherspoon 1980.

11. "Electric Lights and System of Electric Lighting," July 26, 1882 (45: 449–457).

12. Ibid.: 7 (45: 455).

13. Letter, E. M. Marble to TAE, September 13, 1882 (45: 458–459).

14. Letter, R. N. Dyer to Commissioner of Patents, September 13, 1884 (45: 460–461).

15. Letter, B. Butterworth to TAE, November 5, 1884 (45: 462).

16. Letter, R. N. Dyer to Commissioner of Patents, November 3, 1886 (45: 463).

17. Letter, W. V. Montgomery to TAE, November 8, 1886 (45: 463–464).

18. Letter, R. N. Dyer to Commissioner of Patents, February 2, 1888 (45: 466–475).

19. Letter, B. J. Hall to TAE, February 16, 1880 (45: 476–477).

20. Letter, R. N. Dyer to Commissioner of Patents, February 20, 1888 (45: 478–480).

21. Letter, B. J. Hall to TAE, February 27, 1888 (45: 481–482).

22. Letter, C. E. Mitchell to TAE, April 18, 1890 (45: 496–499).

23. Letter, W. E. Scimonds to TAE, April 19, 1892 (45: 502–503).

24. Letter, J. S. Seymour to TAE, May 15, 1894 (45: 513).

25. Letter, R. N. Dyer "Memorandum for Applicant" (45: 514–520).

26. Letter, J. S. Seymour to TAE, April 27, 1895 (45: 532).

27. For the detailed story of the EELC's attempt to enforce its patent victory, see pp. 89–93 of Bright 1949.

28. *Bulletin of The Edison Company for Isolated Lighting*, May 15, 1885–June 7, 1886 (97: 3–129).

29. See e.g. "List of Edison Isolated Plants Installed Prior to October 1, 1885, in the United States" (96: 271–284).

30. This was republished in "A Warning from the Edison Electric Light Co. " (1888), an 84-page pamphlet presenting extensive arguments and facts, documents, editorials, and articles bolstering Edison's position.

31. See e.g. advertisement by Edison Electric Light Company, *Electrical Review*, August 4, 1888: 17 (95: 233).

32. Edison Electric Light Company, Report of the Board of Trustees to the Stockholders at their Annual Meeting, October 27, 1885 (96: 5–14).

33. Opinion of Justice Bradley in *The Consolidated Electric Light Company v. McKeesport Light Co.* (46: 952).

34. Indeed, he had filed designs for vacuum designs as early as April 12, 1879 (patent 227,229), and he had filed many other vacuum-related patents during 1880. The claims section of the patent judged to be crucial (223,898) even uses the phrase "from which receiver the air is exhausted," but only as an ancillary description.

35. The records of both the interference and the McKeesport trial became parts of the New York trial record, available on reels 46–48 of the microfilmed Edison papers.

36. Judge Bradley's decision in *Consolidated Electric Light Company v. McKeesport Light Co.* can be found in *Edison Electric Light v. U.S. Electric Light*, volume 1. *Pleadings, Complainants Prima Facie Proofs and Decisions relating to the Patent in Suit*: 382–398 (46: 952–959).

Chapter 13

1. See Edward Johnson's letter of gratitude to TAE, January 25, 1879 (50: 230–231), Edwin Marshall Fox's, January 26, 1879 (50: 237–238), James Banker's, January 30, 1879 (50: 241), and William Croffut's, February 3, 1879 (50: 242–243).

2. Letter, Otto Moses to TAE, June 1, 1879. (49: 914).

3. Letters, Otto Moses to TAE, January 23, 1882 (63: 140–142), January 27, 1882 (62: 221–222).

4. William J. Hammer as told to Willis J. Ballinger, "Edison—By the Man Who Knew Him Best," NEA Service, 1931, part 2.

5. Ibid., part 4.

6. Letter, Francis Upton to Elijah Wood Upton, March 23, 1879 (95: 514–516).

7. Letter, Francis Upton to Elijah Wood Upton, April 27, 1879 (95: 527–528).

8. Letter, Francis Upton to Elijah Wood Upton, November 30, 1879 (95: 577–579).

9. Letter, Francis Upton to Elijah Wood Upton, December 28, 1879 (95: 586–588).

10. See (63: 146–183; 70: 709–794; 76: 578–641). See also distributions from TAE's account of shares in Edison Ore Milling Co. (54: 449).

11. List of Stockholders (58: 125–132).

12. See letters and telegrams to TAE from Batchelor, Lowrey, and Moses (83: 922–935).

13. Maurice Holland, "Edison—Organizer or Genius?" April 1927. Accession 1630, Box 12, 12-7, Greenfield Village Research Center.

14. The rise of corporate forms during this period when the large modern corporation was developing is documented extensively in Yates 1988.

15. Letter, Charles Tate to Charles Batchelor, September 23, 1884 (89: 591).

16. Ibid.

17. Letter, Charles Tate to Charles Batchelor, September 24, 1884 (89: 593).

18. Letter, Charles Tate to Charles Batchelor, October 9, 1884 (89: 610).

19. Letter, Charles Tate to Charles Batchelor, October 3, 1884 (89: 601).

20. Canvass Book (97: 191–193).

21. Similarly, as Edison operations spread throughout the country, printed reporting forms become impersonal regulators of organizational information flows, such as the printed forms "Estimate for the Erection and Equipment of an Edison Electric Light Central Station" and the monthly report form for local Edison

Electric Illuminating companies. "Estimate for the Erection and Equipment of an Edison Electric Light Central Station," Accession 1630, box 16, folders 57 and 2, Greenfield Village Research Center.

22. "Questions for Central Station Engineers," November 21, 1883 (97: 135–180). Titles appear on (97: 136). The editors of the papers identify the missing author of "Answers Relating to Running of Engine and Boiler" as W. D. Rich (97: 135)

23. T. A. Edison, "Instructions" (97: 177–180).

24. *Instructions for the Installation of Isolated Plants Published Exclusively for the Private Use of the Agents of the Edison Company for Isolated Lighting* (New York: Edison Company for Isolated Lighting, 1881).

25. (53: 4–18). See also files 1881 (57: 3–18); 1882 (60: 5–31); 1883 (66: 392–483); 1884 (73: 25–95); 1885 (77: 539–554); 1886 (79: 306–314).

26. For contemporary accounts of Bergmann & Co., see "Edison's Environments," *Manufacturers' Gazette,* March 10, 1883 (24: 88–89) and "A Representative American Electrical Manufactory," *Electrical World* II: 17 (December 22, 1883): 275.

27. According to a proposed agreement of April 1881 (57: 7–10), Edison and Bergmann each were to receive 44 percent of the profit and Johnson the remaining 12 percent. Edison was to provide $7500 capital, and Bergmann was to receive a salary for managing the company. This agreement was apparently never signed and a revised agreement was prepared the same month, and modified before signing in September 1882, requiring a contribution of $38,290.49 from Edison and $35,000 from Johnson and resulting in an equal partnership among the three. Partnership Agreement, September 4, 1882 (58: 20–25). This combination of documents suggests that the arrangement was carried out on an ad hoc arrangement for the first year until formalized in 1882. A letter from Samuel Insull to Charles Batchelor dated September 28, 1882 (63: 167–179) further specifies that 10 percent of Edison's third was assigned to Batchelor. See also Memorandum "Licensing Bergmann," S. B. Eaton, January 24, 1883 66: 395–406. Edison's annotations suggest that an exclusive license was not granted to Bergmann.

28. Letter, R. G. Dun to TAE, annotated by Edison, February 4, 1881 (56: 768).

29. Memo for Contract between the E. E. L. Co & E. E. Lamp Co., January 1881 (57: 761–765). This contract seems not to have been executed until March 8, 1881; see draft of revised contract letter from Grosvenor Lowrey to TAE, May 12, 1882 (61: 754–767). That revised contract seems not to have been enacted; see letter, Samuel Insull to Charles Batchelor, September 28, 1882 (63: 167–179). Another contract was proposed January 19, 1883; see S. B. Eaton, Memorandum on Proposed Contract (67: 116–138) and contract (67: 280–293). The contract was again amended in 1885 (77: 888–916).

30. (54: 2–32).

31. See files of Edison Lamp Company 1881 (57: 756–1021); 1882 (61: 720–853); 1883 (67: 90–298); 1884 (23: 359–485); 1885 (77: 859–916); 1886 (79: 485–511).

32. Letter, William Curtis to TAE, October 25, 1883 (67: 255–256).

33. See also letter, Francis Upton to TAE, November 8, 1883 (67: 264–266), demanding financial arrangements be regularized, as the company's affairs were in dire disorder, with Upton's own personal finances in danger because of the funds he had advanced to the company to cover expenses.

34. Certicate of Organization (67: 294–298).

35. By-Laws of Edison Lamp Co. (73: 463–485).

36. Letter, Bergmann & Co to Francis Upton, October 12, 1881 (57: 12).

37. Letter, S. Bergmann to TAE, 19 October 1881 (57: 14).

38. Letter, Samuel Insull to Bergmann & Co, October 25, 1881 (57: 15).

39. Letter, C. Goddard to TAE, June 3, 1881 (58: 34).

40. File of Edison Machine Works 1881 (58: 203–346); 1882 (61: 1000–1171); 1883 (67: 353–441); 1884 (73: 548–651); 1885 (77: 921–990); 1886 (79: 512–675).

41. By-Laws Edison Machine Works, March 1881 (58: 246–254).

42. See e.g. Copy of Contract, signed TAE, June 7, 1882 (61: 1072), granting Dean 7 percent of the profits from a major order, and Rocap 3 percent.

43. Letter, TAE to Charles Rocap, February 10, 1883 (67: 360).

44. Letter, Charles Rocap to TAE, February 13, 1883 (67: 361).

45. Letter, William Anderson to Samuel Insull, August 18, 1883 (67: 390–391); however, notice also a receipt near the end of the year for a $1000 settlement to be given to Rocap. John Tomlinson, November 30, 1883 (67: 427).

46. See e. g. letter, Charles Batchelor to Samuel Insull, August 10, 1883 (67: 392) and September 12, 1883 (67: 408).

47. Minutes of the Board of Trustees, Edison Machine Works, February 13, 1884 (73: 552–559).

48. See letter, Charles Batchelor to Edison Construction Department, May 28, 1884 (73: 578), signed by Batchelor as "Gen'l Manager. " Batchelor then took over control of the correspondence for the Machine Works; this was made official shortly thereafter (memorandum, Charles Batchelor, June 1884 (73: 601)).

49. The tube company produced supplies for the laying of cable. Edison served as president, John Kruesi as treasurer, and Samuel Insull as secretary. Kruesi immediately took over control of the operations and a letterhead listing the three officers was soon printed. Kruesi at first passed on information and queries, as well as requests for Edison's signature on checks, either directly to Edison or via Insull. However, soon Kruesi handled almost all matters on his own. Electric Tube

Company 1881 (58: 347–378); 1882 (62: 90–144); 1883 (67: 442–508); 1884 (73: 710–756); 1885 (77: 1024–1057).

50. TAE Construction Department 1883 (68: 302–668); 1884 (74: 383–523).

51. Agreement, July 14, 1884 (73: 653–655); Edison Shafting Company File 1884 (73: 652–709); 1885 (77: 991–1023); 1886 (79: 676).

52. Agreement, July 14, 1884 (73: 657).

53. By-Laws of the Edison Shafting Company (73: 685–691).

54. Agreements, August 28, 1884 (73: 661–675) (duplicate 73: 692–709) The ownership was split Edison 44 percent, Livor 44 percent, and Batchelor 12 percent (73: 683). Edison later assigned three of his shares to Kruesi (letter, John Kruesi to Samuel Insull, December 15, 1884 (73: 680)). Further stock was issued in 1885 (Certificate of Proceedings of Meeting of Stockholders (77: 1019–1021)).

55. It was under Batchelor's auspices at the Machine Works that Tate was sent on the canvassing trip to Michigan and Canada that generated the forms discussed above. Tate was even given a fancy certificate authorizing his role (A. O. Tate, July 19, 1884 (73: 604)). Under Batchelor's auspices as general manager of the Machine Works, arrangements were also made to establish Edison Companies in South America (Agreement with Milton Adams, August 1884 (73: 600–601) and Agreement with Carlos Montiero, August 1884 (73: 602–634); see also list of canvassers (74: 545–548)).

56. Agreement, January 1, 1886 (79: 513–517).

57. See Agreements, May 1886 (79: 537–540; 541–543). Apparently the merger was made official on February 18, according to court documents in another matter (Edison Machine Works against the Standard Stand Rock Drill Company) (79: 568–571).

58. An extended description of the Schenectady works appeared as "A Day with Edison at Schenectady," *Supplement to the Electrical World* 12 (August 25, 1888): 1–12.

59. Edison United Manufacturing 1886 (79: 677–686).

60. TAE, Notes (79: 681–682).

61. "To the Unfortunate Stockholders" (79: 683–685).

62. See chapter 2 of Israel 1992.

63. See Garnet 1985 and Stehman 1925.

64. See files TAE Construction Department 1883 (68: 302-70: 523); 1884 (74: 383-76: 453). We can see some of the dynamics of local enthusiasm that Shaw was working on in a newspaper story from Cythiana, Kentucky. The town had been inspired in by the early plant in Sunbury to seek its own central power plant. "The Electric Light. Shall Cythiana Have It? A Company Formed and a Large Part of the Stock Already Taken," *Cythiana Democrat*, October 12, 1883 (24: 91).

65. The detailed involvement of Edison in the development of Sunbury, Brockton and others is detailed in Hellrigel 1989, especially chapter 3. We can also see the level of Edison's involvement in his vetoing of a proposed station in Danbury, Connecticut. (69: 426–433).

66. Another plant was also contemplated in Louisville, apparently in conjunction with the Southern Exposition, but there is no record of an Edison company being formed or operating in Louisville in the 1880s, despite the successful display at the exposition. The exposition power plant was dismantled at the close of the event. None of the principals in the early electrical companies in Louisville correspond with names appearing in the correspondence with Edison. George Yater, Private Correspondence, October 11, 1997. In the Norvin Green Papers at the Filson Club in Louisville is a letter (dated July 11, 1881) from Norvin Green to Calvin Goddard giving a candid assessment of two groups of investors seeking an Edison franchise. Even though one group included the mayor and a former President of the Board of Alderman, neither group appears to have actually organized a power and light company. Norvin Green, personal letter, Letter Press Copy Book, volume 2, pp. 2–3.

67. For details of the Boston Edison company I rely on Sicilia 1991. See also Toner 1951.

Chapter 14

1. However, in the changing corporate and technological climate of the early 1880s, the ambitions of all participants changed. Major factors in this change were Thomas Edison's leadership; the realignment of the companies through agreements, mergers, and purchases of patents; and the availability of more powerful generators, incandescent lighting, and other products changing the capabilities of systems. For a detailed account of the development of the electrical industries, see Passer 1953.

2. This widely reproduced article is also discussed in chapter 10 in relation to the original exposition and in chapter 15 in relation to aesthetics of the home.

3. The Edison Company for Isolated Lighting, Edison Light (96: 115–124).

4. See e.g. The Edison System of Incandescent Electric Lighting as Applied in Mills, Steamships, Hotels, Theatres, Residences, &c. (New York: Edison Company for Isolated Lighting, 1883); The Edison Electric Light System: Light and Power (London: The Edison Electric Light Company, Limited, 1883); La Lumière Edison, tenth edition (Paris, 1884) (96: 374–402); Deutschen Edison Gesellschaft, Das Edison Glühlicht und seiner Bedeutung für Hygiene und Rettungswesen (Berlin: Julius Springer, 1883) (96: 416–440); Deutschen Edison Gesellschaft, Elektrisch Beleuchtung von Theatern mit Edison-Glühlicht (Berlin: Julius Springer, 1884) (6: 441–474); Compañia Eléctrica de Edison, Luz Eléctrica de Edison (Valparaiso: Imprenta Excelsior, 1885) (96: 491–497).

5. Edward H. Johnson, "Edison Electric Light Stock considered as a speculative holding for the ensuing quarter," September 15, 1881 (58: 37–53): 4 (58: 40).

6. Ibid. 2 (58: 38).

7. Ibid. 9 (58: 45).

8. Ibid. 17 (58: 53).

9. "Edison's Light Turned on," *Sun,* September 5, 1882.

10. "The Edison Light," *New York Times,* October 26, 1881: 8.

11. "Annual Meeting of Stockholders of Light Company," *Fifteenth Bulletin of the Edison Electric Light Company,* December 20, 1882: 37–39 (96: 784–785).

12. "Annual Meeting of the Edison Electric Illuminating Company of New York City," *Fifteenth Bulletin of the Edison Electric Light Company,* December 20, 1882: 39–43 (96: 785–787).

13. *Fifteenth Bulletin:* 41.

14. Ibid.: 43

15. "Annual Meeting of Isolated Company," *Fifteenth Bulletin:* 43–46 (96: 787–789).

16. *Fifteenth Bulletin:* 46.

17. "Annual Meeting of the Isolated Company," *Twenty-Second Bulletin of the Edison Electric Light Company,* April 9, 1884: 23–28 (96: 935–938).

18. "Annual Meeting of the Edison Electric Illuminating Company of New York City," *Twenty-Second Bulletin of the Edison Electric Light Company,* April 9, 1884: 16–23 (96: 932–935).

19. "Annual Stockholder's Meeting, Edison Electric Light Company," *Twentieth Bulletin of the Edison Electric Light Company,* October 31, 1883: 41–50 (96: 887–892).

20. Edison Electric Illuminating Company of New York, Report of the Board of Trustees to the Stockholders at their Annual Meeting, December 9, 1884 (96: 297–303). This is the last report of the Edison Electric Illuminating Company of New York available in the Edison papers.

21. Edison Company for Isolated Lighting, Report of the Board of Trustees to the Stockholders at their Annual Meeting, November 18, 1884. (96: 266–370). This is the last report of the Edison Company for Isolated Lighting available in the Edison papers.

22. Edison Electric Light Company, Report of the Board of Trustees to the Stockholders at their Annual Meeting, October 27, 1885 (96: 5–14).

23. Edison Electric Light Company, Report of the Board of Trustees to the Stockholders at their Annual Meeting, October 28, 1886 (96: 15–23).

24. Another such is The Edison System of Incandescent Lighting as Operated from Central Stations (first edition: New York, 1884; second edition: New York, 1885) (96: 27–40).

25. Edison Electric Light Company, Central Station Catalogue, second edition (New York, 1885) (96: 41–98).

26. W. Jenks, Description of Edison Electric Light Plant, of Brockton Mass. (Brockton, 1885) (96: 286–295).

27. See also "Laramie Letter" (New York: Edison Electric Light Company, 1886) (6: 99–101).

28. Bulletin of the Edison Electric Light Company 1–22 (January 26, 1882–April 9, 1884) (96: 667–840).

29. See e.g. Western Edison Light Company, Bulletin 1 (1882) (96: 305–310); Compagnie Continentale Édison, Bulletin 1–5 (1882–1883) (96: 330–372).

30. *Bulletin of The Edison Company for Isolated Lighting,* May 15, 1885–June 7, 1886 (97: 3–129).

31. Edison United Manufacturing Company, The Edison System of Incandescent Light and Electromotive Power from Central Stations (Schenectady: Edison United Manufacturing Company, 1888).

Chapter 15

1. Catalogue and Price List of Edison Light Fixtures Manufactured by Messrs. Bergmann & Co. (New York: Bergmann & Co., 1883) (96: 185–264).

2. Ibid.: 3.

3. See e.g. the catalogs of seven companies gathered in *Lamps & Other Lighting Devices, 1850–1906* (Pyne, 1972).

4. "A Representative American Electrical Manufactory," *Electrical World* 2 (December 22, 1883): 275.

5. "I Want Something Different," Beardslee Talks, November 1925: 12.

6. "A Representative American Electrical Manufactory," *Electrical World* 2 (December 22, 1883): 276.

7. "Poetry of Electric Light," *Electrical World* 2 (November 24, 1883): 210.

8. "Aesthetics of Electric Lighting," *Electrical World* II: 7 (October 13, 1883): 100.

9. "A Representative American Electrical Manufactory," *Electrical World* 2 (December 22, 1883): 275.

10. "The Incandescent Light and Decorative Art," *Electrical World* 2 (September 8, 1883): 17.

11. For another early attempt to link Edison's light for domestic use with aesthetics, see the opening paragraphs of "The Electric Light in Houses," *Harper's Weekly*, June 24, 1882 (24: 145).

12. See e.g. "A Dining Room Illuminated by Incandescent Lights," *Electrical World* 2 (December 1, 1883): 227; "An Art Room Lighted By a Chandelier of Incandescent Lights," *Electrical World* 2 (September 8, 1883): 17.

13. This illustration was also reproduced in a separate volume extolling the Edison system: du Moncel and Preece 1882, p. 14.

14. "The Novel Electrical Displays at the Cincinnati Centennial Exposition," *Electrical World* 12, October 20, 1888: 212–214; "A Nocturnal Fairyland," *Cincinnati Commercial Gazette*, June 10, 1888: 4.

15. Half a century later, the tenement would be epitomized by the bare flat above a noisy street, illuminated by a garish commercial sign outside the window—a scene immortalized in numerous detective films. This is literally the back side of another electric aesthetic: that of the "public great white way," which electric lighting was also establishing for the cities. Edison's associate William Hammer was as central in creating this other public aesthetic as Johnson and Bergmann were central in creating the domestic.

16. "List of Edison Isolated Plants Installed Prior to October 1, 1885, in the United States" (96: 271–284).

17. "Electricity in the House Beautiful," *Electrical World* 12 (July 16, 1887): 3–4.

18. Sixth Bulletin of the Edison Company for Isolated Lighting, July 25, 1885: 1.

19. Ibid.: 18.

20. "A Revolution in Housework," in *Thirty Years of New York, 1882–1912* (New York: New York Edison Company, 1913): 87–100.

Bibliography

Adams, M. F. 1868. "Automatic Telegraphy." *Journal of the Telegraph,* 1 June.

Adams, W. Grylls. 1882. "The Scientific Principles Involved in Electric Lighting." *Journal of the Telegraph* 15 (1 January and 16 January): 1–3, 17–19.

Allen, Oliver E. 1993. *The Tiger: The Rise and Fall of Tammany Hall.* Addison-Wesley.

Alpern, Andrew. 1975. *Apartments for the Affluent: A Historical Survey of Buildings in New York.* McGraw-Hill.

Altick, Richard D. 1978. *The Shows of London.* Belknap.

American Antiquarian Society. 1991. *Proceedings 100 part 2. Three Hundred Years of the American Newspaper.*

Appleyard, Rollo. 1939. *The History of the Institution of Electrical Engineers (1871–1931).* London: Institution of Electrical Engineers.

Arent, Leonora. 1919. *Electric Francises in New York City.* Columbia University Press.

Association of Edison Illuminating Companies. 1904. *Edisonia: A Brief History of the Early Edison Electric Lighting System.* New York.

Austin, John. 1962. *How to Do Things with Words.* Oxford University Press.

Baldwin, Neil. 1995. *Edison: Inventing the Century.* Hyperion.

Barth, Gunther. 1980. *City People: The Rise of Modern City Culture in Nineteenth Century America.* Oxford University Press.

Bazerman, Charles. 1988. *Shaping Written Knowledge: The Genre and Activity of the Experimental Article in Science.* University of Wisconsin Press.

Bazerman, Charles. 1991. "How Natural Philosophers Can Cooperate." In *Textual Dynamics of the Professions,* ed. C. Bazerman and J. Paradis. University of Wisconsin Press.

Bazerman, Charles, and David Russell. 1994. "The Rhetorical Tradition and Specialized Discourses." In *Landmark Essays in Writing Across the Curriculum*, ed. C. Bazerman and D. Russell. Hermagoras.

Benedict, Burton. 1983. *The Anthropology of World's Fairs*. Scolar.

Benson, Susan Porter. 1986. *Counter Cultures: Saleswomen, Managers, and Customers in American Department Stores, 1890–1940*. University of Illinois Press.

Bijker, Wiebe. 1995. *Of Bicycles, Bakelite, and Bulbs*. MIT Press.

Bijker, Wiebe, Thomas Hughes, and Trevor Pinch. 1987. *The Social Construction of Technological Systems*. MIT Press.

Blondheim, Menahem. 1994. *News over the Wires: The Telegraph and the Flow of Public Information in America, 1844–1897*. Harvard University Press.

Borut, Michael. 1977. The Scientific American in Nineteenth Century America. Ph.D. dissertation, New York University.

Bowers, Frederick. 1989. *Linguistic Aspects of Legislative Expression*. Vancouver: University of British Colombia Press.

Brackett, C. F., and C. A. Young. 1883. "Economy of the Edison Dynamo-Machine." In *The Edison System of Incandescent Electric Light as applied in Mills, Steamships, Hotels, Theatres, Residences, etc.* (Edison Company for Isolated Lighting): 22–25 (96: 146–149).

Breen, Matthew P. 1899. *Thirty Years of New York Politics*. New York: Matthew Breen.

Bright, Arthur Aaron, Jr. 1949. *The Electric Lamp Industry: Technological Change and Economic Development from 1800 to 1947*. Macmillan.

Britton, James E. 1977. *Turning Points in American Electrical History*. IEEE Press.

Brock, W. H. 1981. "The Japanese Connexion: Engineering in Tokyo, London, and Glasgow at the end of the Nineteenth Century." *British Journal for the History of Science* 14: 227–243.

Buchanan, R. A. 1985. "The Rise of Scientific Engineering in Britain." *British Journal for the History of Science* 18: 218–233.

Bugbee, Bruce W. 1967. *Genesis of American Patent and Copyright Law*. Public Affairs Press.

Burke, Doreen Bolger, et al. 1986. *In Pursuit of Beauty: Americans and the Aesthetic Movement*. Metropolitan Museum of Art and Rizzoli.

Butrica, Andrew J. 1987. "The Ecole supérieure de Télégraphie and the Beginnings of French Electrical Engineeering Education." *IEEE Transactions on Education* E-30: 3: 121–128.

Carlson, W. Bernard. 1991. *Innovation as a Social Process.* Cambridge University Press.

Carlson, W. Bernard, and Michael Gorman. 1989. "Thinking and Doing at Menlo Park." In *Working at Inventing,* ed. W. Pretzer. Henry Ford Museum and Greenfield Village.

Carlson, W. Bernard, and Michael Gorman. 1993. Invention as Re-Representation: The Case of Edison's Sketches of the Telephone. Unpublished.

Chandler, Alfred D., Jr. 1990. *Scale and Scope: The Dynamics of Industrial Capitalism.* Belknap.

Chandler, Alfred D., Jr. 1968. In *The Changing Economic Order,* ed. S. Bruchey and L. Galambos. Harcourt, Brace & World.

Churchill, Allen. 1958. *Park Row.* Rinehart & Company.

Conot, Robert. 1979. *A Streak of Luck.* Seaview.

Coontz, Stephanie. 1988. *The Social Origins of Private Life: A History of American Families.* Verso.

A Course of Thirteen Practical Talks to the Working Men of the Edison Electric Light Company of Philadelphia. 1895. Philadelphia: Edison Co.

Cromley, Elizabeth Collins. 1990. *Alone Together: A History of New York's Early Apartments.* Cornell University Press.

Crouthamel, James. 1989. *Bennett's New York Herald and the Rise of the Popular Press.* Syracuse University Press.

Davenport, Neil. 1979. *The United Kingdom Patent System.* London: Kenneth Mason.

de Parville, Henri. 1881. "Exposition d'électricité." *Journal des Debats,* 22 October (94: 667–670); 27 October (94: 680–684).

Dicke, Thomas S. 1992. *Franchising in America: The Development of a Business Method, 1840–1980.* University of North Carolina Press.

Dood, Kendall J. 1983. "Patent Models and the Patent Law: 1790–1880" (Part I) *Journal of the Patent Office Society,* 65: 4: 187–216. Part II 65: 5: 234–274.

Dood, Kendall J. 1991. "Pursuing the Essences of Inventions: Reissuing Patents in the 19th Century." *Technology and Culture* 32, no. 4: 999–1018.

du Moncel, Theodore. 1880. "Some Reflections in Regard to the New Lamp of Mr. Edison." *La Lumière Electrique,* 1 January: 12–13 (48: 603–605).

du Moncel, Theodore. 1881. "Progress of Electrical Science in 1880." *Journal of the Telegraph* 14 (16 March and 1 April): 81–82, 97.

du Moncel, Theodore, and William Henry Preece, 1882, *Incandescent Electric Lights.* D. Van Nostrand.

Dutton, H. I. 1984. *The Patent System and Inventive Activity during the Industrial Revolution 1750–1852.* Manchester University Press.

Dyer, Frank Lewis, and Thomas Commerford Martin. 1910. *Edison: His Life and Inventions.* Harper & Brothers.

Thomas A. Edison. 1868. "Edison's Double Transmitter." *Telegrapher,* 11 April: 265.

Edison, Thomas A. 1879. "The Action of Heat in Vacuo on Metals." *Journal of the Franklin Institute* 108: 333–337.

Edison, Thomas A. 1880a. "The Success of the Electric Light." *North American Review,* October: 295–300.

Edison, Thomas A. 1880b. "The Success of the Electric Light." *North American Review,* October: 295–300; translated and reprinted Thomas A. Edison, "Der Erfolg des electrischen Lichtes." *Frankfurter Presse,* 9 October (89: 14).

Edison, Thomas A. 1976. The Beginning of the Incandescent Lamp and Lighting System. Edison Institute.

Edison, Thomas A. 1984. "Papers, Part I," ed. R. Jenkins. Microfilm edition. University Publications of America.

Edison, Mrs. Thomas A. 1947. "Mrs. Edison Says Inventor Put Home First." Associated Press, *The Sunday Star,* Washington D. C., 2 February C. 2.

Edison, Thomas A., and Charles T. Porter, 1882, "Description of the Edison Steam Dynamo." *Journal of the Franklin Institute* 113: 1–12.

The Electric Lighting Act, 1882. Minutes of Evidence given before the Select Committee of the House of Commons and Full Text of the Act. London: The Scientific Publishing Company, Limited.

Faraday, Michael, 1839. *Experimental Researches in Electricity,* Vol 1. London: Bernard Quaritch.

Federico, P. J. 1929a. "Origin and Early History of Patents" *Journal of the Patent Office Society* 11: 292–305.

Federico, P. J. 1929b. "Colonial Monopolies and Patents." *Journal of the Patent Office Society* 11: 358–365.

Frederico, P. J. 1936. "Operation of the Patent Act of 1790." *Journal of the Patent Office Society* 18, no. 4: 237–251.

Federico, P. J. 1937. "The Patent Office in 1837" *Journal of the Patent Office Society* 19, no. 12: 954–969.

Federico, P. J. 1939. "The Patent Office in 1839." *Journal of the Patent Office Society* 21, no. 18: 786–794.

Finn, Bernard. 1989. "Working at Menlo Park." In *Working at Inventing*, ed. W. Pretzer. Henry Ford Museum and Greenfield Village.

Ford, Henry. 1930. *Edison as I Know Him.* New York: Cosmopolitan Book Corp.

Fox, Edwin M. 1879. "How Edison Disposed of Old Stock in Trade." *Operator* 10: 6 (15 March): 4–5.

Fox, Harold G. 1947. *Monopolies and Patents.* University of Toronto Press.

Fox, Robert. 1986. "Edison et la presse française à l'exposition internationale d'électricité de 1881." In *1880–1980: Un Siecle d'électricité dans le Monde*, ed. F. Cardot. Presses Universitaires de France.

Franklin Institute, 1884. *Official Catalogue of the International Electrical Exhibition.* Philadelphia: Burk & McFetridge.

Frazer, Persifor, Jr. 1878. "Examination of the Phonograph Record under the Microscope." *Journal of the Franklin Institute* 105: 348–350.

Friedel, Robert, and Paul Israel. 1986. *Edison's Electric Light.* Rutgers University Press.

Galambos, Louis, and Joseph Pratt. 1988. *The Rise of the Corporate Commonwealth: U.S. Business and Public Policy in the Twentieth Century.* Basic Books.

Garnet, Robert W. 1985. *The Telephone Enterprise: The Evolution of the Bell System's Horizontal Structure.* Johns Hopkins University Press.

Garrett, Elisabeth Donagy. 1990. *At Home: The American Family 1750–1870.* Abrams.

Giddens, Anthony. 1984. *The Constitution of Society: Outline of the Theory of Structure.* University of California Press.

Gilfillan, F. S. 1935. *The Sociology of Invention.* Follett.

Gomme, Allan. 1946. *Patents of Invention.* Longmans, Green.

Gooday, Graeme. 1991. "Teaching Telegraphy and Electrotechnics in the Physics Laboratory." *History of Technology* 13: 73–111.

Gooding, David. 1990. *Experiment and the Making of Meaning.* Kluwer.

Gooding, David, and Ryan Tweney. 1991. "Introduction." *Michael Faraday's 'Chemical Notes, Hints, Suggestions, and other Objects of Pursuit' of 1822.* Peregrinus.

Green, Harvey. 1983. *The Light of the Home: An Intimate View of the Lives of Women in Victorian America.* Pantheon.

Greene, Theodore. 1970. *America's Heroes.* Oxford University Press.

Greenhaigh, Paul. 1988. *Ephemeral Vistas.* Manchester University Press.

Gruber, Howard. 1974. *Darwin on Man: A Psychological Study of Scientific Creativity.* Dutton.

A Guide to Thomas A. Edison Papers: A Selective Microfilm Edition, parts I (1850–1878) and II (1879–1886). 1985. University Publications of America.

Guralnick, Stanley M. 1979. "The American Scientist in Higher Education, 1820–1910." In *The Sciences in the American Context,* ed. N. Rheingold. Smithsonian Institution Press.

Hammack, David C. 1982. *Power and Society: Greater New York at the Turn of the Century.* Russell Sage Foundation.

Hammer, William J., as told to Willis J. Ballinger. 1931. "Edison—By the Man Who Knew Him Best." *NEA Service,* part 2.

Hammond, John Winthrop. 1941. *Men and Volts: The Story of General Electric.* Lippincott.

Hantmann, Ronald D. 1991. "Prosecution History Estoppel." *Journal of the Patent and Trademark Office Society* 73: 121–135, 248–256.

Hart, H. L. A., and A. M. Honore. 1959. *Causation in the Law.* Clarendon.

Headrick, Daniel R. 1988. *The Tentacles of Progress.* Oxford University Press.

Heap, David Porter, 1884. *Report on the International Exhibition of Electricity Held at Paris August to November 1881.* Government Printing Office.

Hellrigel, Mary Ann. 1989. Creating an Industry: Thomas A. Edison and His Electric Light System. MA thesis, University of California, Santa Barbara.

Hendrickson, Robert. 1979. *The Grand Emporiums: The Illustrated History of America's Great Department Stores.* Stein and Day.

Hickenlooper, A., 1886. *Edison's Incandescent Electric Lights for Street Illumination.* Cincinnati: Robert Clarke.

"Historical Patent Statistics." 1964. *Journal of the Patent Office Society* 46, no. 2: 91–170.

Holland, Maurice. 1927. "Edison—Organizer or Genius?" April, #1630 Box 12, 12–7. Greenfield Village Research Center.

Horstman, I., G. M. MacDonald, and A. Slivinski. 1985. "Patents as Information Transfer Mechanisms." *Journal of Political Economy* 93: 837–858.

Hotchkiss, Horace L. 1905. "The Stock Ticker." In *The New York Stock Exchange,* ed. E. Stedman. New York: Stock Exchange Historical Co.

Hounshell, David A. 1980. "Edison and the Pure Science Ideal in 19th Century America." *Science* 207, no. 8 (8 February): 612–617.

Houston, Edwin. 1876. "The Phenomenon of Induction." *Journal of the Franklin Institute* 101: 59–63.

Howell, John 1881. "Electric Lighting." *Engineering*, 23 December: 637–638.

Howell, John W., and Henry Schroeder. 1927. *The History of the Incandescent Light*. Schenectady: Maqua.

Hoyt, Edwin P. 1966. *The House of Morgan*. Dodd, Mead.

Hughes, Thomas P. 1989. *American Genesis: A Century of Technological Enthusiasm 1870–1970*. Penguin.

Hughes, Thomas P. 1983. *Networks of Power: Electrification in Western Society, 1880–1930*. Johns Hopkins University Press.

Iconography of Manhattan Island. 1928. New York: Robert H. Dodd.

Israel, Paul. 1992. *From Machine Shop to Industrial Laboratory*. Johns Hopkins University Press.

Jehl, Francis. 1929. *The Birth of a Great Invention*. Dearborn: Edison Institute.

Jehl, Francis. 1937. *Menlo Park Reminiscences* (three volumes). Dearborn: Edison Institute.

Jones, Stacy. 1971. *The Patent Office*. Praeger.

Josephson, Matthew. 1959. *Edison: A Biography*. McGraw-Hill.

Keating, Paul W. 1954. *Lamps for a Brighter America*. McGraw-Hill.

Keil, Maurice. 1881. "The Future of Electricity." *Journal of the Telegraph* 14 (1 August): 223–224.

Kevelson, Roberta. 1990. *Pierce, Paradox, Praxis*. Mouton de Gruyter.

Kitch, E. W. 1977. "The Nature and Function of the Patent System." *Journal of Law and Economics* 20: 265–290.

Kline, Richard. 1995. "Construing 'Technology' as 'Applied Science': Public Rhetoric of Scientists and Engineers in the United States, 1880–1945." *Isis* 86, no. 2: 194–221.

Klitzke, Ramon A. 1959. "Historical Background of the English Patent Law." *Journal of the Patent Office Society* 41, no. 9: 615–650.

Kobre, Sidney. 1969. *Development of American Journalism*. Dubuque: Wm. C. Brown.

Koch, Robert. 1971. Louis C. *Tifffany's Glass, Bronzes, Lamps: A Complete Collector's Guide*. Crown.

Kursh, Harry. 1959. *Inside the U.S. Patent Office*. Norton.

Kurzon, Dennis. 1986. *It Is Hereby Performed. . . .* Amsterdam: John Benjamins.

Lamps & Other Lighting Devices, 1850–1906. 1972. Pyne.

Lathrop, George Parsons. 1890. "Talks with Edison" *Harper's New Monthly Magazine* 80 (February): 425–435.

Latour, Bruno. 1987. *Science in Action.* Harvard University Press.

Latour, Bruno. 1988. *The Pasteurization of France.* Harvard University Press.

Lemay, J. A. Leos. 1964. *Ebenezer Kinnersley: Franklin's Friend.* University of Pennsylvania Pres.

Loeb, Lori Anne. 1994. *Consuming Angels.* Oxford University Press.

Lowenthal, Leo. 1984. "The Triumph of Mass Idols." *Literature in Mass Culture.* New Brunswick: Transaction Books.

Lubar, Steven. 1991. "The Transformation of Antebellum Patent Law." *Technology and Culture* 32, no. 4: 932–959.

Luckhurst, Kenneth. 1950. *The Story of Exhibitions.* London: Studio Publications.

Luhmann, Niklas. 1989. *Ecological Communication.* University of Chicago Press.

Machlup, Fritz, and Edith Penrose. 1950. "The Patent Controversy in the Nineteenth Century." *Journal of Economic History* 10, no. 1: 1–29.

MacLeod, Christine. 1988. *Inventing the Industrial Revolution.* Cambridge University Press.

Maltbie, Milo. 1911. *Franchises of Electrical Companies in Greater New York.* New York: Public Service Commission for the First District.

Mandelbaum, Seymour J. 1990. *Boss Tweed's New York.* Chicago: Ivan R. Dee.

Manson, George A. 1886. "An Electrical Engineer." *St. Nicholas,* February: 300–302.

Marks, William D. 1895. "History of Edison Electric Light Co. of Philadlphia." In *A Course of Thirteen Practical Talks to the Working Men of the Edison Electric Light Company of Philadelphia.* Philadelphia: Edison Co.

Marvin, Carolyn. 1988. *When Old Technologies Were New.* Oxford University Press.

McClure, J. B., 1879. *Edison and His Inventions.* Chicago: Rhodes & McClure.

McMahon, A. Michal. 1976. "Corporate Technology: The Social Origins of the American Institute of Electrical Engineering." *Proceedings of the IEEE* 64, no. 9: 1383–1390.

McMahon, A. Michal. 1984. *The Making of a Profession: A Century of Electrical Engineering in America.* IEEE Press.

Merton, Robert King. 1973. *The Sociology of Science.* University of Chicago Press.

Michels, John. 1880. "Historical Notes on Gas Illumination." *Science* 1: 23 (December 4): 275–277.

Millard, André. 1989. "Machine Shop Culture at Menlo Park." In *Working at Inventing*, ed. W. Pretzer. Henry Ford Museum and Greenfield Village.

Millard, André. 1990. *Edison and the Business of Invention.* Johns Hopkins University Press.

Miller, Raymond C. 1957. *Kilowatts at Work: A History of the Detroit Edison Company.* Wayne State University Press.

Mokyr, Joel. 1990. *The Lever of Riches: Technological Creativity and Economic Progress.* Oxford University Press.

Mole, Charles C. 1978. "The Expanding Scientific Role of the Federal Government in the Nineteenth Century: The Patent Office as A Case Study." *Journal of the Patent Office Society* 60, no. 5: 328–347.

Morton, Henry. 1880. "To the Editor." *Sanitary Engineer,* 1 January (48: 593).

Morton, Henry, A. M. Mayer, and B. F. Thomas. 1880. "Some Electrical Measurements of One of Mr. Edison's Horseshoe Lamps, *Scientific American,* 17 April: 241 (48: 630–633).

Mott, Frank Luther. 1950. *American Journalism: A History of Newspapers in the United States through 260 Years.* Revised edition. Macmillan.

Myers, Gustavus. 1968. *The History of Tammany Hall.* New York: Burt Franklin.

Myers, Gustavus. 1974. *History of Public Franchises in New York City.* Arno.

Nerney, Mary Childs. 1934. *Thomas A. Edison: A Modern Olympian.* Smith & Haas.

Norris, James D. 1990. *Advertising and the Transformation of American Society, 1865–1920.* Greenwood.

Norwig, E. A. 1954. "The Patents of Thomas A. Edison." *Journal of the Patent Office Society* 36, no. 3: 213–296.

Nye, David E. 1994. *American Technological Sublime.* MIT Press.

Nye, David E. 1990. *Electrifying America: Social Meanings of a New Technology.* MIT Press.

Nye, David E. 1983. *The Invented Self: An Anti-biography from the Papers of Thomas A. Edison.* Odense University Press.

O'Brien, Frank M. 1968. *The Story of the Sun.* New Edition. Greenwood.

Outerbridge, A. E. 1880. "The Edison Electric Light." *Journal of the Franklin Institute* 109: 145–156 (48: 619–625).

Passer, Harold C. 1953. *The Electrical Manufacturers.* Harvard University Press.

Pershey, Edward Jay. 1989. "Drawing as a Means to Invention." In *Working at Inventing*, ed. W. Pretzer. Henry Ford Museum and Greenfield Village.

Perks, Sidney. 1905. *Residential Flats of All Classes*. London: B. T. Batsford.

Platt, Harold. 1991. *The Electric City: Energy and the Growth of the Chicago Area, 1880–1930*. University of Chicago Press.

Plush, S. M. 1878. "Edison's Carbon Telephone Transmitter, and the Speaking Phonograph." *Journal of the Franklin Institute* 105: 265–271.

Polanvyi, M. 1944. "Patent Reform." *Review of Economic Studies* 11: 61–76.

Pollard. William. 1790. Patent Application, 29 June. In *Restored U.S. Patents*, volume 1 (Research Publications), reel 1, frames 16 and 17.

Preece, William. 1879. "The Criteria of the Electric Light." *Telegraphic Journal* 7 (15 February): 59–60 (48: 556–557).

Preece, William. 1881. "Electric Lighting at the Paris Exposition." *Electrician*, 17 and 24 December: 76, 91–92.

Pretzer, William S., ed. 1989. *Working at Inventing: Thomas A. Edison and the Menlo Park Experience*. Henry Ford Museum and Greenfield Village.

Priestley, Joseph. 1775. *The History and Present State of Electricity*. Bathurst (1966 edition: Johnson).

Proceedings of the Board of Alderman of the City of New York, volumes 151–152, 1881. New York: Martin B. Brown.

Reader, W. J. 1991. " The Engineer Must Be a Scientific Man": The Origin of the Society of Telegraph Engineers." *History of Technology* 13: 112–118.

Restored U.S. Patents, volume 1: 1790–1803. New Haven: Research Publications.

Rich, Giles S. 1982. "P. J. (Pat) Federico and His Works." *Journal of the Patent Office Society* 64, no. 1: 3–11.

Riordan, William L. 1948. *Plunkitt of Tammany Hall*. Knopf.

Rose, Mark. 1993. *Authors and Owners*. Harvard University Press.

Rosenberg, Robert. 1983. "American Physics and the Origins of Electrical Engineering." *Physics Today*, October: 48–54.

Rosenberg, Robert. 1984. "The Origins of EE Education." *IEEE Spectrum*, July: 60–68.

Rowland, H. A., and G. F. Barker. 1880. "On the Efficiency of Edison's Light." *American Journal of Science* 19 (April): 337–339 (48: 628–630; 53: 745–747).

Rydell, Robert. 1984. *All the World's a Fair*. University of Chicago Press.

Ryder, John D., and Donald G. Fink. 1984. *Engineers & Electrons: A Century of Electrical Progress.* IEEE Press.

Satterlee, Herbert L. 1939. *J. Pierpont Morgan: An Intimate Portrait.* Macmillan.

Schaffer, Simon. 1983. "Natural History and Public Spectacle in the Eighteenth Century." *History of Science* 21: 1–43.

Schivelbusch, Wolfgang. 1988. *Disenchanted Night: The Industrialization of Light in the Nineteenth Century.* University of California Press.

Schroeder, Fred E. H. 1986. "More 'Small Things Forgotten': Domestic Electrical Plugs and Receptacles, 1881–1931." *Technology and Culture* 27, no. 3: 525–543.

Schudson, Michael. 1978. *Discovering the News: A Social History of American Newspapers.* Basic Books.

Schudson, Michael. 1995. *The Power of News.* Harvard University Press.

Scientific American Reference Book, The. 1881. Munn.

Schwarzlose, Richard A. 1989. *The Nation's Newsbrokers.* Northwestern University Press.

Scientific American, 1876. *Hand Book.* New York: Munn & Co.

The Scientific American Reference Book, 1881. New York: Munn & Co.

Searle, John. 1969. *Speech Acts.* Cambridge University Press.

Searle, John. 1979. *Expression and Meaning,* Cambridge University Press.

Shenton, James, ed. 1977. *Free Enterprise Forever! Scientific American in the 19th Century.* New York: Images Graphiques.

Sherwood, Morgan. 1983. "The Origins and Development of the American Patent System." *American Scientist* 71: 500–506.

Sicilia, David B. 1991. Selling Power: Marketing and Monopoly at Boston Edison, 1886–1929. Dissertation, Brandeis University.

Smith, Adam. 1978. *Lectures on Jurisprudence.* Clarendon.

Smith, James G. 1864. "President's Report" *Telegrapher* 1: 1, 26 September: 1–2.

Smithsonian Institution Libraries. 1992. *The Books of the Fairs: Materials about World's Fairs, 1834–1916.* American Library Association.

Sprague, Frank J, 1884. Report on the Exhibits at the Crystal Palace Electrical Exhibition, 1882. Government Printing Office.

Stedman, E. C., and A. N. Easton. 1905. "History of the New York Stock Exchange" In *The New York Stock Exchange*, ed. E. Stedman. New York: Stock Exchange Historical Co.

Stehman, J. Warren. 1967. *The Financial History of the American Telephone and Telegraph Company.* Boston; Houghton Mifflin, 1925; New York: Augustus M. Kelley.

Stotz, Louis. 1938. *History of the Gas Industry.* New York: Stettiner Bros.

Strange, P. 1985. "Two Electrical Periodicals: The Electrician and The Electrical Review 1880–1890." *IEE Proceedings* A 132A, no. 8: 574–581.

Stringham, Emerson. 1941. *Semiotic of Patent Interference Count.* Pacot Publications.

Takahashi, Yuzo. 1986. "Institutional Formation of Electrical Engineering in Japan." In *1880–1980: Un Siecle d'électricité dans le Monde,* ed. F. Cardot. Presses Universitaires de France.

Terman, Frederick E. 1976. "A Brief History of Electrical Engineering Education." *Proceedings of the IEEE* 64, no. 9: 1399–1407.

Thirty Years of New York, 1882–1912. 1913. New York Edison Company.

Toner, James V. 1951. *The Boston Edison Story.* New York: Newcomen Society.

Trachtenberg, Alan. 1982. *The Incorporation of America: Culture and Society in the Gilded Age.* Hill and Wang.

Tweney, Ryan D. 1991a. "Faraday's 1822 'Chemical Hints' Notebook and the Semantics of Chemical Discourse." *Bulletin for the History of Chemistry* 11: 51–55.

Tweney, Ryan D. 1991b. "Faraday's Notebooks: The Active Organization of Creative Science." *Physics Education* 26, no. 5: 301–306.

Upton, Francis R. 1880. "Edison's Electric Light." *Scribner's Monthly,* February: 531–544.

Upton, Francis, C. F. Brackett, and C. A. Young. 1880. "The Edison Light." *Scientific American,* 15 May: 308 (24: 19).

U.S. Department of Commerce. 1981. *The Story of the United States Patent and Trademark Office.* Government Printing Office.

van Nostrand, A. D. 1997. *Fundable Knowledge.* Erlbaum.

Vaughan, Floyd L. 1956. *The United States Patent System: Legal and Economic Conflicts in American Patent History.* University of Oklahoma Press.

Veblen, Thorstein, 1899. *The Theory of the Leisure Class.* Macmillan.

Vygotsky, Lev. 1978. *Mind in Society: The Development of Higher Psychological Processes.* Harvard University Press.

Wachhorst, Wyn. 1984. *Thomas Alva Edison: An American Myth.* MIT Press.

Wallace, William III, and Russell J. Christensen. 1986. *Ebasco Services Incorporated.* New York: Newcomen Society.

Warshow, Robert Irving. 1929. *The Story of Wall Street.* New York: Blue Ribbon Books.

Weber, Max. 1994. *Sociological Writings,* ed. W. Heydebrand. Continuum.

Weil, Vivian, and John W. Snapper, eds. 1989. *Owning Scientific and Technical Information: Value and Ethical Issues.* Rutgers University Press.

Werner, M. R. 1968. *Tammany Hall.* Greenwood.

Williams, L. P. 1991. "Michael Faraday's Chemical Notebook" *Physics Education* 26, no. 5: 278–283;.

Wingate, Charles F., ed. 1875. *Views and Interviews on Journalism.* New York: F. B. Patterson.

Wise, George. 1985. *Willis R. Whitney, General Electric, and the Origins of U.S. Industrial Research.* Columbia University Press.

Witherspoon, John J., ed. 1980. *Nonobviousness—The Ultimate Condition of Patentability.* BNA Books.

Woodmansee, Martha, and Peter Jaszi, eds. 1994. *The Construction of Authorship.* Duke University Press.

Wright, Brian D. 1983. "The Economics of Invention Incentives: Patents, Prizes, and Research Contracts." *American Economic Review* 73, no. 4: 691–707.

Wyckoff, Peter. 1972. *Wall Street and the Stock Markets: A Chronology (1644–1971).* Chilton.

Yates, Joanne. 1988. *Control Through Communication.* Johns Hopkins University Press.

Young, Owen D. 1941. "Appendix." In John Winthrop Hammond, *Men and Volts.* Lippincott.

Zunz, Olivier. 1990. *Making America Corporate, 1870–1920.* University of Chicago Press.

Index

Adams, James, 171
Adams, M. F., 123
Adams, W. Grylls, 135
Aesthetic movement, 326, 328
Aesthetics, 217, 313–315, 319–331, 349
American Association for the
    Advancement of Science, 130–131,
    178, 210–211
American Electrical Society, 129, 134
American Gas Light Association, 155,
    164
American Institute of Electrical
    Engineering, 113–114, 120, 137, 139
American Society of Mechanical
    Engineers, 131
Andrews, William, 287
Apartments, 328–330
Association of Edison Illuminating
    Companies, 289, 311
Ashley, James, 114, 122–125, 134
Austin, John, 101, 168
Ayrtoun, William, 117

Bailey, Francis, 93
Bailey, Joshua, 201–210, 266
Ball, Clinton, 17–18
Banker, J. H., 171, 175
Barker, George, 13–15, 18, 134, 139,
    145, 167, 188, 192–193, 210–212, 262
Barrett, William, 151–153
Batchelor, Charles, 48, 60–66, 69,
    73–75, 83, 112, 171, 179, 197,
    201–208, 211, 214, 241, 266,
    269–274, 278–282, 288

Beardslee Company, 319
Bennett, James Gordon, Jr., 35, 178,
    195, 229
Berggren, E. J., 281
Bergmann, Sigmund, 112, 197, 278,
    331
Bergmann & Company, 197, 278–279,
    282, 288, 314–324
Berlin Exhibition, 199
Bernhardt, Sarah, 196
"Black boxes," 344–346
Bliss, George, 288
Boehm, Ludwig, 48
Boston Edison Company, 288
Brackett, C. F., 139, 191–192
Bradley, Justice, 250–251
Bright, Sir Charles Tilson, 119, 208
Broadway Transit Company, 228
Brockton Edison Company, 287–288,
    305–306
Brooklyn Bridge, 221
Brush, Charles, 9, 190, 291
Butler, Howard, 174

Cables, submarine, 117, 120
Campbell, Allan, 223
California Electric Light Company, 291
Canadian Edison Manufacturing
    Company, 283
Carman, William, 48, 61–62, 65, 73,
    76, 78
Central Station Construction
    Department, 282
Charisma, 251–268, 286–289, 334, 339

Chicago, 220, 288–289
Chicago Edison, 289
Chicago World's Fair, 217
Cincinnati, 220, 327
Civil War, 24, 120
Claims, as speech acts, 160, 165–169
Clarke, Charles, 115, 197, 275, 287
Claudius, Hermann, 230
Cognitive institutions, 347
Collaboration, 4, 60, 75–76, 108, 278,
    348–349
Commonwealth Edison, 289
Compagnie Continentale Edison, 307
Competitive analogy, 150–156, 298, 310
Consolidated Electric, 248
Consumption, and affluence, 315,
    319, 323–331
Communicative systems, 334–340,
    344, 348–350
Cooper, Edward, 222–223, 226–227
Cornell University, 112
Craig, Hugh, 15–17
Croffut, William, 33, 163, 171–173
Crowley, David, 35
Crystal Palace Exhibition, 119, 139,
    199–201, 214–216, 233
Cummings, Amos J., 163

Dana, Charles, 35
Davy, Humphrey, 9
Dean, Charles, 281
DeLong, George Washington, 178
Demonstrations, 15, 43, 45, 178, 226,
    338
  for backers, 174–178
  at Menlo Park, 131, 137, 147, 180–189,
    196–199, 233, 297, 308
  for press, 161–164
  for technical community, 188–195
Department stores, 327
de Parville, Henri, 205
DeSaguliers, Jean, 115
Detroit, 24, 274, 289
Discourse
  politics as, 220
  technology and, 2–3, 18–19,
    159–160, 340

Discursive systems, 334–340, 344,
    348–350
Documentary records, 3–4, 220, 223,
    268–274
du Moncel, Theodore, 117–118, 135,
    137, 139, 186, 203–205, 212–215,
    292–294, 325
Dungeness lighthouse, 9
Dyer, Richard, 245

Eaton, S. B., 147, 226
Ebasco, 287
Economics, and persuasion, 141,
    155–156
Edgar, Charles, 288
Edison Company for Isolated Lighting,
    146–147, 197, 229, 248, 282, 285,
    287, 292, 296–302, 307–308, 330
Edison effect, 131
Edison Electric Illuminating
    Company of New York, 197, 248,
    292, 296, 299, 301–302
Edison Electric Lamp Company, 197,
    278–279, 282
Edison Electric Light Company,
    43–44, 146–147, 164, 171–174, 197,
    229–232, 253, 256–257, 265,
    277–278, 281–283, 286–292, 296,
    299–308
Edison Electric Light Company of
    Europe, 265
Edison Machine Works, 197, 281–282,
    288
Edison Shafting Manufacturing
    Company, 282
Edison United Manufacturing
    Company, 276, 282–283, 309–311
Edison Pioneers, 262
Edson, Franklin, 227
Edson, Tracy, 164, 171, 175, 178, 223
Electrical engineering, 112–119, 350
Electrical Society, 115–116
Electric Bond and Share Company, 287
Electric generators, 10–11, 145–146,
    201, 213, 215, 232
Electric lighting, arc, 9–10, 148, 150,
    199

Electric lighting, incandescent
  Edison's earliest interest in, 10
  fixtures for, 315–322
  and gas lighting, 150–155, 199
  history of, 9–10
  as marketing strategy, 147–149
  as ordinary, 234
  as patentable novelty, 253
  as public sensation, 181–198
Electric power, 1, 145–148. *See also*
  Power stations
Electric Railway Company of America,
  146
Electric Tube Company, 198, 281–282
Enlistment, 343–344, 348
Etheric force, 27–28, 128
European craftsmanship and taste,
  324–326
Exhibitions, 199–217. *See also names
  and sites of particular exhibitions*

Fabbri, Egisto, 176, 180, 183, 197, 226
Fall River, Mass., 229, 288
Faraday, Michael, 55, 57–58
Farmer, Moses, 167, 248
Filaments
  bamboo, 196
  platinum, 176–178
Financial markets, 17, 40
Finsbury Technical College, 117
Fire Insurance Underwriters, 306
Floral design, 313–317, 322–324,
  330–331
Force, Martin, 48
Fox, Edwin Marshall, 136, 161,
  171–173, 185
Fox, Robert, 203
Franchises, 285–286, 303, 310–311
Franklin, Benjamin, 115
Franklin Institute, 131, 192
Frazer, Persifor, Jr., 130
Friedel, Robert, 1, 50

Garfield, James, 196
Garrison, William Lloyd, Jr., 287
Gas utilities, 10, 143, 148–155, 171,
  189–190, 199, 220, 298, 306–307, 316

General Edison, 253, 276, 289
General Electric, 248, 253, 257, 283,
  286, 289
Goddard, Calvin, 171, 175, 280
Goerck Street Works, 223, 281
Gold and Stock Telegraph Company,
  41, 53–54, 56, 123
Gooding, David, 344
Gouraud, George, 201
Grace, William, 222–223, 226–227
Greeley, Horace, 35
Green, Norvin, 197
Griffin, Stockton, 45, 48, 83, 171, 176

Haid, Alfred, 115
Hammer, Charles, 112, 201, 263–264
Harrisburg, Pa., 219
Hastings, Frank, 288
Havana, 228
Hickenlooper, Major A., 220
Hospitalier, Edmond, 203–206
Hauksbee, Francis, 115
Heterogeneous symbolic engineering,
  334–336, 338
Holland, Maurice, 267
Hotels, 328–330
Hounshell, David, 139
Houston, Edwin, 129, 291
Howell, John, 294–295
Hughes, David Edward, 119, 208
Hughes, Thomas, 1, 50, 155

Imperial College of Engineering, 117
Institute of Electrical and Electronics
  Engineers, 113
Institution of Civil Engineers, 116
Insull, Samuel, 112, 263, 265,
  281–282, 288–289
Intellectual property, 85, 109
International Telegraph Treaty,
  117–118
Inventor, as status, 108–109, 114
Investment, rhetoric of, 291–312, 337
Investors, in Edison companies,
  15–17, 42–45, 164–167, 173–180,
  188–189, 197, 219, 234, 277,
  291–292, 337

Irving Hall, 221, 227
Ivrey-sur-seine, 215
Israel, Paul, 1, 50, 114

Jablochoff, Paul, 9, 202
Jehl, Francis, 48, 165, 179, 211, 224,
    262–263
Johnson, Edward, 28, 112, 130, 138,
    171, 197, 211, 266, 278–279, 288,
    296–298, 322–324, 328, 331

Kabath, N., 203
Kansas City, Mo., 138
Keil, Maurice, 135
Kelly, "Honest John," 221–222
Kalakaua, King, 134
Kruesi, John, 48, 60, 69, 72–78, 83, 266,
    282

Lahure, A., 206
Lane-Fox, St. George, 135, 203
Latour, Bruno, 90, 343–344
Law, as symbolic system, 240, 253
Lawrence, Mass., 288
Lecture, as genre, 13
Leisure, 326–328, 331
Leonard, H. Ward, 283, 288
Leonard and Izard, 283
Livor, Harry, 282
Louisville, Ky., 27, 138, 217, 219
Lowrey, Grosvenor, 42–43, 164,
    173–176, 197, 211–212
Luhmann, Niklas, 349

Machine shop culture, 60, 83,
    114–115
Man, Albon, 45, 187, 239–244,
    248–249, 308
Marketing, 143, 147–149
Marvin, Carolyn, 1
Massachusetts Institute of Technology,
    112
Masson, George, 203
Maxim, Hiram, 111, 202–210, 248
McCormick Reaper Company,
    285–286
McKeesport Light Company, 249–253

Menlo Park, N.J., 29–32, 47–49, 61,
    74, 114–115, 131, 134, 137, 147,
    160–163, 166–168, 175–199, 226,
    261–264, 277, 287–288, 308
Michels, John, 132–134
Michigan, 269, 273–274
Middle class, 325–331, 338
Military Telegraphic School, 117
Miller, W. C., 17
Mills, D. O., 328
Moore, A. T., Jr., 288
Morgan, J. Pierpoint, 328
Morris, Joseph, 223–224
Morton, Henry, 15, 134–135, 155,
    186–187, 193
Moses, Otto, 138, 201–208, 214, 262,
    266–267
Mott, Samuel D., 48
Munich, exhibitions at, 201, 217, 324
Munn & Company, 94–95, 128
Myers, Gustavus, 227

National Telegraphic Union, 121
Newark, N.J., 226
Newburgh, N.Y., 288
Newspapers,
    British, 208, 213–215
    changes in, 23, 35–36, 339
    Edison and, 23–29, 36–38
    French, 203–213
    power of, 18
New York City, 149–151, 196, 219–234,
    326
New York Electric Lines Company,
    227
Nichols, Edward, 115
Notebooks
    function of, 48–84
    Menlo Park, 61–66, 69–70
    Newark Shop, 60, 63
    Unbound, 64–74
Nye, David, 1

Ohio, 287
Ontario, 270–271
Ordinary, rhetoric of, 231–234, 278
Outerbridge, A. E., 131

Ott, Fred, 48, 60, 83
Ott, John, 48, 60, 83, 264

Painter, Uriah H., 171
Palais de l'Industrie, 200
Paris Exposition, 119, 135–139,
    199–212, 216, 228, 233, 262, 266,
    292–295, 325
Paris Opera House, 202, 213
Patent agents, 53, 128
Patent claims, 95–98
Patent drawings, 76, 81
Patent examination process, 85–86,
    106–108
Patent laws, 92–94
Patent litigation, 248–251, 345
Patent models, 91–92, 95
Patent Office, U.S., 114, 237–246
Patents, 9–10, 85–109, 237–257, 298
    appeal of, 239–240
    challenges to, 239
    concept of, 90–91, 109
    crucial, 251
    as genre, 92–98, 102–108
    history of, 92–94
    infringement of, 247
    interference with, 240, 247
    as rhetorical form, 92–98, 104–108,
        239, 339
    texts of, 88–89
    value of, 15, 39, 85, 94, 109,
        237–239, 248, 337, 346
Patent specifications, 95–96
Pennsylvania, 287
Persuasion, 141, 155–156, 341–342.
    See also Rhetoric
Philadelphia, 289
Philadelphia Electrical Exhibition,
    113, 139, 199–201, 217, 288
Phonograph, 28–35, 130
Plunkitt, George Washington, 221
Politics, as discourse, 220
Pollard, William, 92–93
Pope, Edison, and Company, 122
Pope, Franklin, 114, 122–124, 134
Port Huron, Mich., 24, 51
Porter, Charles T., 131.

Porter, Rufus, 128
Power lines, underground, 150, 153,
    224–227, 306
Power stations, 112, 135, 147, 199,
    215–219, 223, 228–233, 274–275,
    283–289, 296–300, 303–305,
    308–311
Promise, as a speech act, 159–160,
    168–169, 298
Preece, William, 118–119, 139, 186,
    208, 214–215, 294–295
Prescott, George, 124
Presence, 333–340
Priestley, Joseph, 57–58, 115
Puskas, Theodore, 201–210, 266

Railroads, 23–24, 39–41, 146, 148, 285
Reid, Whitelaw, 35, 328
Representation, 2–5, 105–108,
    336–340, 343, 348, 350
Resting points, representational,
    344–346
Rhetoric, 3–5, 141–143, 339–342
Rio de Janeiro, 228
Rocap, Charles, 281
Roselle, N.J., 287
Rowland, Henry, 139, 192–193

San Francisco, 27, 219
Saussure, Ferdinand de, 101
Sawyer, William, 45, 111, 167, 174,
    187, 239–244, 248–251, 308
Schenectady, N.Y., 138, 282, 308
Science, ideal of, 127–128, 139, 188
Scrapbooks, Menlo Park, 61–62
Searle, John, 99, 101, 105, 168, 234, 342
Serrell, Lemuel, 53
Sewing machine, 146–147
Shaw, Charles A., 15
Shaw, P. B., 287
Siemens, Charles, 119, 215, 294–295
Singer, Isaac, 285
Singer Sewing Machine Company,
    285–286
Smith, Adam, 141
Smith, Gerrit, 17
Social facts, 342, 346–347

Smithsonian Institution, 130
Society for Civil Engineering, 116
Society of Telegraphic Engineers, 116–119
Soldan, Gustav, 281
Speech acts, 98–108, 168–169, 239, 250–251, 334, 341–347
Sprague, Franklin, 287
Sprague Electric Railway Company, 283
Stager, Ansons, 288
Stanley, William, 248
Stephenson, George, 27
Stucker, N., 17
Sturgeon, William, 115
Structuration, 344, 347–349
Sunbury, Pa., 287
Swan, Joseph, 111, 135, 139, 186, 203, 208–214, 307

Talk, and writing 3, 220, 223
Tasimeter, 130, 192
Tate, Alfred O., 269–274
Telegraphy, 27, 115–127, 284
Telephone industry, 284
Thomson, Elihu, 291
Thomson, William, 186
Thomson Houston Company, 283, 291
Tiffany, Louis D., 327–328
Tissandier, Gaston, 203–206
Tweed, William Marcy, 221
Twombly, Hamilton, 42, 164
Tyndall, John, 186
Typification, 144–145, 337, 343, 350

Unger, William, 60
Upton, Francis, 48, 61–64, 73, 112, 115, 134, 137, 165, 174–183, 187, 190, 192, 230, 264–266, 278–279, 282, 288
U.S. Electric Light Company, 138, 202–203, 241, 248–249

Veblen, Thorstein, 327–328
Vienna Exposition, 199–200, 217
Villard, Henry, 146, 195, 197, 229, 287–288
Vygotsky, Lev, 347

Walker, Charles, 116
Wallace, William, 10–11, 145, 160
Weber, Max, 259, 276–277, 283
West Orange, N.J., 276
Western Edison, 288–289, 307
Western Electric, 288
Western Union, 41–43, 115, 123–127, 164, 180, 284
Westinghouse, 248–249, 257

Yates, Joanne, 40, 222
Young, Charles, 139, 191–192